Donald Ary, Northern Illinois University
Lucy Cheser Jacobs, Indiana University

INTRODUCTION TO STATISTICS
Purposes and Procedures

HOLT, RINEHART AND WINSTON
New York   Chicago   San Francisco
Atlanta   Dallas   Montreal
Toronto   London   Sydney

Figure 6.3 is from Henry E. Garrett, *Statistics in Psychology and Education*, p. 104. Copyright 1926, 1937, 1947, 1953, © 1958, 1966 by David McKay Co., Inc. Reprinted by permission of the publisher.

Figure 6.5 is reprinted with permission of Macmillan Publishing Co., Inc. from Anne Anastasi and John P. Foley, Jr., *Differential Psychology*, © Macmillan Publishing Co., Inc., 1958.

Table A6 is reprinted by permission from *Statistical Methods* by George W. Snedecor and William G. Cochran, sixth edition © 1967 by Iowa State University Press, Ames, Iowa.

Table A7 is from R.A. Fisher: *Statistical Methods for Research Workers*, 14th edition. Copyright © 1972 by Hafner Press, a division of Macmillan Publishing Co., Inc.,

Table A8 is reprinted by permission of Prentice-Hall from Gene V. Glass and Julian C. Stanley, *Statistical Methods in Education and Psychology*, © 1970.

Library of Congress Cataloging in Publication Data

Ary, Donald.
   Introduction to statistics.

   Includes index.
   1. Statistics. I. Jacobs, Lucy Cheser, joint author. II. Title.
HA29.A78   519.5   75-38831

ISBN 0-03-088412-8

Copyright © 1976 by Holt, Rinehart and Winston
All rights reserved
Printed in the United States of America

890 038 20 19 18 17 16 15 14 13 12 11

PREFACE

This text provides an introduction to statistical procedures for students in education and psychology. The book is based on four assumptions: (1) The field of statistics is a highly organized structure built around a few basic concepts; (2) The really important concepts in statistics are straightforward and reasonable ideas; (3) If the basic concepts can be mastered the whole structure can be comprehended; (4) The decision-making processes in statistics can be defined and charted.

Many people who aren't particularly fond of the idea of learning statistics find themselves in a position where they have to do so. However, it is our impression that most authors of statistics texts write to impress their colleagues, not to communicate with people who need to learn statistics. The result has been statistics texts that explain everything in the most complex possible way and consequently confuse and frustrate their readers. We have attempted to explain everything in a way that will be reasonable and meaningful to students. We have tried to use a style of writing which we hope beginning students in statistics will find readable and which will maintain their interest and increase their comprehension. Whenever possible, we have used vivid examples to illustrate the statistical concepts. It has been our experience that students are often able to recall a vivid example that was used for a particular concept and are then able to recall the concept itself.

We have emphasized the rationale and applications of statistics and the role of statistics in practical decision-making processes. We first explained the concepts thoroughly and then dealt with the mathematical computations related to the concepts. We do not presume extensive mathematical training on the part of the reader. Complex proofs and formula derivations which are too often not understood by students have been eliminated. Only computations related to the basic concepts and procedures are required. Whenever mathematical computations are introduced, we explain the process. The examples chosen for illustration purposely use small numbers because this eliminates laborious and time-consuming calculations while, at the same time, illustrating the statistical procedure just as effectively as more complicated examples. As a further help to students, simple problems along with the answers are included throughout each of the chapters. Working these problems provides readers with a way to check their grasp of the concepts and procedures before proceeding to a new section. In addition, a review set of problems is presented at the end of each chapter.

We have provided a list of objectives at the beginning of each chapter which expresses the content of each chapter in terms of what the student should expect to learn. Instructors may also find the objectives helpful as a basis for organizing lecture material and for writing test items. In addition, a flow chart has been included at the end of each chapter to show how the statistical procedures described in that chapter become part of the decision-making process.

We are indebted to the Literary Executor of the late Sir Ronald A. Fisher, F.R.S. for permission to reprint Table A7 from the book *Statistical Methods for Research Workers*. We are also indebted to the aforementioned and to Dr. Frank Yates, F.R.S., for permission to reprint Tables A4 and A5 from their book *Statistical Tables for Biological, Agricultural and Medical Research*. We are especially indebted to our many students whose fears and frustrations with statistics have led us to attempt a student-oriented presentation of statistics.

De Kalb, Illinois *DONALD ARY*
Bloomington, Indiana *LUCY CHESER JACOBS*
December 1975

CONTENTS

PREFACE v

**PART ONE
DESCRIPTIVE STATISTICS** 1

CHAPTER ONE
BASIC CONCEPTS 3
Objectives 3
Purpose of Statistics 3
Descriptive Statistics 4
Inferential Statistics 5
Variables 9
Scales of Measurement 12
Summary 18

CHAPTER TWO
FREQUENCY DISTRIBUTIONS AND GRAPHS 23
Objectives 23
Purpose of Frequency Distributions and Graphs 24
Frequency Distributions 24
Graphing Frequency Distributions 34
Forms of a Frequency Polygon 42
Summary 45

CHAPTER THREE
MEASURES OF CENTRAL TENDENCY 51
Objectives 51
Averages 52
Mode 52
Median 54
Mean 63
How To Mislead with Measures of Central Tendency 70
Comparison of the Mean, Median, and Mode 70
Selecting a Measure of Central Tendency 74
Summary 79

CHAPTER FOUR
MEASURES OF VARIABILITY 82
Objectives 82
Introduction 83
The Range 84
Quartile Deviation 86
Variance 90
Standard Deviation 94
A Comparison of the Measures of Variability 105
Summary 107

CHAPTER FIVE
INTERPRETATION OF INDIVIDUAL SCORES 112
Objectives 112
Percentile Rank 113
Percentiles 118
Quartiles and Deciles 123
Standard Scores 125
Summary 134

CHAPTER SIX
THE NORMAL CURVE 138
Objectives 138
The Concept of the Normal Curve 138
The Standard Normal Curve Table 146
Using the Normal Curve Table 148
Summary 157

CHAPTER SEVEN
CORRELATION 162
Objectives 162

Scattergrams 164
Bivariate Frequency Distribution 170
Calculation of the Pearson Correlation Coefficient 172
Factors Influencing the Correlation Coefficient 184
Interpreting a Correlation Coefficient 186
Summary 187

CHAPTER EIGHT
PREDICTION 195
Objectives 195
Prediction 196
Predicting z-Scores 198
Regression toward the Mean 199
Least Squares Criterion 203
Standard Error of Estimate 214
Summary 220

CHAPTER NINE
OTHER MEASURES OF RELATIONSHIP 226
Objectives 226
Spearman Rank Correlation Coefficient: ρ_s 227
Point-Biserial Correlation Coefficient: ρ_{pbi} 232
Biserial Correlation Coefficient: ρ_{bi} 236
Tetrachoric Correlation: r_t 241
The Phi Coefficient: r_ϕ or Simply ϕ 244
Coefficient of Contingency 248
Summary 249

PART TWO
INFERENTIAL STATISTICS 257

CHAPTER TEN
THE LOGIC OF INFERENTIAL STATISTICS 259
Objectives 259
Introduction 259
Binomial Probability 269
Aids Available for Determining Binomial Probabilities 277
Normal Curve Approximation of the Binomial 278
Summary 281

CHAPTER ELEVEN
RANDOMNESS AND SAMPLING ERROR IN HYPOTHESIS TESTING 287
Objectives 287

Introduction 287
Random Sampling 288
Random Assignment 290
Concept of Sampling Error 291
The Lawful Nature of Sampling Errors 292
Using the Normal Curve To Test Hypotheses about a Population Mean 295
The Principal and the Superintendent 299
One-tailed and Two-tailed Tests 303
Areas of Rejection and Retention for Raw Score Means 306
The Effect of Numbers on the Likelihood of Type II Errors 307
Steps in Testing a Hypothesis about a Population Mean 308
Summary 308

CHAPTER TWELVE
THE t-TESTS 313

Objectives 313
Degrees of Freedom 315
Using the t-Distribution for Hypothesis Testing 317
The t-Test for Significance of the Difference between Means 322
The Null Hypothesis 326
The Logic of the t-Test 327
Assumptions Necessary for a t-Test 329
Significance of the Difference between Two Means: Independent Samples 330
Significance of the Difference between Two Means: Correlated or Nonindependent Samples 334
The t-Test for Two Correlated Samples 336
Comparison of the Power of Tests Based on Independent and Correlated Samples 340
The t-Tests for Pearson r Correlation Coefficients 342
Summary 344

CHAPTER THIRTEEN
ONE-WAY ANALYSIS OF VARIANCE 349

Objectives 349
Introduction 349
Partitioning the Sum of Squares 354
The F-Ratio 356
Using the F-Table 357
Computational Formulas for Sums of Squares 359
Comparison of Group Means Following the F-Test 364

Assumptions Underlying Analysis of Variance   365
Summary   365

CHAPTER FOURTEEN
TWO-WAY ANALYSIS OF VARIANCE   371
Objectives   371
Factorial ANOVA   371
Computation for Two-Way Analysis of Variance   374
K-Factor Analysis of Variance   386
Summary   388

CHAPTER FIFTEEN
NONPARAMETRIC STATISTICS   394
Objectives   394
Introduction   394
Chi Square   395
Summary, Goodness-of-Fit Chi Square   400
Chi Square Test of the Independence of Categorical Variables   402
Restrictions in the Use of the Chi Square   409
Correlation Coefficients Derived from Chi Square   411
Sign Test   413
Median Test   416
Comparison of Parametric and Nonparametric Statistics   419
Summary   420

APPENDIX   427

INDEX   455

**INTRODUCTION
TO STATISTICS
Purposes and
Procedures**

PART ONE
DESCRIPTIVE STATISTICS

CHAPTER ONE
BASIC CONCEPTS

OBJECTIVES

Chapter One is intended to provide an introduction to some basic statistical concepts. After studying this chapter the reader should be able to:

1. Describe how statistics are useful to the scientist.
2. List the purposes of both descriptive and inferential statistics.
3. Distinguish between population and sample, and parameter and statistic.
4. Explain the purpose of sampling in research.
5. Define "variable" and list some of the ways that variables are classified.
6. Distinguish between continuous and discrete variables and identify examples of each.
7. Distinguish between independent and dependent variables.
8. Describe the role of measurement in research.
9. Describe the basis for the different levels of measurement and the meaning that numbers have at each level.
10. Name the four different types of scales of measurement, list their respective characteristics, and recognize each in a measurement situation.

Purpose of Statistics

Statistics constitute a body of scientific methods that are used for the quantitative analysis of data. Statistical procedures serve two principal functions: (1) they aid the scientist in organizing, summarizing, interpreting, and communicating quantitative information obtained from observations; (2) they permit the scientist to go beyond the data gathered from a small number of subjects to reach tentative conclusions about the large group from which the smaller group was derived. The statistical procedures which deal with the first function are referred to as *descriptive statistics*; the statistical procedures concerned with the second function are called *inferential statistics*. This text presents an introduction to both types of statistics, since they are intimately related processes.

It may not be overstating the case to say that a knowledge of basic statistical procedures is indispensable to the educator and the psychologist in many aspects of their work. For example, statistics enable us to understand and interpret the scores from the myriad of educational and psychological tests that are administered. Even reading a test manual in order to evaluate the adequacy of the test for a particular purpose requires statistical knowledge. More importantly, if the educator and the psychologist are to be able to read, understand, and benefit from the research literature, an understanding of the statistical concepts and methods used is essential. Most of the journal articles in education and psychology report findings in statistical form or apply statistical concepts in their discussion. Consequently, we need at least a rudimentary knowledge of statistics in order to be an intelligent consumer of research. Furthermore, if we intend to do research in our area of interest, we must understand statistics in order to plan appropriate procedures and to interpret and communicate the findings from our research.

Statistics are utilized in all fields of research—in the social sciences, the physical and biological sciences, business and economics, agriculture, and so on. Although the actual procedures employed may differ from one field to another, they are all based on the same underlying theory of statistics. This text is written primarily for the teacher and the psychologist, however; the concepts and methods discussed are those considered basic to statistical theory. Thus the student who grasps the basic concepts included in this text should be prepared to apply statistics in any field of research.

Descriptive Statistics

When one has a mass of numerical data to interpret, it is first necessary to organize and summarize these data in such a way that they can be meaningfully understood and communicated. As indicated above, descriptive statistical procedures are employed for this purpose. Frequently the most useful first step is to arrange the data in some logical order. For example, the test scores from the administration of a standardized reading test could be arranged from high to low. Inspection of the data can then provide information about the general pattern of the performance of the particular group. One can see the highest score, the lowest score, and whether the scores tend to cluster at certain points. However, numerical indexes are generally employed to extract further information from the data. Such indexes represent characteristics of the distribution of scores and their meaning can be interpreted without having to see the entire distribution. The indexes most often calculated include the most frequently occurring score, the score which divides the distribution into two equal parts, the average score, and the variability or spread of the scores about the average. Statistical techniques are also available for determining whether a relationship exists between two sets of data. In our example, we might wish to know whether or not there is a relationship between the group's performance on the standardized reading test and their performance in everyday class work. Using yet other statistics, we could use the standardized test scores to make predictions of the future reading levels of this particular group of students. Thus, through the use of appropriate descriptive statistics, concise statements can be made about characteristics of a set of data. Descriptive statistics, however, are always concerned with the particular set of data that has been gathered by the researcher.

Inferential Statistics

Information about particular small groups is often of little intrinsic interest. The researcher is interested in determining if the findings from a small finite group are likely to be true for larger groups or for all the potential observations that could be made. Typically the aim of research is to produce broad generalizations and principles that permit explanation and prediction of events. For example, a researcher may compare the effectiveness of two methods of teaching reading to first graders at Central Elementary School in Columbus, Ohio. His real interest is not which method is more effective with just these first graders, but rather his interest is in arriving at conclusions about the differential effectiveness of the two methods which could be generalized to other first graders throughout the United States. He uses the small group in order to obtain the data which will enable him to draw such inferences. It is at this point that inferential statistics make a powerful contribution to the work of the researcher. It is inferential statistics that provide the methods needed for making inferences about a large group on the basis of small group findings. Of course, there is always some uncertainty involved in such a process, but inferential statistics enables the researcher to calculate the risk he takes in going from specific data to general conclusions. It has been said that inferential statistics are the science of making reasonable decisions with incomplete information.

6 DESCRIPTIVE STATISTICS

In order to fully understand this distinction between descriptive and inferential statistics, it is necessary to discuss the concepts of population and sample.

Populations

By definition a *population* is an entire group of people, objects, or events all having at least one characteristic in common. A population is not restricted to people, but may be defined to include laboratory rats, brands of cigarettes, types of industrial accidents, highway fatalities, and so on. All freshmen students enrolled at state universities at a certain time or all the cars produced by a given manufacturer during a particular year could be the population of interest. In research, the term population refers to the entire group of individuals or events to which generalizations are to be made. The nature of the research project determines the scope of the population to be considered. A population may vary in size from a single classroom of students to several million persons or events (for example, the voting population in the United States) or even to an infinite number of cases.

As was stated above, the final aim in inferential statistics is to arrive at inferences about characteristics of populations. Assume we are interested in determining the average height of male high school seniors in the United States. In this case, the population would include all male students classified as seniors in all high schools in the United States. It would refer to a defined aggregate of persons having the characteristics of maleness and high school senior status in common, all of whom would be measured for height. If we were interested in ascertaining the average score made by Indiana high school seniors on the SAT test, the population would be restricted to all high school seniors in the state of Indiana who take the SAT test. A study might be concerned with the attitudes of university students in the United States toward the legalization of marijuana. In this case, the population would include all university students, both male and female, enrolled in all the universities in the United States at the time of the study.

When measures, such as an average, are obtained for defined populations, they are called *parameters*. A parameter is defined as a numerical characteristic of a population. If we had the average height of all male high school seniors in this country, the average would be a parameter of that population.

Samples

Because of the large size of most populations, it is not usually possible or practical to measure every single person or element in that population. For example, it would not be feasible to measure the height of every male high school senior in the United States. It would be too expensive and time-consuming to measure the attitudes of all U.S. college students toward the legalization of marijuana and to produce parameters based on the population. The pre-election polls cannot contact every eligible voter in the population. Fortunately, it is usually not necessary to measure every person or element in a given population. It is sufficient to select and measure what is known as a sample of the persons or objects and then make inferences about the entire population on the basis of the sample data. A *sample* is defined as a subgroup selected from the complete population. For example, if there are 60 million voters in the United States, then one million or ten million drawn from that number would represent samples. Descriptive measures are obtained from the sample and these data are used to make statements about the characteristics of the entire population.

The usual procedure in a scientific study is first to specify the population with respect to the characteristic of interest, and then, by an appropriate method, select a sample from this population. If sample data are to be used as the basis for generalizations about the population, it is essential that the sample be representative of that population. A representative sample is one that is like the population with respect to the characteristics under consideration in the study, that is, it is a true replica of the population from which it is drawn. To be representative the composition of a sample must reflect accurately the proportion or frequency of relevant characteristics in the entire population. For example, to be representative, a sample of voters in a pre-election poll should be like the total population of voters on the distribution of factors relevant to voting behavior such as age, occupation, race or ethnic membership, social class, and geographic factors. Only on the basis of such a representative sample could reasonable inferences be made about the total population of voters. A sample of faculty members at a university selected for an attitude survey should be representative of the total faculty population on such factors as age, rank, department, school, and so on.

Representativeness is often approximated by means of random selection, a procedure in which every object or individual in the population has an equal probability of being included in the sample. Such a random procedure ensures that chance alone determines the members chosen for the sample. Of course, any number of different samples could be drawn from a given population.

Measures obtained from a sample, such as average test scores, proportions agreeing or disagreeing with an opinion statement, and so on, are called *statistics*. A statistic is any numerical characteristic descriptive of a sample. It indicates something about a sample as a parameter indicates something about a population. Lower case italic letters are generally used to symbolize statistics and lower case Greek letters are generally used to symbolize parameters.

8 DESCRIPTIVE STATISTICS

Exercises

1. Which sentences below indicate the use of descriptive statistics and which indicate the use of inferential statistics?
   (a) The teacher identified the highest and lowest scores in this week's spelling test.
   (b) The superintendent had every tenth student on his list of freshmen take a reading test in order to get an idea of how well all freshmen read.
   (c) A college placement officer sent a questionnaire to a random sample of his graduating seniors in order to assess the job-finding success of all graduating seniors.
   (d) A principal administered spelling tests to all his sixth graders and recorded the number of pupils who scored above the national norm on the test.
2. Mathematical values which describe population data are:
   (a) statistics
   (b) parameters
3. An investigator has a list of (i) all third graders in the city school system. From this list he drew a (ii) subgroup of 50 children for his study.
   (a) Which is a population?
   (b) Which is a sample?
4. A researcher is interested in the effect of special inquiry training on the ability of elementary school children to solve mathematical problems. He selects the third, fourth, and fifth graders at Elmwood Elementary School for his study. After the study, the mean score of these students on a math problem-solving test is compared with a group of control students. In this study, identify the (a) populations, (b) samples, and (c) statistics.
5. The average height of the population of seventh grade boys in Bloomington, Indiana is called a _____. The average height of a sample of seventh grade boys in this city is called a _____.
6. A researcher is interested in studying the personality characteristics of children in Columbus, Ohio, who have IQ's of 150 and above. He locates 30 of these children and then selects 10 of them for intensive study. Identify (a) the population, (b) the sample in this study.

Solutions

1. (a) descriptive   (b) inferential   (c) inferential   (d) descriptive
2. (b) parameter
3. (a) all the third graders   (b) the subgroup of 50
4. (a) elementary school children with or without inquiry training   (b) third, fourth, and fifth graders at Elmwood Elementary School assigned to the two groups   (c) the mean scores on the math test
5. parameter; statistic
6. (a) the 30 children with IQ's of 150 and above   (b) the 10 included

Variables

The characteristics that researchers observe and measure are called *variables*. The term variable refers to a characteristic that can take on more than one value and that shows variation from person to person or case to case. One may observe a group and measure the age, height, weight, intelligence, achievement, and attitudes of the members of that group. Such characteristics are called variables because one would expect to find variation from person to person of these attributes. Variables may be classified as *quantitative* or *qualitative*. Quantitative variables are those which vary in quantity and are thus recorded in numerical form, as for example, age, the scores on a spelling test, or the time required to solve arithmetic problems; qualitative variables vary in quality and may be recorded in other than numerical form. Qualitative variables, such as sex, color of hair, color of eyes, handedness, and so on, may be indicated by means of a verbal label or may be recorded through the use of code numbers.

Characteristics that serve as variables in one study may be kept constant in another study by selecting members of a sample on the basis of similarity in this characteristic. The term *constant* is used to refer to characteristics that do not vary for the members of a particular group. For example, if a research study is concerned only with fourth graders, then grade level would be a constant. If one were studying only ten-year-olds, then in this instance, age would be a constant and not a variable.

In research a distinction is made between *independent* and *dependent* variables. Variations in an independent variable are presumed to result in variations in the dependent variable. In an experimental study, the independent variable is the one that is manipulated by the experimenter in order to see what effect changes in that variable have on the other variable, which is hypothesized to be dependent upon it. The latter is thus named the dependent variable because its value is thought to depend on, or vary with, the value of the independent variable. In an experiment designed to determine the effectiveness of a new method of teaching reading on the reading achievement of first graders, the method of teaching is the independent variable and the reading achievement would be the dependent variable. In nonexperimental studies the independent variable is not manipulated by the experimenter but is a pre-existing variable which is hypothesized to influence a dependent variable. For example, if social class is hypothesized to influence school attendance patterns, social class is the independent variable and school attendance is the dependent variable.

Another important classification of variables is made on the basis of whether they are *discrete* or *continuous*. A discrete variable is one that can take on only a specific set of values. For example, the number of foreign-born students in a school is a discrete variable. There might be 14 or 15 of these individuals but never $14\frac{1}{2}$. Family size is another discrete variable. A family may be composed of 3, 4, 6, or more people, but values between these numbers would not be possible. Discrete variables may, of course, be either qualitative (sex, marital status, handedness, state of birth) or quantitative (number of books in a library, number of children in a classroom).

A continuous variable is one that can assume any value, including fractional ones, within a range of values. Thus, continuous variables are measured in both whole and fractional units. Age, height, weight, intelligence, achievement test scores, daily caloric intake, and so on, are examples of continuous variables. An individual can be described as $10\frac{1}{2}$ years old and weighing $75\frac{1}{2}$ pounds. Continuous variables are always quantitative in nature. We can think of continuous variables as existing along a continuum from the smallest amount of the variable at one extreme to the largest amount possible at the other end. To go from one unit to the next on a continuous variable, one must pass through a large number of fractional parts. For example, as a child grows in height he does not suddenly grow from 47 to 48 inches or even from 47.0 to 47.1 but rather passes gradually through a succession of points as he grows the one inch.

Measurement of continuous variables is always an approximation of the true value. No matter how accurately one measures it is not possible to measure and record all the possible values of a continuous variable. Therefore, continuous variables are measured to the nearest convenient unit. If a child's height is recorded as 47 inches on his school record chart, this does not mean that he is exactly 47 inches tall. Height measured in inches is regarded as representing an interval from half a unit of measurement below to half a unit above the reported value. A height of 47 inches would be considered as representing an interval ranging from 46.5 to 47.5, but for the school's purposes height reported to the nearest whole inch is precise enough. An IQ of 120 on a school record represents an approximation of a score between 119.5 and 120.5. If an insurance company records a child's age as 10 years, this means that he is somewhere between nine years, six months to one day less than ten years, six months of age. This illustrates the fact that, although continuous variables exist along a continuum, the recorded values of these variables are, in practice, discrete. In other words, definite predetermined points along the continuum are chosen to report the variables. Convenience and lack of refinement in the measuring instruments account for this practice.

In research studies, we often find variables that are inherently continuous being treated as though they were discrete. For example, age is a continuous variable, but subjects in a study may be placed into age categories, such as under 30, 30–50, 50–70, and so on, rather than using actual ages. Test scores may be categorized as pass-fail instead of using the raw scores, or intelligence scores will be categorized simply as high-average-low. On the other hand, we sometimes find discrete variables being treated as though they were continuous. You may have seen statistical studies reporting that the average family in the United States consists of 4.5 people. Of course, a family would not consist of 4.5 people. The average, 4.5 is an index resulting from treating a discrete variable as if it were a continuous one.

Exercises

1. Match each definition with the term on the right.
   (a) a value that is the same for all members of a population
   (b) a characteristic that classifies a subject into a labeled category
   (c) a value that is known to or believed to change as another value changes
   (d) a characteristic that is best described by a number rather than by a verbal label
   (e) a characteristic that can take certain specific values but not intermediate values
   (f) a value that is known or believed to influence another value
   (g) a characteristic that can assume any value within a range of values

   quantitative variable
   constant
   qualitative variable
   independent variable
   discrete variable
   continuous variable
   dependent variable

2. Which variables are discrete and which are continuous?
   (a) length of hair
   (b) incidents of disruptive behavior
   (c) amount of anxiety a person manifests
   (d) number of tables in a college library
   (e) intelligence
   (f) number of students who passed an exam
   (g) reading ability
   (h) items passed on an achievement test
   (i) number of doctors in a town

3. A study investigates the relationship of prior kindergarten experience on the amount of anxiety shown by first graders in the classroom. Identify the (a) discrete variable, (b) continuous variable, (c) independent variable, (d) dependent variable, and (e) constant in this study.

4. Study A investigates the effect of intelligence on the speed of completion of a problem-solving task. Study B investigates the effect of an early stimulating environment on the intelligence of children. In Study A, intelligence is the _____ (independent, dependent) variable. In Study B, intelligence is the _____ (independent, dependent) variable.

12 DESCRIPTIVE STATISTICS

Solutions
1. (a) constant   (b) qualitative variable   (c) dependent variable   (d) quantitative variable   (e) discrete variable   (f) independent variable   (g) continuous variable
2. (a) continuous   (b) discrete   (c) continuous   (d) discrete   (e) continuous   (f) discrete   (g) continuous   (h) discrete   (i) discrete
3. (a) kindergarten experience   (b) anxiety   (c) kindergarten experience   (d) amount of anxiety   (e) grade level
4. (a) independent   (b) dependent

Scales of Measurement

A fundamental step in the conduct of research is measurement. Measurement is the process through which observations are translated into numbers. S. S. Stevens stated, "In its broadest sense, measurement is the assignment of numerals to objects or events according to rules."[1] Researchers begin with variables, then use rules to determine how these variables will be expressed in numerical form. The variable *religious preference* may be measured according to the numbers indicated by students who are asked to select one among (1) Catholic, (2) Jewish, (3) Protestant, or (4) Other. The variable *weight* may be measured as the numbers observed when subjects step on a scale. The variable *social maturity* may be measured as the scores of subjects who have taken the Vineland Social Maturity Scale.

The nature of the measurement process that produces the numbers determines the interpretation that can be made from them and the statistical procedures that can be meaningfully used with them. The most widely quoted taxonomy of measurement procedures is Stevens'[2] "Scales of Measurement" in which he classifies measurement as nominal, ordinal, interval, and ratio.

Nominal Scale

The most primitive scale of measurement is *nominal* measurement. Nominal measurement involves the placing of objects or individuals into categories which are qualitatively rather than quantitatively different. Measurement at this level only requires that one be able to distinguish two or more relevant categories and know the criteria for placing individuals or objects into one or another category. The required empirical operation at

---
[1] S. S. Stevens, "Mathematics, Measurement, and Psychophysics," in S. S. Stevens (ed.), *Handbook of Experimental Psychology* (New York: Wiley, 1951), p. 1.
[2] *Op. cit.*, pp. 1–49.

this level involves recognizing that a given object or individual belongs in a given mutually exclusive category or that it does not. The only relationship between the categories is that they are *different* from each other; there is no suggestion that they represent "more" or "less" of the characteristic being measured. Classifying students according to sex, male or female, would constitute nominal measurement.

Numbers are often used at the nominal level, but only in order to identify the categories. The numbers arbitrarily assigned to the categories serve merely as labels or names. All the members of a category are assigned the same number and no two categories are assigned the same number. For example, in preparing data for a computer, the numeral 0 might be used to represent a male and the numeral 1 to represent a female. There is no empirical relationship among the numbers used in nominal measurement that corresponds to the mathematical relation between the numbers. The 1 does not indicate more of something than the 0. The numbers could be interchanged without affecting anything but the labeling scheme used. The numbers assigned to the categories may not be ordered, added, or divided; however, it is permissible to count the number of cases in each category.

The purist may not regard this process of categorization as measurement at all. Some statistics textbooks do not include nominal procedures in their discussion of levels of measurement. However, if measurement is defined as "the assignment of numbers to objects or events according to rules," then the process of categorization and numbering as described above indicates measurement.

The numbers used in a nominal scale do not represent absolute or relative amounts of any characteristic. They merely serve to identify the members of a given category. For example, the numbers assigned to football players constitute a nominal scale. We would not say that the player with number 48 on his jersey is necessarily a better player than the one with 36 on his jersey. Nor could we say that the difference in playing ability between players with the numbers 40 and 48 is equal to the difference between players with the numbers 50 and 58. The statement that the player with number 36 is three times as good as the player with number 12 is likewise meaningless.

The identifying numbers in a nominal scale can never, of course, be arithmetically manipulated through addition, subtraction, multiplication, or division. One may use only those statistical procedures based on mere counting, such as reporting the number of observations in each category.

Ordinal Scale

The next highest scale of measurement is *ordinal.* In ordinal measurement, one determines the relative position of objects or individuals with respect to some attribute, but without indicating the distance between positions. The essential requirement for measurement at this level is an empirical criterion for ordering objects or events with respect to the attribute—that is, some procedure for determining, for each thing being measured, whether that individual or object has more, the same amount, or less of the attribute in question. Ordinal measurement occurs, for example, when teachers rank students on certain characteristics, such as their social maturity, leadership abilities, cooperativeness, and so on. Students are frequently ranked according to their academic achievement or according to their performance in a music or athletic contest.

In ordinal measurement, the empirical procedure used for ordering objects must satisfy a criterion known as the *transitivity postulate*. This postulate is written: If $(a > b)$ and $(b > c)$, then $(a > c)$; it means that the relationship must be such that if object a is greater than object b, and object b is greater than object c, then object a is greater than object c. Of course, other words may be substituted for "greater than"; these might include "stronger than," "precedes," "has more of some attribute," and so on.

The empirical operation in ordinal measurement involves only direct comparison of the objects or individuals in terms of the extent to which they possess the attribute in question. Thus, when numbers are assigned to the objects, the only information considered is the *order* of the objects. Consequently, the only characteristic of the numbers having meaning is their order. The numbers assigned in ordinal measurement indicate only the order of position and nothing more. Neither the difference between the numbers nor their ratio has meaning. When the numbers 1, 2, 3, and so on, are used in ordinal measurement there is no implication that rank 1 is as much higher than rank 2 as 2 is than 3, and so on. The distance between the child with a rank of 1 and the child with a rank of 2 may be the same, less than, or greater than the distance between children with rankings of 2 and 3. There is simply no basis for interpreting the magnitude of differences between the numbers or the ratio of the numbers. In an untimed footrace, we may know who came in first, second, third, and so on, but we would not know how much faster one runner was than another. The difference between the first and second would not necessarily be the same as the difference between the second and third, or the third and fourth. Neither could one say that the runner who came in second was twice as fast as the runner who came in fourth.

A good example of an ordinal scale is the scale of hardness of minerals. Minerals are arranged according to their ability to scratch one another. If mineral A can scratch mineral B, then mineral A is said to be harder than mineral B. On this basis a diamond is ranked as the hardest, since it can scratch all other known minerals but cannot be scratched by any others. A set of ten minerals ranging in hardness from the softest to the hardest was selected as a standard and assigned the numbers from 1 to 10, with 1 indicating the softest mineral and 10 the hardest. Other minerals are assigned numbers on the basis of the scratch test. Thus, we know the order of hardness of minerals, but we do not know how much harder one mineral is than another. We may not assume that a mineral assigned a value of 4 is twice as hard as a mineral with a value of 2, or that the difference in hardness between minerals 2 and 4 is the same as the difference in hardness between minerals with values 1 and 3.

The arithmetical operations of addition, subtraction, multiplication, and division cannot be used with ordinal scales. The statistics appropriate for an ordinal scale are limited. Since the size of the interval between the categories is unknown, one cannot use any statistical procedure that assumes equal intervals. Statistics that indicate the points below which certain percentages of the cases fall are appropriate with an ordinal scale. Statistics appropriate for the nominal scale may also be used with an ordinal scale.

Interval Scale

An interval scale is one which provides equal intervals from an arbitrary origin. An interval scale not only orders objects or events according to the amount of the attribute they represent, but also establishes equal intervals between the units of measure. Equal differences in the numbers represent equal differences in the attribute being measured. The Fahrenheit and Centigrade thermometers are examples of interval scales.

On an interval scale, both the order and distance relationships among the numbers have meaning. We may assert that the difference between 50 and 51 degrees Centigrade is equal to the difference between 30 and 31 degrees Centigrade. We could not say, however, that 50 degrees is twice as hot as 25 degrees. This is because there is no true zero point on an interval scale. A zero point is established by convention, as in the Centigrade scale which assigns the value 0 degrees to the freezing point of water.

Likewise, the zero point on a psychological or educational test is arbitrary. For example, there is no zero intelligence; there is no way in our standardized intelligence tests to identify an individual of zero intelligence. A student may occasionally receive a score of zero on a statistics test, but this does not mean that he has zero knowledge of statistics. If we had three students who made scores of 15, 30, and 45 on a statistics quiz, we could not say that the score of 30 represents twice as much knowledge of statistics as the score of 15 or that the score of 45 represents three times as much knowledge as the score of 15. To understand the reason why this is so, let us assume that 15 very simple items are added to the quiz so that all three students are able to answer them correctly. The three scores would now become 30, 45, and 60 for the three students. If we attempted to form ratios between the values on this interval type scale, we would mistakenly report that the student with a score of 60 had twice as much knowledge of statistics as the student with a score of 30, whereas in the earlier ratio we had incorrectly assumed that the same student had three times as great a knowledge of statistics as the other student.

Thus, because the zero is arbitrary, multiplication and division of the numbers are not appropriate; as we have seen, ratios between the numbers on an interval scale are meaningless. However, the difference between positions on an interval scale may be reported or the numbers may be added. Any statistical procedures based on adding may be used with this level scale along with the procedures appropriate for the lower level scales. These include most of the common statistical procedures.

In the measurement of certain psychological variables, such as achievement, intelligence, attitudes, and personality traits, one is not always certain about the equality of the intervals. It is often necessary to make assumptions about the variable and the meaning of the numerical scores obtained from the measurements. A close correspondence between the units of measurement and the underlying attribute is assumed in the better standardized intelligence and achievement tests. For example, it is conventional to assume that a given intelligence test score represents a standard amount of intelligence; that is, a score of 130 represents more intelligence than a score of 125. It is also assumed that the intervals between score categories are equal; that is, the difference between 125 and 130 is equal to the difference between 120 and 125. In teacher-made achievement tests, however, where two students may receive the same score by passing items of widely varying difficulty levels, one may not be dealing with interval measurement at all. Consider a spelling test with the words: dog, book, tree, pneumatic, psoriasis, xanthophyll. The distance on the score scale between 1 and 3 correct and between 3 and 5 correct is the same. However, when considered in terms of spelling ability, the difference between 3 and 5 correct suggests a greater difference in ability than does the difference between 1 and 3 correct. Unless one can say that the distance between 3 and 5 on the spelling test represents the same amount of spelling ability as does the distance between 1 and 3, then these scores indicate only the rank order of the students and as such represent ordinal and not interval measurement. In the measurement of psychological variables, it is necessary to aim at the closest possible correspondence between the intervals observed between the scores and intervals of the attribute being measured.

Ratio Scale

A ratio scale, the highest type, is one which provides a true zero point as well as equal intervals. Ratios can be formed between any two given values on the scale. A yardstick used to measure length in units of inches or feet is a ratio scale, for the origin on the scale is an absolute zero corresponding to no length at all. Thus, it is possible to state that a stick 6 feet long is twice as long as a stick 3 feet long. With a ratio scale, it is possible to multiply or divide each of the values by a certain number without changing the properties of the scale. For example, we can multiply 2 pounds by 16 to change the unit of measurement to 32 ounces or we can multiply 6 feet by 12 to change the unit to inches. We can multiply and maintain the same ratio as before the multiplication. For example, we can multiply 4 quarts of milk and 2 quarts of milk by 2 and change the unit of measurement to pints. In pints, 8 pints is still twice as much as 4 pints.

Ratio scales are found primarily in the physical sciences. Few educational and psychological characteristics can be measured as ratio scales. Few measures in education and psychology can be interpreted as ratio scales. While we may say that a person 6 feet tall is twice as tall as a person 3 feet tall, we cannot say that a person with an IQ of 150 is twice as intelligent as a person with an IQ of 75.

All types of statistical procedures are appropriate with a ratio scale.

Exercises

1. Match the following examples with the appropriate measurement level listed on the right:
    (a) Jennifer is the second most popular girl in the class   nominal
    (b) A coding procedure: blue eyes = 1, brown eyes = 2   interval
    (c) A wrestler weighs 110.5 pounds   ratio
    (d) Jennifer was born in 1967 and Mary was born in 1973   ordinal

2. What kind of scale is used in each of the following statements?
    (a) *Recipe #21, Dutch chocolate cake*, is baked at (b) *350 degrees* for (c) 30 minutes. It requires (d) *more* chocolate than recipe #20 but less than recipe #22.

3. Identify the measurement indicated in each of the following examples as nominal, ordinal, interval, or ratio.
    (a) Classifying female students as blondes, brunettes, or redheads for a study of personality characteristics.
    (b) Recording the heights of kindergarten students.
    (c) Obtaining reading readiness scores for kindergarten students.
    (d) Ranking students in a music contest.
    (e) Classifying elementary students according to their eating habits as: (1) "hot lunchers," (2) "sack lunchers," (3) "goes home for lunch."

Solutions

1. (a) ordinal (b) nominal (c) ratio (d) interval
2. (a) nominal (b) interval (c) ratio (d) ordinal
3. (a) nominal (b) ratio (c) interval (d) ordinal (e) nominal

SUMMARY

Statistics serve many useful purposes. Statistical procedures help us reduce vast quantities of information to manageable form (descriptive statistics) and to reach reasonable decisions with limited information (inferential statistics). In the statistician's vocabulary population is used to refer to an entire aggregate of people, things, events, behaviors, and so on, all of which have a given characteristic in common. Sample is used to refer to a selected subgroup of the population. Samples must be representative of the population in order that the researcher may have reasonable confidence in the generalizations which he makes about the population.

Scientists are interested in discovering the relationships between variables. Variables are characteristics which show variation from individual to individual or object to object. Variables may be classified as: (a) quantitative or qualitative; (b) discrete or continuous; and (c) independent or dependent.

Variables and the relationship between them may be more precisely described through the process of measurement. Measurement is defined as the assignment of numbers to objects or individuals according to some rule. Measurement is possible because of the correspondence between the empirical relations among the objects to be measured and properties of the numbers used. On the basis of the empirical relationships and the rule for assigning numbers, four different scales of measurement may be distinguished: nominal, ordinal, interval, and ratio. Flowchart 1 summarizes the steps for distinguishing these four scales. The levels differ in the interpretation that can be made of the numbers used.

Nominal scales categorize observations into groups. Numbers are employed merely as labels. Ordinal scales not only categorize observations but place them in rank order on the basis of some quantifiable characteristic. Numbers are used to indicate rank order only; one must not assume that the equally spaced numbers represent equal differences in the attribute being measured. Interval scales have equal intervals. They not only allow the researcher to categorize and rank order his observations, but they also allow him to assess the magnitude of difference between observations. Interval numbers are used to represent the degree to which a given characteristic is present, but since this scale is lacking an absolute zero point, multiplication and division are inappropriate, although addition and subtraction are justified. Ratio scales have both equal intervals and absolute zero points. Numbers in this type of scale represent the highest level of measurement. Ratio measures are rare in educational and psychological research.

There are different methods of statistical analysis available which are appropriate for the data derived from the four different scales of measurement. We discuss this more extensively in subsequent chapters.

FLOWCHART 1   SCALES OF MEASUREMENT

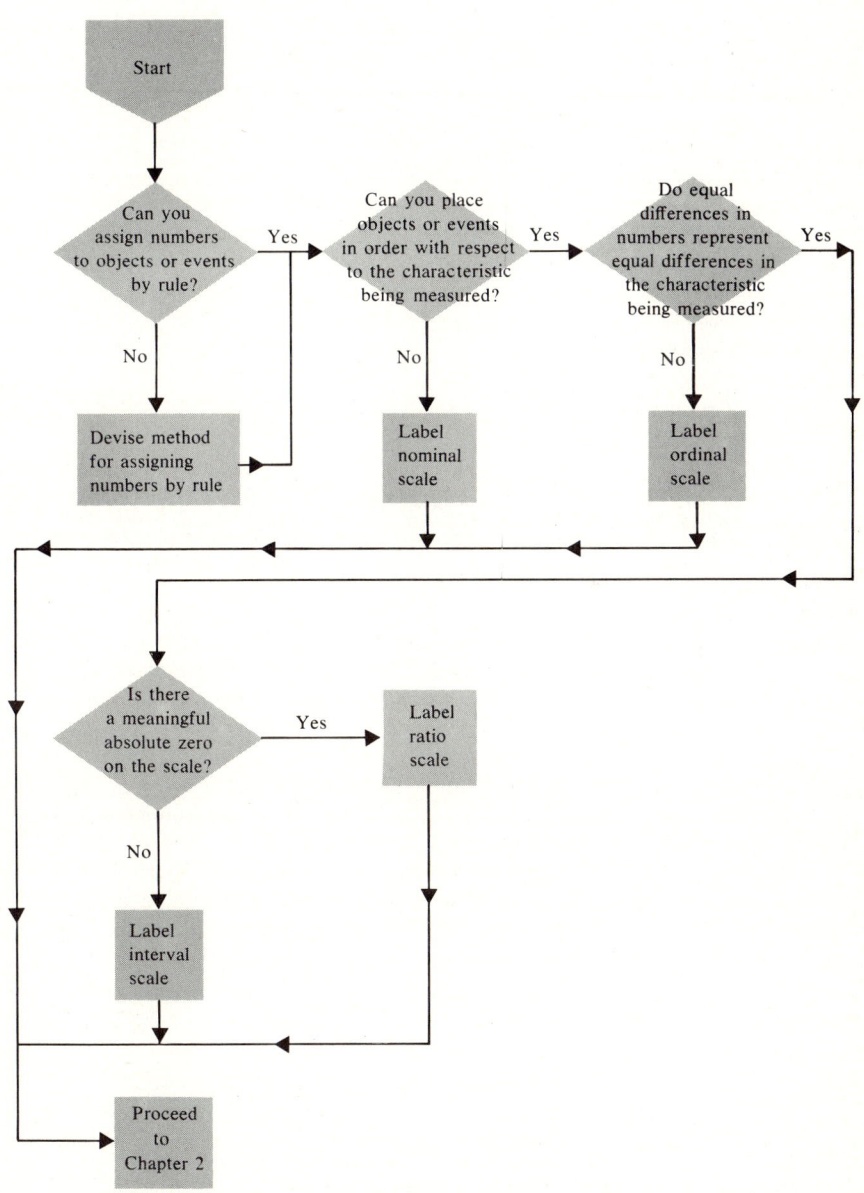

EXERCISES

Questions 1 to 7 concern the same narrative. Use the keys preceding each paragraph to indicate which term best describes each italicized phrase in the text.
Questions 1–3:
   (a) frequency distribution
   (b) population
   (c) sample
   (d) quantitative variable
   (e) qualitative variable

A head nurse suspected that (1) *the 200 nurses in her hospital* were getting rusty on antiseptic procedures. She did not want to give an exam to all 200 of them in order to test this hypothesis. She instructed her secretary to place every nurse's name on a slip of paper, place these slips in a bowl, mix well, and blindfolded, draw 30 names from the bowl. (2) *The 30 nurses drawn* were given a test on antiseptic procedures. (3) *The scores on the test* represented the number of items answered correctly.

Questions 4–5:
   (a) statistic
   (b) parameter

The head nurse would like to know (4) *the typical examination performance of all her nurses.* She can only compute the (5) *average performance of the 30 nurses who have taken the exam.*

Questions 6–7:
   (a) independent variable
   (b) dependent variable

She might ask each nurse in her sample to indicate (6) *how many years have elapsed since the completion of this training* to see if this is related to (7) *scores on the test.*

8. Statistics are useful in (i) organizing, summarizing, interpreting, and communicating quantitative information and (ii) enabling the researcher to go beyond the data gathered from a sample to reach tentative conclusions about a population from which the sample was drawn.
   (a) Which statement describes the purpose of inferential statistics?
   (b) Which statement describes the purpose of descriptive statistics?

9. What kind of measurement scale is indicated in each of the following statements? (i) nominal (ii) ordinal (iii) interval (iv) ratio
   (a) John finished the math test in 35 minutes, while Jack finished the same test in 25 minutes.
   (b) Jack speaks French, but John does not.
   (c) Jack is taller than John.
   (d) John is 6 feet, 2 inches tall.
   (e) John's IQ is 120, while Jack's IQ is 110.

10. Distinguish between the following and give examples of each:
    (a) sample and population
    (b) statistic and parameter

(c) independent and dependent variable
(d) continuous and discrete variables
11. A phone company serves one million households. The president of the company wishes to know the extent to which the company's customers are satisfied with their service. A consultant suggests that the company contact a random sample of clients listed in the phone directories to obtain their opinion about the services. Indicate the population under consideration and the sample. Why is sampling useful in this situation?
12. Classify the following as discrete or continuous variables.
    (a) the temperature during the month of May.
    (b) the number of children who ride the bus at Central Elementary School
    (c) the distance a car can travel on a gallon of gasoline
    (d) the number of books a student checks out of the library during a six-week grading period
13. Indicate whether each of the following is a parameter or a statistic.
    (a) the average SAT score of all freshmen students at State University
    (b) the average SAT score of 100 students whose names were drawn from a list of all freshmen at State University
    (c) the proportion of students in the entire student body at North High School who favor an open-campus policy
    (d) the average height of 25 females whose names are selected from a list of all female students enrolled in the senior class at North High School
14. Given the scale of measurement involved, which of the following statements is/are legitimate?
    (a) Mary made 90 on the mathematics test. Thus she knows twice as much math as John, who scored 45.
    (b) There are 25 students in category 1 (right-handed) and 5 students in category 2 (left-handed).
    (c) John, the heaviest boy in the class, weighs twice as much as Mary, the smallest girl.
    (d) Benford's band ranked second in the state music contest. Central's band ranked fourth. We can conclude that Benford's band is twice as good as Central's.
    (e) Paul finished second in a race; John finished fourth; and Mark finished sixth. The difference between John's performance and Paul's is the same as the difference between John's performance and Mark's.

ANSWERS

1. (b) 2. (c) 3. (d) 4. (b) 5. (a) 6. (a) 7. (b) 8. (a)—(ii) (b)—(i)
9. (a)—(iv), ratio (b)—(i), nominal (c)—(ii), ordinal (d)—(iv), ratio (e)—(iii), interval
10. (a) A *population* refers to an entire group (*all*) of people or objects having one or more characteristics in common. In research, it commonly refers to the entire group in which one is interested. For example, all the graduating seniors at State University.

22 DESCRIPTIVE STATISTICS

A *sample* is a subgroup drawn from the entire population. Most research deals with samples rather than populations. For example, a group of graduating seniors selected from *all* of the seniors.

(b) A *statistic* is a numerical characteristic of a sample. For example, the proportion of graduating seniors in the sample who had obtained a job. A *parameter* is a numerical characteristic of a population. For example, the proportion of *all* graduating seniors at State University who had obtained a job.

(c) The *independent variable* is the one that is manipulated by the experimenter in order to determine its effect on another variable called the *dependent variable* The value of the dependent variable varies with the values of the independent variable. For example, a method of teaching could be an independent variable; student achievement the dependent variable.

(d) A *discrete* variable is one that can take on only certain values. A *continuous* variable can assume any value including fractional ones. For example, number of children is a discrete variable; their age, height, weight, and so on, are continuous variables.

11. The *population* is the one million households who are customers of the phone company. The sample is the smaller number (50,000) selected from one million. Sampling would be useful because it would be too expensive and time-consuming to contact every customer. The opinions of 50,000 would be sufficient to determine the extent of customer satisfaction.

12. (a) continuous (b) discrete (c) continuous (d) discrete

13. (a) parameter (b) statistic (c) parameter (d) statistic

14. (a) Not legitimate, the scale is interval so multiplication and division are not justified.

(b) Legitimate, counting cases within categories is appropriate with a nominal scale.

(c) Legitimate, multiplication and division are justified with a ratio scale.

(d) Not legitimate, multiplication and division are not justified when the scale is ordinal.

(e) Not legitimate, with an ordinal scale the distance between ranks cannot be assumed to be equal.

CHAPTER TWO
FREQUENCY DISTRIBUTIONS AND GRAPHS

OBJECTIVES

Chapter Two deals with procedures that are used for organizing and presenting data. After completing this chapter, the reader should be able to:

1. State the purpose of the frequency distribution in statistics.
2. List the advantages and disadvantages of the grouped frequency distribution as compared with the ungrouped frequency distribution.
3. Construct both an ungrouped and a grouped frequency distribution for a given set of data.
4. Distinguish between the real and the apparent limits of a class interval in a grouped frequency distribution.
5. Convert a frequency distribution into a cumulative frequency distribution or a cumulative percentage distribution.
6. List the types of graphs used to present frequency distributions and the situation in which each is preferred.
7. Construct a histogram and a frequency polygon for a given frequency distribution.
8. Construct a cumulative frequency polygon and a cumulative percentage polygon (ogive) from given data.
9. Define skewness and correctly label curves as positively or negatively skewed.
10. Recognize data that might logically be expected to result in normal or skewed distributions.
11. Label curves as unimodal, bimodal, or multimodal.
12. Recognize ways that graphs may be drawn to distort data.

Purpose of Frequency Distribution(s) and Graphs

The teacher or researcher often is confronted with large masses of data, usually scores of some type, which require interpretation if they are to be useful. For example, the scores in Table 2.1 are hypothetical reading scores for 100 fourth grade students in an elementary school. The scores are listed according to the order in which the test papers were scored.

TABLE 2.1 Reading Scores of 100 Fourth Grade Students

91	95	78	60	59	99	54	90	94	77
75	79	72	85	84	74	60	74	78	71
75	98	80	83	82	95	62	74	97	79
82	72	78	66	65	61	78	81	71	77
72	67	74	84	83	66	74	71	66	73
89	93	76	93	93	81	83	88	92	75
98	86	83	75	75	80	87	97	85	82
84	94	87	82	55	77	82	93	82	83
89	86	69	90	60	97	96	86	85	68
72	92	82	81	56	65	59	71	90	80

It would be difficult to make any "sense" of all these scores beyond possibly identifying the two extreme scores in the group. The task of interpretation would be an even more formidable one if there were several hundred scores to consider. Large collections of scores need to be arranged in some systematic way so that generalizations can be reached concerning the characteristics of the group of data represented. It is basically for this purpose of summarizing data that descriptive statistics are employed by the teacher or researcher. Methods are available for presenting data in compact, yet comprehensible and accurate form so that the essential characteristics of the data are readily discernable. One of the most efficient ways to summarize and organize data is to arrange them into what is called a *frequency distribution*.

Frequency Distributions

A frequency distribution is an organization of measures or observations into score classes along with the frequency in each class. A frequency distribution presents data in a concise and orderly form that is easy to interpret. Table 2.2 presents the scores from Table 2.1 arranged into a frequency distribution. Notice that the frequency distribution includes two columns: a column labeled X which lists all the scores in order of size from

TABLE 2.2 Frequency Distribution of 100 Reading Test Scores

X	f	X	f	X	f
99	1	83	5	67	1
98	2	82	7	66	3
97	3	81	3	65	2
96	1	80	3	64	0
95	2	79	2	63	0
94	2	78	4	62	1
93	4	77	3	61	1
92	2	76	1	60	3
91	1	75	5	59	2
90	3	74	5	58	0
89	2	73	1	57	0
88	1	72	4	56	1
87	2	71	4	55	1
86	3	70	0	54	1
85	3	69	1	Total (N) = 100	
84	3	68	1		

highest to lowest and a column labeled f which gives the frequency of occurrence of each score. In preparing a frequency distribution, we first list all the possible scores and then go through the collection of scores making a tally mark next to each score in the list every time it occurs. Table 2.3 illustrates this procedure. To prepare the final form as shown in Table 2.2 we simply total the tally marks to show the frequency of occurrence of each score. Finally, add the frequency column to show the total number of cases (N).

It is easier to interpret the data in a frequency distribution than it was with the unorganized collection of scores. We can see the highest and lowest scores easily and can also identify the score which occurs the greatest

TABLE 2.3 Procedure Followed in Making a Frequency Distribution

X	TALLY MARKS
99	1
98	11
97	111
96	1
95	11
94	11
93	1111

number of times; in Table 2.2 the most frequent score is 82. Other frequently occurring scores are also obvious, such as 83, 75, and 74. In addition, we can observe how the scores are distributed along the entire scale; that is, it can easily be determined whether the scores are distributed uniformly or whether gaps exist at certain points.

Table 2.2 still involves considerable detail since it lists every possible score value between 54 and 99, the highest and lowest scores, even including scores that have a frequency of zero. These scores, such as 64 and 70, are included in order to make the list of scores consecutive as well as to communicate that there were no scores at these values. Since all possible score values are listed one at a time, the distribution in Table 2.2 is called an *ungrouped frequency distribution*. When there is a wide range of scores, the ungrouped frequency distribution may become a cumbersome way of presenting the data. In such a situation, instead of listing each possible score separately, we may condense the data by setting up intervals which contain a range of possible scores. When score values are grouped to form intervals of scores called *class intervals*, the resulting frequency distribution is known as a *grouped frequency distribution*.

Exercises

1. Prepare an ungrouped frequency distribution for the following scores: 8, 10, 11, 6, 8, 10, 8, 5, 14, 13, 7, 6, 9, 10
    (a) How many score values must be represented in the distribution?
    (b) What is N in this distribution?
2. What is the purpose of a frequency distribution?

Solutions

1. | X | f | | X | f |
   |----|---|---|---|---|
   | 14 | 1 | | 9 | 1 |
   | 13 | 1 | | 8 | 3 |
   | 12 | 0 | | 7 | 1 |
   | 11 | 1 | | 6 | 2 |
   | 10 | 3 | | 5 | 1 |

   (a) 10
   (b) 14
2. To organize and summarize data.

Grouped Frequency Distributions

Grouping scores into class intervals each containing several score values is frequently done when one has a large number of cases spread out over a large range. As a rule-of-thumb, grouping is done when the range (the difference between the highest and lowest score) is 20 or greater. Instead of dealing with the individual scores, the score values are assigned to mutually exclusive and exhaustive classes defined in terms of the particular grouping interval chosen. All values lying within a particular range of score possibilities are grouped together to form a single class interval. For example, if we set up a class interval 81–83, then all scores originally recorded as 81, 82, 83, would be grouped in that interval. The classes are said to be mutually exclusive and exhaustive because all of the scores must fall into one and only one class. Table 2.4 shows the data from Table 2.1 arranged into a grouped frequency distribution.

TABLE 2.4 Grouped Frequency Distribution of Reading Scores

CLASS INTERVAL	f	CLASS INTERVAL	f
99–101	1	75–77	9
96–98	6	72–74	10
93–95	8	69–71	5
90–92	6	66–68	5
87–89	5	63–65	2
84–86	9	60–62	5
81–83	15	57–59	2
78–80	9	54–56	3
			$N = 100$

In Table 2.4 we have a more concise and a more easily interpreted picture of the distribution of reading scores for these 100 students than we had in Table 2.1. It is readily apparent that the greatest concentration of scores is from 72 through 86; in fact, over half of the 100 scores fall between these two scores. The most frequently occurring interval was 81–83; thus, we could say that a large number of scores occur in the low 80's. The scores are scattered throughout the distribution but there is some decrease in frequency at the extremes.

Although the grouped frequency distribution provides an economical and easily communicated method of presenting data, nevertheless some information is lost as a result of the grouping process. For example, from the grouped frequency distribution, we cannot tell exactly how many students received a score of 75; we can see only that nine students received scores in the class interval 75–77, which contains the score of 75. Or it can be seen that one person scored in the class interval 99–101, but we cannot determine the exact score of that person. Individual scores lose their identity in the

28 DESCRIPTIVE STATISTICS

grouping. As a result, some error will inevitably be involved in the statistics calculated on grouped scores. This is because calculations must be made on the basis of the class interval rather than the value of the individual scores. In most cases, however, such error (called grouping error)[1] is slight and does not present a serious drawback to grouping.

Grouping into Class Intervals The first question that arises when organizing data into a grouped frequency distribution is "How does one know how many intervals to use?" The choice of the number of intervals is a somewhat arbitrary matter, however there are some considerations that should be kept in mind when making this decision. On the one hand, if there are more than 20 intervals, one still has a cumbersome and uneconomical arrangement of data. The original purposes of grouping are defeated. Furthermore, it may be difficult to ascertain the characteristics of the distribution when the number of intervals is very large. For example, when we look at Table 2.2 with its 46 class intervals it is difficult to get a mental picture or "feel" of the distribution of the scores. In Table 2.4 with 16 class intervals we can see the general pattern of the distribution much more clearly.

On the other hand, if too few class intervals are used, then each interval must contain a greater number of scores and there is a greater loss of information about the original individual scores with a resulting loss in accuracy in any calculated statistic. For example, we could group the scores in Table 2.1 into two intervals—those below 80 and those above 80—but the loss of information about the original scores would be great and we would have a very incomplete picture of their distribution. Thus, we want to choose the number of class intervals that would serve the purpose of convenience, compactness, and economy of presentation while at the same time avoiding a loss of information about the identity of the original individual scores. In other words, we should strive for a balance between presentation of sufficient information and the need for presenting the data in a compact, comprehensible form. There is a rule-of-thumb that suggests the appropriate number of class intervals for most data in the behavioral sciences is between 10 and 20 with 15 being a popular number. If there are fewer than 10 intervals, one runs the risk of losing too much information and increasing the error in calculations; if there are more than 20 intervals, one still has a bulky arrangement of data whose characteristics may not be readily apparent. Furthermore, the original advantages of grouping, economy, and convenience, are lost.

[1] *Grouping error is defined as the difference between the value of a statistic when computed from a grouped frequency distribution and the corresponding value when computed from the original raw score.*

To determine how many scores to include in a class interval (interval size), one first subtracts the lowest score from the highest score to find the range of the scores. For example, in Table 2.4 where the highest score is 99 and the lowest score is 54, the range is 45. Next the range is divided by 15, the recommended number of intervals; the resulting quotient is called the interval size, i. It indicates the number of scores to be placed in each interval. In our example we get exactly 3 when the range of 45 is divided by 15; thus, $i = 3$. Usually the quotient is not a whole number; however, in such cases, the quotient is generally rounded to the nearest odd number. If the range in our example had been 61, then the quotient 4.07 would be rounded to 5 (the nearest odd number) for the class interval size. If the range had been 102, an interval size of 7 would be recommended.

The advantage of an odd number for the interval size is that, in such a case, the center or midpoint of each class interval, which is often used to stand for all the scores, will be a whole number. If the interval size is an even number, the midpoint will be halfway between two numbers, that is, a fraction. For example, in Table 2.4 where $i = 3$, the midpoint of the interval 54–56 is 55. If $i = 4$ and the interval is 54–57, then the midpoint would be 55.5. An exception to the general rule of using an odd number for class interval size is when the range is between 20 and 30. In such cases, a class interval size of two is used since an interval of one would yield too many classes and three would yield too few. When the range is below 20 there is no advantage to grouping the data.

Having established the class interval size, the next step is to identify the classes into which the data would be grouped. Although it is completely arbitrary, the authors recommend that the class intervals be started at a value that is a multiple of the size of the interval, that is, the lower limit of the interval should be a multiple of i. For example, if the size of the interval is 3, then the intervals are started with some multiple of 3 such as 9, 12, 15, 18, and so on. With a class interval size of 5, the intervals are started with some multiple of 5 such as 10, 15, 20, 25, and so on.

Of course, the lowest class interval is started at a value which would include the lowest observed score. In our example the lowest observed score is 54. Since 54 is a multiple of 3, we begin the lowest interval with score 54. Thus, the lowest class interval would contain score values from 54–56, the next interval from 57–59, the next from 60–62, and so on. Notice that the next higher interval begins with the integer following the maximum score of the interval below it. Successively higher class intervals are set up until the highest score has been included. Tally marks are made for each score in the appropriate interval and these are then totaled to indicate the frequency for each interval. The frequency column is then added to obtain the total number of scores, which should correspond to the number of raw scores. Table 2.4 shows the completed product.

Exercise

1. A grouped frequency distribution is shown below. Identify (a) the number of class intervals shown, (b) the size of the interval, (c) the midpoint of the interval with the highest frequency, (d) the total N.

X	f	X	f	X	f
48–50	1	33–35	3	18–20	5
45–47	0	30–32	7	15–17	5
42–44	1	27–29	8	12–14	4
39–41	2	24–26	10	9–11	3
36–38	2	21–23	6	6–8	4

Solution

1. (a) 15 (b) 3 (c) 25 (d) 61

Real Limits of a Class Interval At first glance the interval 54–56 in Table 2.4 may seem to include only two score values, but actually it includes three: 54, 55, 56. Recall from Chapter 1 that with continuous data, a score is thought of as representing a midpoint in a range or interval of possible values and the value of the midpoint is used for recording purposes. When a child's height is recorded as 61 inches, this represents 60.5 to 61.5 inches recorded to the nearest whole inch. When a filling station attendant tells a motorist that the distance between two towns is 78 miles, he really means between 77.5 and 78.5 miles rather than exactly 78 miles. Even when fractional measures are not possible it is useful to think of each possible score as representing an interval rather than a discrete point. A student who receives a score of 51 on an achievement test can be considered as having a score somewhere between 50.5 and 51.5. Achievement is a continuous variable, but our test only measures it in whole numbers. If a more detailed measuring instrument has been used the score obtained would have fallen within the limits 50.5–51.5.

The same is true for values arranged in a grouped frequency distribution. Thus, we write down the limits for the various class intervals, as for example 75–79, but these are only the *apparent limits*. The class interval 75–79 actually includes all values from 74.5 to 79.5 while the interval 80–84 includes all scores between 79.5 and 84.5, and so on. It is important to remember this distinction between the apparent limits and the *real limits* of the interval, which correspond to values .5 above and below the recorded limits. The class interval size is actually the difference between the real limits of the interval rather than the difference between the apparent limits. For example, in the class interval 75–79, the interval size is $79.5 - 74.5 = 5$, whereas if you subtract the apparent limits 79–75, you would get 4 which does not represent the interval size. We know that there are five scores included in that interval: 75, 76, 77, 78, and 79.

Although it is helpful to know how to arrange data into a grouped frequency distribution, we will not discuss in this text the calculation of the various statistics from grouped data. Older textbooks taught these methods because they represented a saving of time and effort when calculating statistics from a mass of data. However, since the development of inexpensive calculators and electronic computers, calculations can be performed easily with the original data. This has the added, important advantage that there is no loss of information about the original values. Calculations with grouped data involve a loss of accuracy which we no longer need to tolerate.

Thus far in our discussion of real limits we have considered cases where the unit of measurement was one. This is usually but not always the case. For example, most school achievement tests yield grade equivalent scores with a unit of measurement of one tenth of a school year. A score of 5.4 represents 5.4 school years and 5.8 represents 5.8 school years. In such cases the real limits are located midway between adjacent scores. For example, the real limits of a grade equivalent score of 5.7 are 5.65 to 5.75.

Exercises

1. In a set of scores, the lowest score is 14 and the highest score is 62.
   (a) What is the range?
   (b) What size class interval should be used for grouping the data?
   (c) What are the apparent limits of the lowest class?
   (d) What are the real limits of the highest class?
2. (a) With the lowest score equal to 4 and the highest score equal to 30, what is the range?
   (b) What size class interval should be used for grouping the data?
   (c) What are the real limits of the lowest class?
   (d) What are the real limits of the highest class?
3. List the class intervals that would be used in a grouped frequency distribution where the raw scores range from 22 to 89.
4. In the above frequency distribution (Exercise 3), (a) what is the midpoint of the lowest interval; (b) what are the apparent limits of the highest interval?

Solutions

1. (a) $62 - 14 = 48$
   (b) $48 \div 15 = 3.2$, rounded to 3
   (c) $12 - 14 =$ interval is 3; begin interval with a multiple of 3
   (d) 59.5–62.5

2. (a) 30 − 4 = 26
   (b) two, since the range is between 20 and 30
   (c) 3.5–5.5
   (d) 29.5–31.5
3. 85–89   60–64   35–39
   80–84   55–59   30–34
   75–79   50–54   25–29
   70–74   45–49   20–24
   65–69   40–44
4. (a) 22   (b) 85–89

Cumulative Frequency Distributions

For certain statistical purposes, one is interested not in the frequency occurring within individual class intervals, but rather in the number of scores in the distribution which fall below particular values. Such information can be shown easily in a *cumulative frequency distribution* which is prepared by a process of adding successively, from the bottom, the number of cases (f) in each class interval. Thus, the cumulative frequency (cf) of any interval is the *total* number of scores in the distribution which are lower in value than the upper real limit of that interval; that is, the cf of an interval indicates the number of scores up to and including that interval. Table 2.5 shows a cumulative frequency distribution of 50 arithmetic scores.

TABLE 2.5 Cumulative Frequencies for a Distribution of 50 Arithmetic Scores

CLASS INTERVAL	FREQUENCY (f)	CUMULATIVE FREQUENCY (cf)
99–101	1	50
96–98	3	49
93–95	4	46
90–92	3	42
87–89	2	39
84–86	5	37
81–83	7	32
78–80	5	25
75–77	4	20
72–74	5	16
69–71	3	11
66–68	2	8
63–65	1	6
60–62	2	5
57–59	1	3
54–56	2	2
	$N = 50$	

Column 3 in Table 2.5 shows the cumulative frequencies. The cumulative frequency of the fourth class interval 63–65 was found by successively adding the frequencies: $2+1+2+1$, which indicates that a total of six scores were lower than the upper real limit of that interval, which is 65.5. The cumulative frequency of the interval 66–68 was found by successively adding $2+1+2+1+2$ to get 8. Note that the topmost entry in the cumulative frequency (cf) column is always equal to N, the total number of cases. If you do not come out with N as the last entry in the (cf) column, you will know that an error has been made in adding the frequencies.

Cumulative Percentage Distributions

Sometimes it is useful to show the percent of scores that fall below certain values. For this purpose the cumulative frequencies are converted into cumulative percents by dividing each cumulative frequency (cf) by N, the total number, and multiplying by 100 ($cf/N \times 100$). The fourth column in Table 2.6 shows the cumulative percents based on the data from Table 2.5. The cumulative percent for interval 66–68, for example, was found by

TABLE 2.6 Cumulative Percentages for the Distribution of 50 Arithmetic Scores

CLASS INTERVAL	FREQUENCY	cf	CUMULATIVE PERCENT
99–101	1	50	100
96–98	3	49	98
93–95	4	46	92
90–92	3	42	84
87–89	2	39	78
84–86	5	37	74
81–83	7	32	64
78–80	5	25	50
75–77	4	20	40
72–74	5	16	32
69–71	3	11	22
66–68	2	8	16
63–65	1	6	12
60–62	2	5	10
57–59	1	3	6
54–56	2	2	4
	$N = 50$		

34 DESCRIPTIVE STATISTICS

dividing 8(cf) by 50(N) and multiplying by 100 (8/50·100). The advantage of the cumulative percentage distribution is that it readily provides information about the percent of scores falling below given score values. It is generally more meaningful to know the percent of a group below certain points than it is to know just the number. It can be seen in Table 2.6 that 50 percent of the cases fall below 80.5, the upper real limit of the interval 78–80; 84 percent of the cases fall below 92.5. Notice that the top entry in the cumulative percent column is 100. This will always be 100, since all cases will fall below the upper real limit of the highest interval.

Exercise

1. A portion of a frequency distribution is shown below with the lowest interval 30–31. Add a cumulative frequency (cf) and a cumulative percent column to the f distribution.

X	f	cf	c %
38–39	3		
36–37	5		
34–35	3		
32–33	4		
30–31	10		

$N = 25$

Solution

1.

X	f	cf	c %
38–39	3	25	100
36–37	5	22	88
34–35	3	17	68
32–33	4	14	56
30–31	10	10	40

Graphing Frequency Distributions

After organizing data into frequency distributions, it is often helpful to present the data graphically. Graphs communicate the essential characteristics of a frequency distribution in pictorial form so that one can readily identify these characteristics and can compare one frequency distribution with another. Graphs of frequency distributions have as essential characteristics: (1) one axis which represents all the possible classes or scores and

(2) the other axis which represents the frequency of occurrence of these possible scores.

There are two common ways of representing frequency distributions graphically: the *histogram* and the *frequency polygon.* In a histogram, the frequencies of each score or class are represented by areas in the form of vertical bars. Figure 2.1 illustrates a histogram of the data in Table 2.4.

Notice the characteristics of the histogram as illustrated in Figure 2.1:

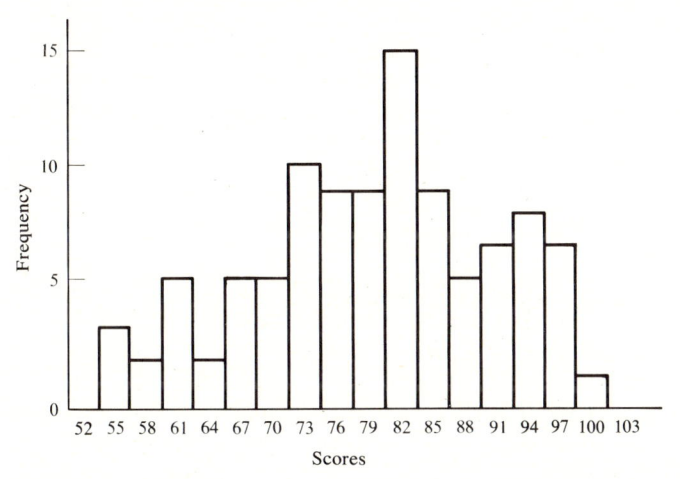

Figure 2.1 Histogram of 100 Reading Scores (Data of Table 2.4)

1. The horizontal axis or *abscissa* represents the score possibilities. These may be single scores or class intervals. If the data are grouped, the abscissa is usually marked off by the midpoints of the class intervals (see above) but sometimes the limits of the intervals are used instead. It is customary to start with the lowest values on the left and proceed to the right with as many intervals as are necessary to include all the scores. It is not necessary to extend this axis to zero, unless scores of zero or near zero actually have been observed. By convention, an empty interval should be left at the lower end of the scale and also at the upper end to show that the frequency reaches zero in those intervals.
2. The vertical axis or *ordinate* represents the frequencies. The vertical axis is marked off by beginning with zero at the bottom and proceeding upward to the greatest frequency.
3. The selection of the distance along either axis to serve as a scale unit is arbitrary. However, it is conventional among statisticians when graphing frequency distributions to follow the "three-quarter-high rule" which states that the vertical axis should be laid out so that the height of the maximum point (highest frequency) is approximately three quarters of the length of the horizontal axis. Following this convention has some

36 DESCRIPTIVE STATISTICS

aesthetic advantage as well as the advantage of preventing personal biases from influencing the choice of the relative unit size of the abscissa and the ordinate.

4. A bar or rectangle is raised above each score interval on the horizontal axis. The width of the bar extends from the lower real limit to the upper real limit of the interval. Its vertical height corresponds to the frequency within that interval. Successive bars are placed adjacent to each other to show the continuity of the scores in continuous data. If there is an interval where the frequency is zero, then an empty space should be left.

5. The vertical axis should be labeled f, or frequency, and the horizontal axis labeled to show what is being measured (scores, height in inches, weight in pounds, reaction time in seconds, and so on). In addition, a descriptive title, indicating what the graph is showing, is always placed below the graph. (See Figure 2.1.)

Exercise

1. Construct a histogram based on the data in the following frequency distribution.

X	f
15–17	1
12–14	5
9–11	2
6–8	4
3–5	3

Solution

1.

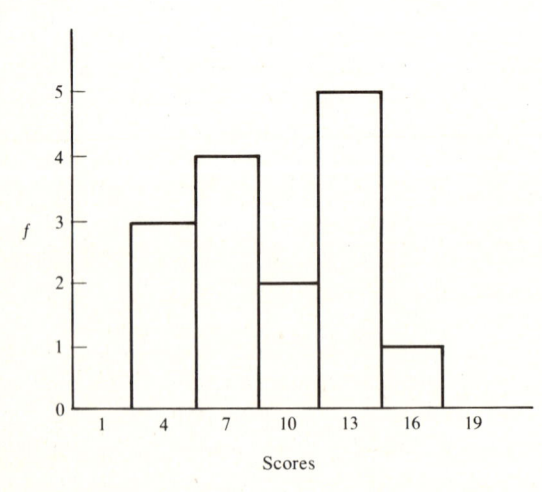

Frequency Polygons

Another method used to represent a frequency distribution graphically is the *frequency polygon.*

To prepare a frequency polygon, we mark off the axes exactly as we did for a histogram. However, instead of drawing a bar above each class interval to represent its frequency, a point is plotted on the graph directly above the midpoint of each interval and at a height corresponding to the frequency of that interval. The points are connected by straight lines and the result is a polygon, from which the graph derives its name.

It is customary to plot an additional interval at the top and bottom of the distribution so that the polygon line can be brought down to the horizontal axis (point of zero frequency) at the midpoint of the interval below the lowest interval and also at the midpoint of the interval placed above the highest one in the distribution. If this attachment to base at each end is not carried out, the figure will not be complete. Figure 2.2 shows a frequency polygon based on the data of Table 2.4.

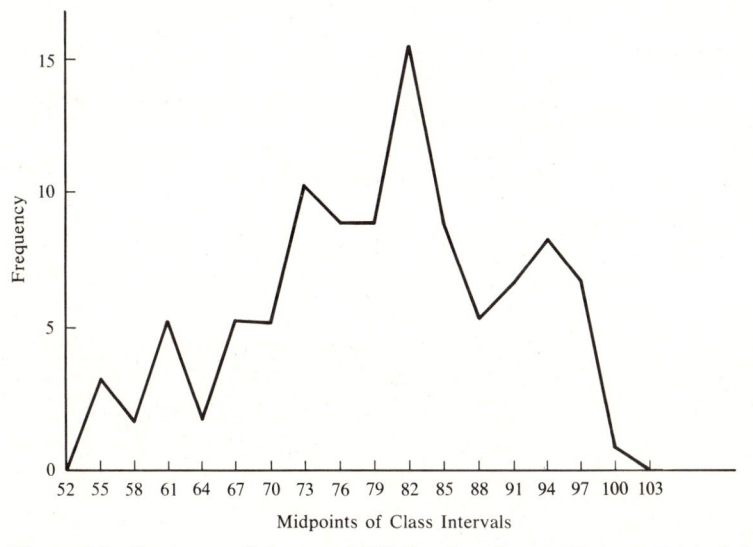

Figure 2.2 Frequency Polygon of 100 Reading Scores (Data of Table 2.4)

Both histograms and frequency polygons convey information about a frequency distribution in pictorial form. We choose the graphic model which seems to provide the clearest representation of the data. The frequency polygon may be preferred for distributions in which underlying continuity is explicit or assumed because the continuous line of the polygon suggests continuity more than do the contiguous bars of the histogram. The latter is

38 DESCRIPTIVE STATISTICS

preferred for discrete distributions. The frequency polygon perhaps has an advantage when the purpose is to compare two distributions; this is especially true when the two distributions are superimposed on the same graph. Polygons may be easily superimposed and the essential features of each are still clearly distinguishable. On the other hand, superimposed histograms present a very confusing picture. Figure 2.3 shows two superimposed frequency polygons of distributions of test scores for a group. The contrast between the group's performance in reading and in mathematics is clearly shown by the polygons. The frequency polygon for the reading scores is less spread out than for the mathematics scores.

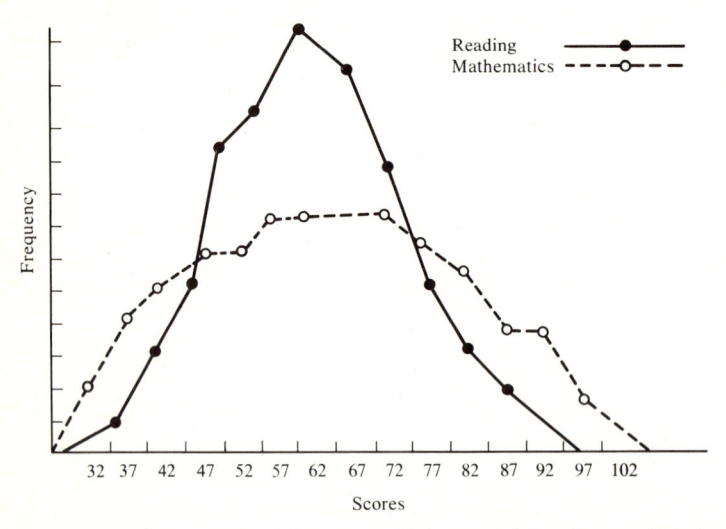

Figure 2.3 Superimposed Frequency Polygons of Reading and Mathematics Scores

Exercises

1. Construct a frequency polygon from the following data:

X	f
50–54	2
45–49	4
40–44	10
35–39	12
30–34	16
25–29	14
20–24	8
15–19	6

2. Construct a frequency polygon for the group of spelling test scores shown below:

15, 13, 15, 14, 12, 13, 14, 15, 11, 10

Solutions

1.

2.

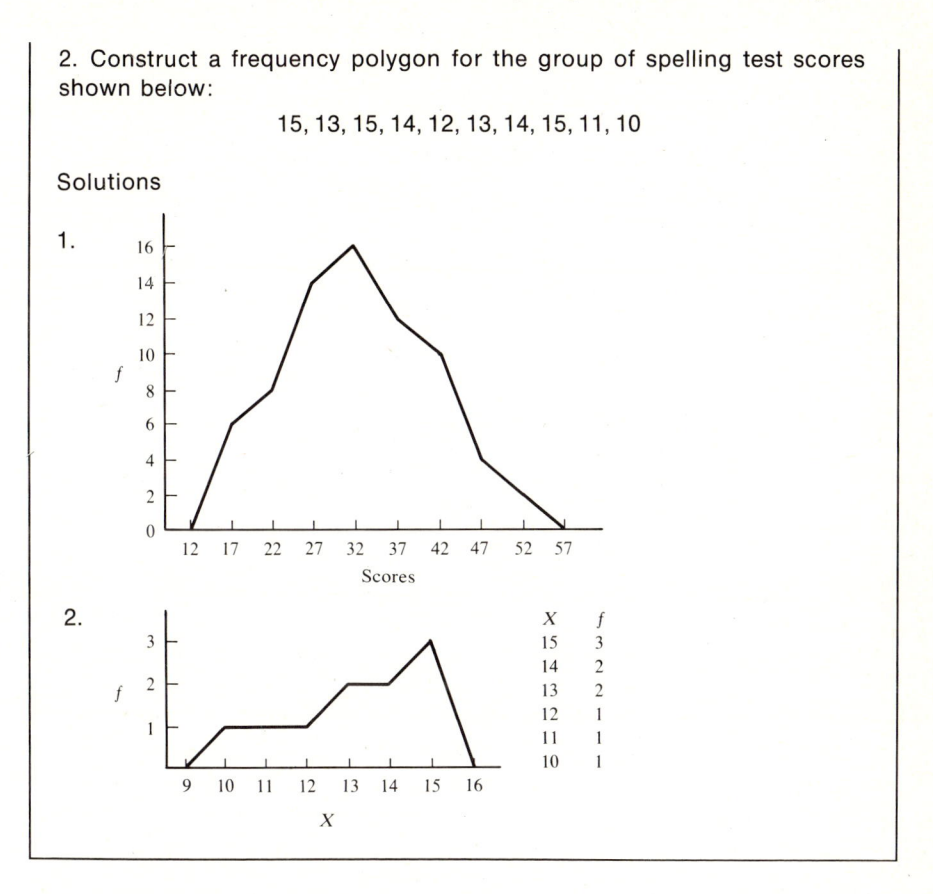

X	f
15	3
14	2
13	2
12	1
11	1
10	1

Cumulative Frequency Polygons

A distribution of measures can also be presented as a *cumulative frequency polygon* or *ogive*. It is constructed from a cumulative frequency distribution as illustrated in Table 2.5. In constructing a cumulative frequency polygon, the *cumulative frequencies* are plotted on the vertical axis and the *real upper limits of the class intervals* are indicated on the horizontal axis. Note this distinction in the cumulative frequency polygon: the real upper limits of each interval are used instead of midpoints because the graph is to indicate the number of cases falling below the upper limit of a certain interval rather than below the midpoint. Thus, when we plot the point indicating cumulative frequency, it is placed at the upper limit of the interval rather than at the midpoint, as was done with the frequency polygon. The cumulative frequency distribution of Table 2.5 is shown as a cumulative frequency polygon in Figure 2.4. The cumulative frequency 50 (total N) is plotted against the top of the real upper limit of the highest

40 DESCRIPTIVE STATISTICS

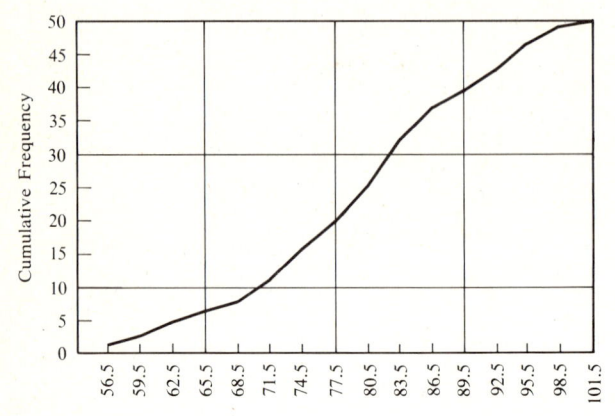

Figure 2.4 Cumulative Frequency Polygon of 50 Arithmetic Scores (Data of Table 2.5)

interval (101.5), the *cf* 49 is plotted against the upper limit (98.5) of the next lowest interval, and so on. Once all points have been plotted the points are connected to reveal a shallow S-type curve.

We note another difference between the cumulative frequency polygon and the frequency polygon. The cumulative frequency always increases as the score values increase because we are successively adding the frequencies in an interval to the frequencies of all lower intervals. Therefore, the cumulative frequency polygon will always go only in an upward direction; the graph will end with the line at the top of the right-hand side. The line is *not* brought back down to the horizontal axis as we did with the frequency polygon.

Cumulative Percentage Polygon In some situations, we may wish to use cumulative percentages rather than cumulative frequencies in constructing a polygon. The cumulative frequencies are converted to percentages and these cumulative percentages are plotted against the upper real limits of the class intervals. Again, the plotted points are connected and the shallow S-curve appears. In the cumulative percentage polygon, the vertical axis represents cumulative percentages, from zero at the bottom to 100 at the top, the horizontal axis again shows the upper real limits of the intervals with the plotted points located above the upper real limits. When cumulative percentages are graphed in this way, the result is called a *cumulative percentage polygon*, or *ogive*. Figure 2.5 shows a cumulative percentage polygon based on the data shown in Table 2.7.

The advantage of the cumulative percentage curve is that one can very quickly approximate the percentage of the total number of cases which fall below certain scores. For example, if one wanted to know the median of the distribution shown in Figure 2.5, which is the point below which 50 percent of the cases fall, one could draw a line from the 50 percent point on the vertical axis out to where this line intersects the curve. The point of intersection can

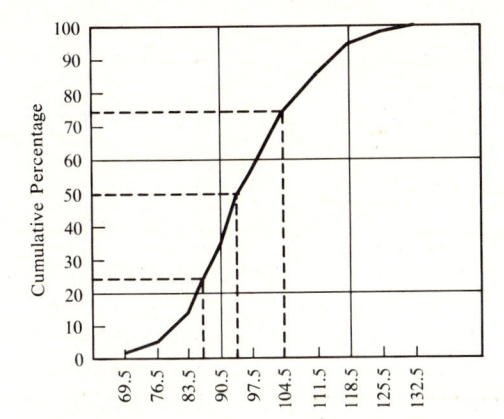

Figure 2.5 Cumulative Percentage Polygon of 72 Achievement Test Scores (Data of Table 2.7)

then be related to the horizontal axis by dropping a perpendicular line to that axis and the value of the median reported. In Figure 2.5 the median is approximately 95. Similarly, one could locate points below which any other percentage of the group would fall. For example, the lines in Figure 2.5 indicate that 75 percent of the group fall below 104.5, while 25 percent fall below approximately 87.

Another advantage of the cumulative percentage curve is that it can be used to compare two groups when the N is different in the groups since it converts both distributions to a common base—a percentage scale. The cumulative frequency curve cannot be used to compare two groups with different N's because a confused picture would result.

TABLE 2.7 Achievement Test Scores for 72 Fourth Grade Students

X	f	CUMULATIVE FREQUENCY	CUMULATIVE PERCENTAGE
126–132	1	72	100
119–125	3	71	99
112–118	6	68	94
105–111	8	62	86
98–104	13	54	75
91–97	16	41	57
84–90	15	25	35
77–83	6	10	14
70–76	3	4	6
63–69	1	1	1

$N = 72$

Forms of a Frequency Polygon

The main reason for constructing histograms and frequency polygons is that they reveal readily how scores are distributed along the score scale. That is, they show the *form* of the distribution. Graphs of frequency distributions may assume many different shapes or forms. One of the most immediately obvious characteristics of the form of a graphed frequency distribution is its symmetry or lack of symmetry (balance). A polygon is *symmetrical* in shape if one side is a mirror image of the other. When one side is not a mirror image of the other it is *asymmetrical*. If an asymmetrical curve is characterized by a high point or hump that is off-center and by tails of distinctly unequal length it is called a *skewed curve*. The hump indicates the scores with the highest frequencies. The areas away from the hump where the frequencies decrease are called the *tails* of the distribution.

Skewed curves are labeled according to the direction in which the longer tail is pointing. When the right tail of the distribution is distinctly longer than the left, the distribution is said to be skewed to the right or to be positively skewed. This is because the longer tail goes in the direction of the high scores or the positive values in the group. This means that while the preponderance of scores tend to be at the lower end of the score values, the extreme scores are at the high end, hence the long tail to the right.

When a distribution is negatively skewed or skewed to the left, the opposite situation exists. In negative skewness, the left tail is longer than the right. The tail extends toward the low end of the score scale. This means that while the higher frequencies are concentrated toward the high scores, the extreme scores are at the low end. Both types of skewed curves are illustrated in Figure 2.6. The label of a skewed curve indicates where the exceptional scores are found, that is, the direction of the longer tail. A very easy test could result in scores with a distribution like polygon A in Figure 2.6. Most scores are high with a few lower scores trailing off in the left-hand or negative side of the distribution. A difficult test could result in a distribution such as Polygon B with most scores on the low side and a few higher scores trailing off to the right or positive direction.

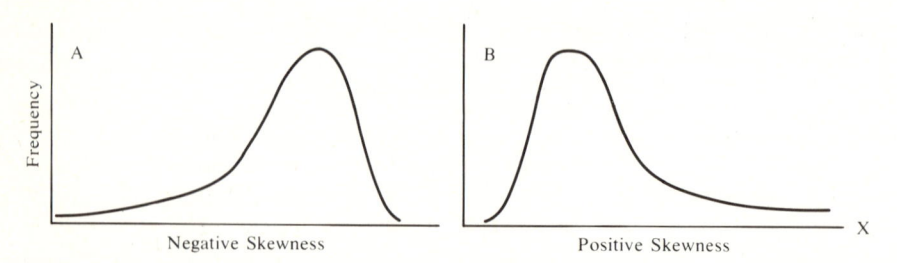

Figure 2.6   Illustration of Skewness in Frequency Polygons

Frequency Distributions and Graphs 43

> Exercises
>
> Tell whether the following polygons would be symmetrical, negatively skewed, or positively skewed:
> 1. A distribution of ages of a faculty with many young teachers and a few middle-aged and elderly teachers.
> 2. The distribution of IQ scores where half the class are in the average range, one-fourth are below the average range and another one-fourth are above the average range.
> 3. The distribution of batting averages for a baseball team where the majority are very good hitters and the remainder are mediocre or poor hitters.
> 4. The distribution of number of dates for girls in a dorm where the majority do not date at all, some girls have a few dates and some have many.
>
> Solutions
>
> 1. positive   2. symmetrical   3. negative   4. positive

Modes

Another noticeable feature of a polygon is the number of highest points or humps in the distribution. When a distribution has only one highest point, it is said to be *unimodal.* This means that there is one score or one score interval that contains more cases than any other; it is the most frequently occurring. If a distribution has two or more humps or peaks, it is said to be *multimodal.* Multimodal covers a range of possibilities; for example, a curve that has only two humps is labeled bimodal, one with three peaks or high points is trimodal, and so on. See Figure 2.7.

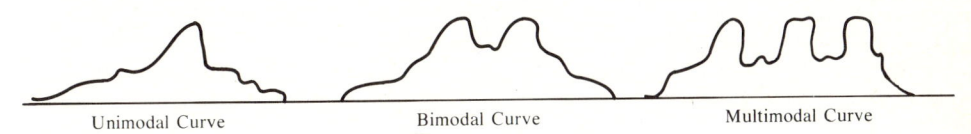

Unimodal Curve   Bimodal Curve   Multimodal Curve

Figure 2.7 Modality in Curves

Keeping Honest with Histograms and Polygons

Like any other tools histograms and polygons can be used to mislead us as well as to inform us. It is possible to draw graphs for the same group of data which give entirely different impressions. For example, an incumbent governor running for re-election might show the state expenditures during his term as illustrated in Figure 2.8. His opponent might show the same

44 DESCRIPTIVE STATISTICS

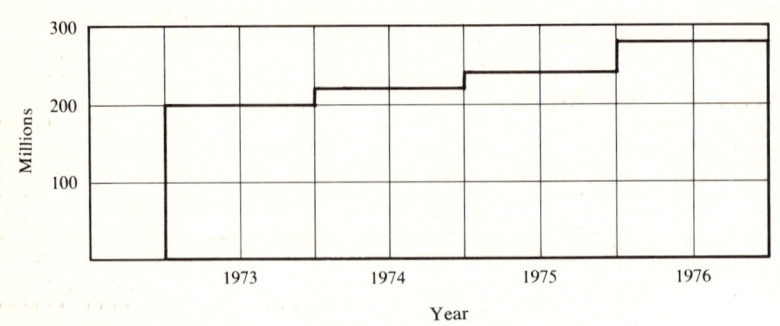

Figure 2.8 Incumbent Governor's Histogram of State Expenditures

expenditures as illustrated in Figure 2.9. Both histograms convey exactly the same information but the governor's histogram gives an impression of gradual, minor increases while his opponent's histogram suggests steeply rising expenditures. The impression conveyed by a graph depends largely on the proportions chosen for the abscissa and ordinate. This illustrates why it is suggested that as a general rule a histogram or polygon should be about three-fourths as tall as it is wide, because this gives relatively undistorted proportions. Figure 2.10 shows the state expenditures in a histogram following this rule.

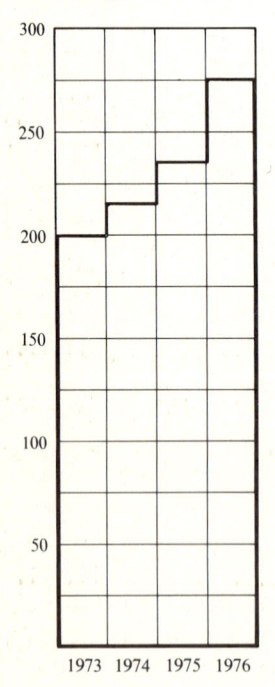

Figure 2.9 Opponent's Histogram of State Expenditures

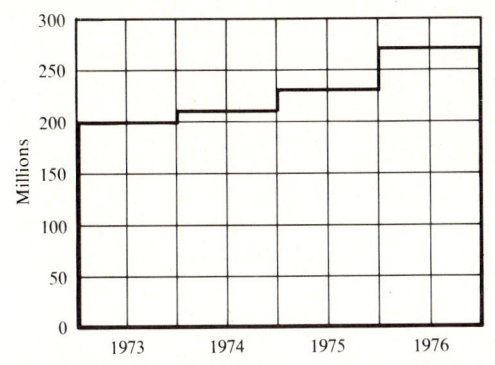

Figure 2.10 State Expenditures in Histogram Following Three-Fourths Rule

It is also recommended that the units on the abscissa and ordinate begin at the zero point. However, this is not always possible especially with score values on the abscissa. If a part of the abscissa or ordinate has been omitted, then the reader's attention should be called to this fact by means of slash marks drawn to cut the axis.

SUMMARY

Frequency distributions and graphs are useful for ordering data and presenting it in an easily interpreted form. An ungrouped frequency distribution contains all possible scores (X) from highest to lowest and the number of occurrences of each score (f). Sometimes a frequency distribution is more readily interpreted if the data are grouped into class intervals. A rule-of-thumb is that one can preserve reasonable definition yet present a readily interpreted distribution with about 15 class intervals.

In Flowchart 2 we see that one first decides whether a frequency distribution would be useful for interpreting one's data. If the answer is yes one next decides whether the distribution is to be grouped or ungrouped. With a range of greater than 20 a grouped distribution is often useful for organizing the data.

The real limits of an interval are half a unit above and below the recorded units. Cumulative frequency distributions record the total number of scores which are lower in value than the upper real limit of each interval. Cumulative percentage distributions record the percent of scores below the upper real limit of each interval.

Graphs are often useful for communicating a frequency distribution. A histogram has rectangular bars which extend up to the frequency on the vertical axis and extend from the lower limit to the upper limit of the interval on the horizontal axis.

FLOWCHART 2

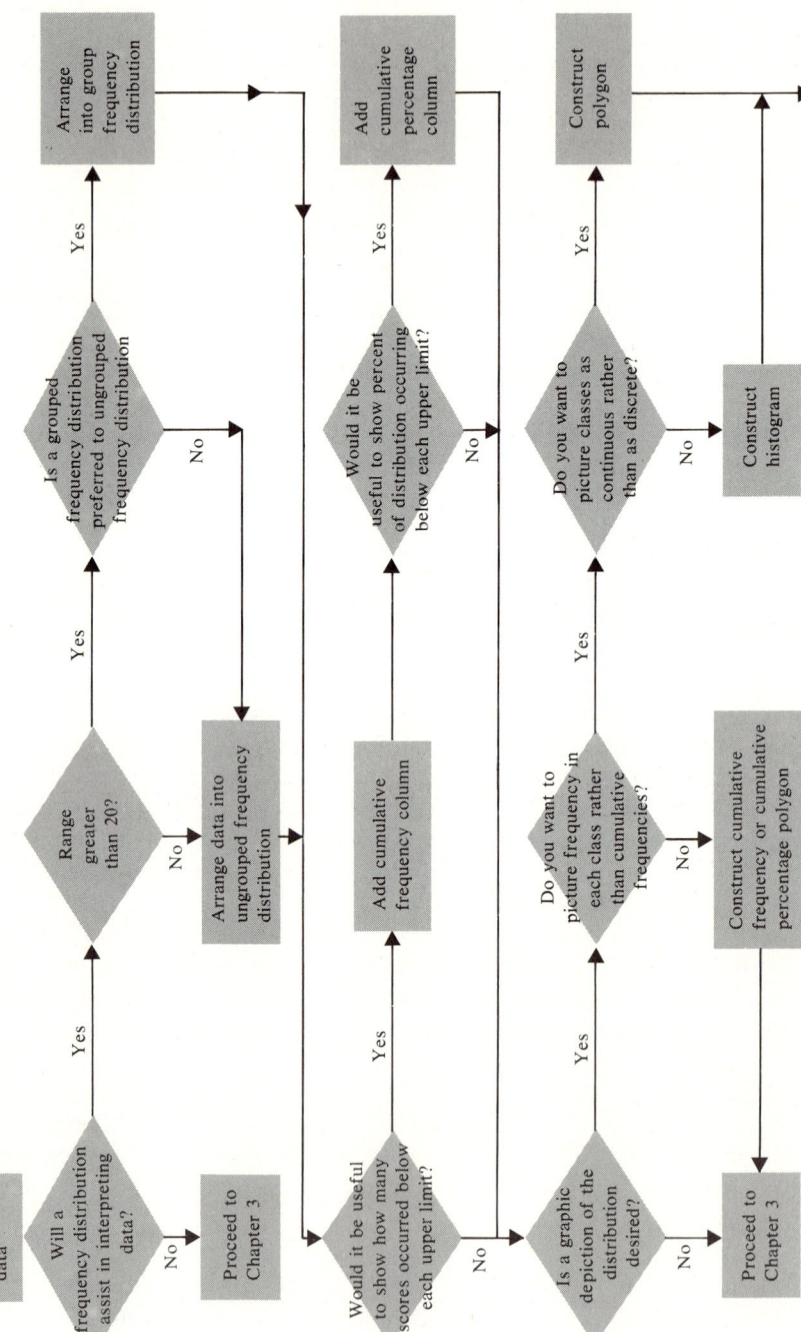

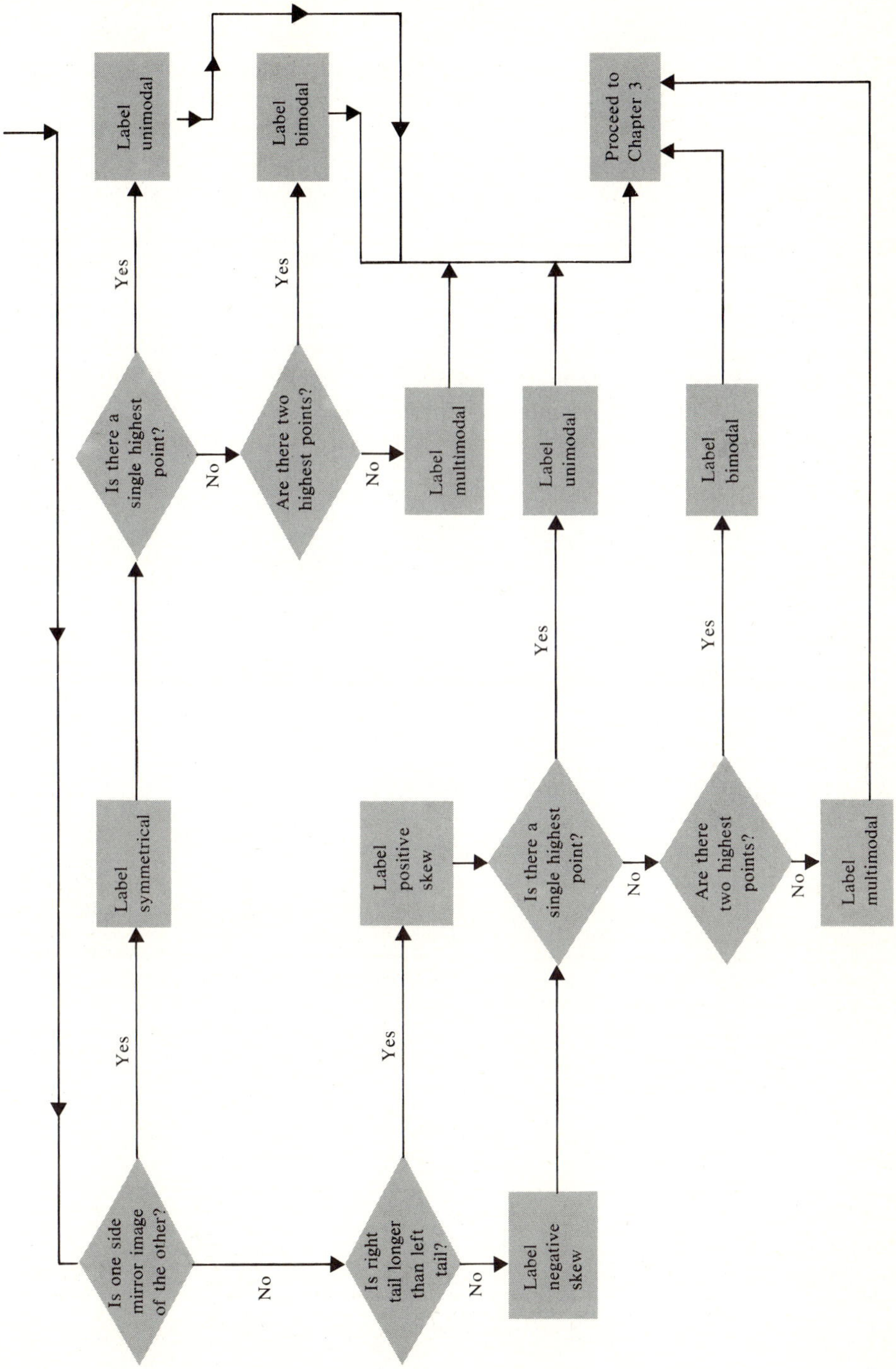

Is one side mirror image of the other?
Yes
No
Label symmetrical
Is there a single highest point?
Yes
No
Label unimodal
Are there two highest points?
Yes
No
Label bimodal
Label multimodal
Is right tail longer than left tail?
Yes
No
Label positive skew
Label negative skew
Is there a single highest point?
Yes
No
Label unimodal
Are there two highest points?
Yes
No
Label bimodal
Label multimodal
Proceed to Chapter 3

48 DESCRIPTIVE STATISTICS

In a frequency polygon, the frequency values are plotted above the midpoint of the interval and the points joined by lines and the ends of the line brought down to zero. The histogram pictures data in a manner which suggests discreteness whereas the polygon suggests continuity.

Cumulative frequency polygons provide a pictorial description of a cumulative frequency distribution. Cumulative percentage polygons do the same for cumulative percentage distributions.

Certain terms have proven useful for describing the shapes of distributions. If one side of a polygon is the mirror image of the other it is described as symmetrical. If this is not the case, the polygon is asymmetrical. An asymmetrical polygon with the right tail longer than the left is positively skewed. One with the left tail longer than the right is negatively skewed.

A polygon with a single highest point is unimodal, one with two highest points is bimodal, one with several points is multimodal.

Polygons, histograms, and other graphs can be used to mislead as well as to inform. In general an honest polygon or histogram should have the base of the ordinate zero and its height should be three-fourths of its width.

EXERCISES

1. What are the real limits, the midpoints, and the interval size for each of the following class intervals?
   (a) 15–19
   (b) 0–5
   (c) 20–21
   (d) 70–79

2. The following are scores obtained by a group of 40 students on a statistics examination.

61	78	70	83	92	67	66	83
76	68	79	84	82	86	81	60
78	77	86	77	81	92	80	77
70	40	75	60	74	82	77	87
63	94	76	87	81	77	87	84

   Prepare a frequency distribution and a cumulative f distribution for these data using a class interval = 3.

3. Using the data in Problem 2, set up a frequency distribution with $i = 1$ (ungrouped frequency distribution) and $i = 10$. Discuss the advantage and disadvantages of employing these intervals.

4. Given the following frequency distribution, draw a histogram.

X	f	cf	X	f	cf	X	f	cf	X	f	cf
80–84	1	116	60–64	6	100	45–49	20	57	30–34	7	11
75–79	2	115	55–59	15	94	40–44	17	37	25–29	3	4
70–74	3	113	50–54	22	79	35–39	9	20	20–24	1	1
65–69	10	110									

Frequency Distributions and Graphs 49

5. Given the following frequency distribution of 75 scores draw a frequency polygon, a cumulative frequency polygon, and a cumulative percentage polygon.

X	f	cf	X	f	cf
95–99	2	75	65–69	10	21
90–94	3	73	60–64	6	11
85–89	6	70	55–59	3	5
80–84	9	64	50–54	1	2
75–79	16	55	45–49	1	1
70–74	18	39			

6. Describe the types of distributions you would expect if you were to graph each of the following:
   (a) a test designed for grades 4–8 is given to a group of fourth graders
   (b) the heights of adult American females
   (c) a distribution of intelligence test scores for a sample of advanced graduate students in physics
7. Which graphic representation of data would be recommended in the following situations?
   (a) one wishes to graph a distribution where data are discrete
   (b) one wishes to graph a distribution where data are continuous
   (c) one wishes to compare two groups where the numbers are not the same in the two groups
   (d) one wishes to show two simple frequency distributions in one graph
   (e) one wishes to show the total number falling below the upper real limits of each interval

ANSWERS

1.
	Real Limits	Midpoint	Interval Size
(a)	14.5–19.5	17	5
(b)	−.5–5.5	2.5	6
(c)	19.5–21.5	20.5	2
(d)	69.5–79.5	74.5	10

2.
X	f	cf	X	f	cf
93–95	1	40	60–62	3	4
90–92	2	39	57–59	0	1
87–89	3	37	54–56	0	1
84–86	4	34	51–53	0	1
81–83	7	30	48–50	0	1
78–80	4	23	45–47	0	1
75–77	8	19	42–44	0	1
72–74	1	11	39–41	1	1
69–71	2	10			
66–68	3	8			
63–65	1	5			

50 DESCRIPTIVE STATISTICS

3.

X	f	X	f	X	f	X	f	X	f
94	1	81	3	68	1	55	0	42	0
93	0	80	1	67	1	54	0	41	0
92	2	79	1	66	1	53	0	40	1
91	0	78	2	65	0	52	0		
90	0	77	5	64	0	51	0		
89	0	76	2	63	1	50	0		
88	0	75	1	62	0	49	0		
87	3	74	1	61	1	48	0		
86	2	73	0	60	2	47	0		
85	0	72	0	59	0	46	0		
84	2	71	0	58	0	45	0		
83	2	70	2	57	0	44	0		
82	2	69	0	56	0	43	0		

$i = 1$

Advantage—can see the value of each individual score.
Disadvantage—involves too much detail; all score values must be listed, even those with zero frequency.

X	f
90–99	3
80–89	15
70–79	14
60–69	7
50–59	0
40–49	1

$i = 10$

Advantage—compact presentation of data; quickly set up.
Disadvantage—too much information lost about the individual scores.

6. (a) positively skewed
   (b) symmetrical
   (c) negatively skewed
7. (a) histogram
   (b) frequency polygon
   (c) cumulative percentage curve
   (d) frequency polygon
   (e) cumulative frequency polygon

CHAPTER THREE
MEASURES OF CENTRAL TENDENCY

OBJECTIVES

In Chapter Two, we considered the frequency distribution as a means for organizing and summarizing data. In this chapter, we discuss certain descriptive statistics which may be calculated and used to communicate characteristics of the frequency distribution in a still more compact way. After studying this chapter, the reader should be able to:

1. Identify the reason for using measures of central tendency.
2. List the three commonly used measures of central tendency.
3. Define mode and determine the mode of a distribution of measures.
4. List the characteristics of the mode as a measure of central tendency.
5. Define median.
6. Determine the value of the median in a set of measures and in a frequency distribution.
7. List the characteristics of the median as a measure of central tendency.
8. Define mean.
9. Calculate the mean of both raw scores and simple frequency distributions.
10. List the characteristics of the mean as a measure of central tendency.
11. Know the guidelines for selecting a measure of central tendency for use in a particular situation.
12. State the relationship between the three measures in a symmetrical and in a skewed distribution.

Averages

We often need more concise ways of summarizing information than the frequency distributions described in Chapter Two. Certain summary scores are calculated in order to provide a description of the characteristics of the entire distribution. One of these characteristics is the general location of the scores. The most convenient method for indicating the general location of a set of scores is to report an *index of central tendency*, commonly known as an *average*. There are three commonly used indexes of central tendency: the mode, the median, and the mean. All three of these measures are used to convey information about the average scores in a distribution. As we will see, each of these interprets average in a slightly different way.

Mode

The mode is the simplest measure of central tendency and is easily determined merely by inspection. The mode is defined as the *score* value which occurs most frequently in a group of scores. We identify the mode merely by looking at the scores and locating the one that occurs most often. Examine the following collection of scores: 4, 5, 5, 5, 6, 7, 7, 8, 8, 8, 8, 9, 10. In this example, the score 8 occurs more frequently than any of the other score values and is the mode. The most frequently occurring score is usually somewhere near the center of a distribution. In such a case the mode is a legitimate index of central tendency. However, the mode does not always occur near the center of a distribution and therefore we cannot rely on it to accurately reflect the center of a set of scores. The mode is an unreliable average or index of central tendency.

The distribution in the above example would be described as *unimodal*, which means that there is only *one* score which occurs with a greater frequency than any other scores. In some distributions no score occurs with greater frequency than any other. For the set of observations: 10, 11, 13, 16, 18, 19 there is no mode. In other distributions there may be two or more modes. A distribution with two or more modes is said to be *multimodal*. Multimodal covers a range of possibilities. In the example: 4, 5, 6, 6, 6, 6, 7, 7, 8, 8, 8, 8, 9, 10, the scores 6 and 8 both occur four times, a frequency which is greater than the frequency of occurrence of any other value. In this case, the distribution would be described as bimodal with both 6 and 8 considered as modes. If a distribution has three modes, it would be labeled trimodal, and so on. (See p. 43.)

The mode has limited usefulness as a measure of central tendency. It is especially unreliable with small groups, because it will fluctuate in response to changes in the value of a few scores. Consider the reported catches in a fishing contest: 40, 36, 36, 36, 36, 21, 21, 21, 5, 0; the mode is 36. However, if two of the fishermen reporting 36 were discovered to have included an underweight fish which should have been thrown back, their catch would have been 35, and the mode would shift right down to 21. Because of its instability the mode is seldom used to indicate central tendency. Its use is largely restricted to that of a quick inspection kind of "average." The mode cannot be manipulated mathematically and thus has very limited use in statistical calculations. A mode may be reported for any of the scales of measurement but is the only measure of central tendency that may be legitimately used with nominal data.

As we have seen in Chapter Two it is sometimes useful to describe a distribution according to whether it has a single mode or not. To give an example, suppose the professor of an advanced statistics class gives a pretest the first day of class and finds that a polygon of the scores has the following shape:

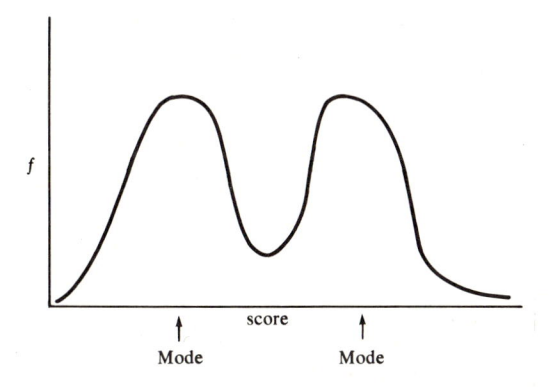

The bimodality of the polygon suggests that the class is divided into two subgroups. He checks the students' records and finds that most of the students in the lower part of the distribution have not taken the introduction to statistics class.

Exercises

1. What is the mode in the following distribution?
    85, 92, 96, 96, 98, 99, 99, 99, 100, 101, 101, 103, 105, 110
2. Half of the inhabitants of a village are midgets and the other half are not. Would we expect a polygon showing the heights of the inhabitants to be unimodal?

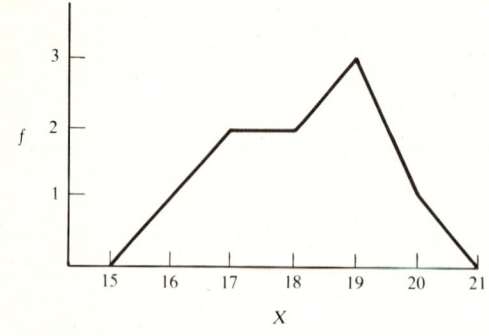

3. What is the mode in the distribution shown above?
4. What are the advantages of the mode as a measure of central tendency?

Solutions

1. 99
2. No, we would expect a bimodal distribution.
3. 19
4. It is easy to compute. It can be used with nominal data.

Median

The median is the middle point in a distribution. Half of the distribution is located above the median and half is below the median.

The first step in locating the median is to arrange the scores in order, thus enabling us to locate the middle point. Consider the following observations: 3, 8, 12, 18, 19, 21, 25. The median is 18 since it is the middle point; there are three observations above the median 18, and three observations below it.

If the median falls between two adjacent scores, as it will when the total number of cases is *even*, the median is the point that is the upper real limit of the lower of the two middle scores and the lower real limit of the higher score. If we add a score of 30 to the above distribution we get 3, 8, 12, 18, 19, 21, 25, 30; the median is then 18.5. That is, the median is the point that separates the real limits of the two middle scores 18 and 19.

To summarize, if there is an odd number of cases, the median is the middle case in the distribution. With an even number of cases, the median is the point halfway between the *two* middle cases.

Exercises

1. What is the median of each of the following groups of scores?
(a) 6, 4, 9, 13, 7

(b) 6, 4, 9, 7
(c) 1, 7, 5, 3, 2

Solutions

1. (a) 7 (N is odd and 7 is the middle score)
   (b) 6.5 (N is even and 6.5 is halfway between the two middle scores)
   (c) 3

Computation of the Median

When there is a repetition of the same score in the middle of the distribution, the location of the median requires computation. Consider the following distribution: 4, 5, 6, 7, 7, 7, 8, 9, 9, 10. Since there are ten cases in the distribution, the median is the point below which one-half, or five, of the cases fall. If we count from the bottom of the distribution, we find that five of the cases fall between two of the 7's; that is, between the fifth and sixth cases in the distribution. Since there is an additional 7 falling below the fifth case, we cannot report the median as simply 7. Looking at the distribution we can see that there are three scores below 7 but four scores above 7; thus, 7 would not fit the definition of the median. We need to locate the point below which one-half, or five cases, fall. Again, counting from the lower end of the score values, we find that three cases fall below the interval 7. There are three cases in the interval 7; however, if we include all three of those cases, we will have more than the five cases we need for the median. Thus, we know that the median falls somewhere *within* the interval 7. To locate the median point within the interval we engage in a process known as *interpolation*. In order to interpolate within an interval it is necessary to assume that the three 7's are evenly spread throughout the interval 6.5–7.5. This is consistent with the statistical assumption that scores are continuous and that a score represents a range of possible values in an interval.

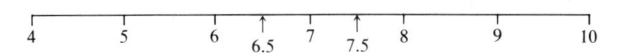

```
|────|────|─↑─|─↑─|────|────|
4    5    6 6.5 7 7.5 8    9    10
```

The score of 7 is thought of as ranging from midway between 6 and 7 (6.5) on the lower side to midway between 7 and 8 (7.5) on the upper side. We want to locate the median point which falls somewhere between 6.5 and 7.5.

Three cases fall below 6.5, the lower limit of the interval 7. We need two more cases from the interval 6.5–7.5 in order to reach the fifth case or the median. Since there are three 7's and we need two of them, we will only go $\frac{2}{3}$ of the way into the interval 6.5–7.5. The 2 in the numerator of the fraction represents the number of scores we need from the interval and the 3 in the denominator of the fraction represents the number of scores in the interval.

56 DESCRIPTIVE STATISTICS

Thus, the median would be: $6.5 + \frac{2}{3}$ or $6.5 + .67 = 7.17$. The median of the distribution is 7.17, the point below which exactly one-half of the cases fall. See Figure 3.1 for an illustration. It can be seen that the area of the histogram is divided in half at the median 7.17.

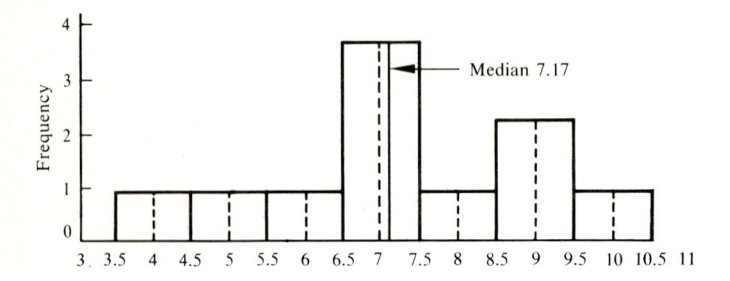

Figure 3.1 Histogram Showing Interpolated Median

Now consider a distribution with scores: 3, 3, 4, 5, 6, 6, 8, 9, 10. Again, we wish to locate the point that will divide the distribution into two equal parts. Since there are nine cases in the distribution, we want the point below which and above which there will be 4.5 cases ($\frac{1}{2} \times 9$). Counting from the lower end of the distribution, we find that four cases fall below 5.5 (the lower limit of interval 6). Thus, we need only .5 more in order to get to the 4.5th case which is the median. We must interpolate within the interval 5.5–6.5 in order to reach this point. Since four cases fall below interval 5.5–6.5 and we need .5 more, we will go .5/2 of the way into this interval. Again, the .5 in the numerator represents the number of scores we need from the interval and the 2 in the denominator represents the number of scores in the interval. We take the lower limit of the interval, 5.5, and add .5/2 to it. Thus, the median would be $5.5 + .5/2$ or $5.5 + .25 = 5.75$. Exactly half of the cases fall below 5.75 in this distribution. (See Figure 3.2.)

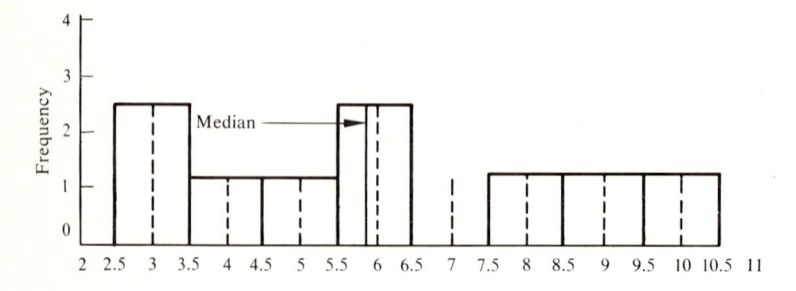

Figure 3.2 Histogram Showing Interpolated Median

Measures of Central Tendency   57

Exercises

1. What is the median of each of the following groups of scores?
   (a) 8, 9, 10, 11, 11, 11, 12
   (b) 3, 4, 5, 6, 4

Solutions

1. (a) 10.75 $[10.5 + \frac{1}{4}]$
   (b) 4.25 $\left[3.5 + \frac{1.5}{2}\right]$

Let us find the median of scores that have been arranged in a frequency distribution. The distribution is shown in Table 3.1. There are 20 scores in the distribution; therefore, the median is the point below which 10 ($\frac{1}{2} \times 20$) cases fall. Counting from the bottom, we find that there are nine cases below

TABLE 3.1 A Frequency Distribution of Scores

X	f
10	1
9	2
8	3
7	5
6	4
5	2
4	2
3	1
	20

interval 7; the lower limit of which is 6.5. Thus, we need to go into interval 7 for the additional one case. There are five cases in interval 7; thus, we will go $\frac{1}{5}$ or .20 unit into that interval. The lower limit of the interval is 6.5 and we add .20 to that and find the median to be 6.7.

It is not necessary to go through this tedious process in every case. The above procedure can be summarized as a formula which simplifies the location of the median.

$$\text{Mdn} = L + \left(\frac{N/2 - cf_b}{f_w}\right)i \quad \text{where: Mdn} = \text{the median}$$

N = the total number of cases in the distribution

58 DESCRIPTIVE STATISTICS

L = the lower real limit of the interval containing the median

cf_b = the total frequency in all intervals below the interval containing the median (cumulative frequency below)

f_w = the frequency of cases within the interval containing the median

i = interval size[1]

Formula (3.1)
The Median

Let us apply Formula (3.1) to locate the median in the first example that we considered above: 4, 5, 6, 7, 7, 7, 8, 9, 9, 10. In this example, the value of L is 6.5, N is 10, cf_b is 3, f_w is 3, and the interval size, i, is 1. Substituting in the formula, we have:

$$\text{Mdn} = 6.5 + \left(\frac{5-3}{3}\right)1 = 6.5 + .67 = 7.17$$

This formula provides us with the most efficient method for determining the median of a frequency distribution. Table 3.2 shows a frequency distribution and the computation of the median of the scores using the formula.

The first step is to locate the interval within which the median score ($N/2$) lies. This is done by referring to the cumulative frequency (cf) column, which shows the frequencies of the scores up to and including any given interval in the distribution, and proceeding from the lowest value up until the first value equal to or greater than $N/2$ is located.

In the distribution in Table 3.2, the total number of scores is 36 and we need to locate the $\frac{36}{2}$ or the 18th score. Looking up the cf column, we see that there are 17 cases up to and including the score 13, and 24 cases up to and including the score 14; thus, the median is located within the interval represented by the score of 14. We then apply the formula for the median. The value of L is 13.5 because that is the lower limit of the interval that contains the median; recall that the interval represents a range from 13.5 to 14.5. The value of cf_b is 17 because 17 scores fall below the interval containing the median. The value of f_w is 7 because the distribution indicates that there are 7 scores within the interval represented by the value 14. The interval size, i, again is 1 because each score represents an interval width of 1. Substituting in the formula we find that the median of the distribution is 13.64.

[1] The value i is necessary only when the size of the interval is other than 1. In the examples considered thus far, each of the intervals represented one score point, therefore, $i = 1$.

TABLE 3.2 Computation of the Median of a Frequency Distribution of Scores

SCORES (X)	FREQUENCY (f)	CUMULATIVE FREQUENCY (cf)
20	1	36
19	2	35
18	3	33
17	1	30
16	1	29
15	4	28
14	7 f_w	24
13	7 ⎫	17
12	5 ⎬	10
11	3 ⎬ f_b	5
10	1 ⎬	2
9	1 ⎭	1
	N = 36	

$$\text{Mdn} = 13.5 + \left(\frac{36/2 - 17}{7}\right) 1 = 13.5 + \left(\frac{1}{7}\right) 1 = 13.64$$

Exercise

1. Compute the median of the following frequency distribution:

X	f
10	5
9	3
8	8
7	4
6	3
5	1

Solution

1. 8 $\left(7.5 + \frac{4}{8}(1) = 8\right)$

Computing the Median when Interval Size Is Other than One In some cases we need to deal with data where the interval size is other than one. For example, elementary achievement test results are frequently reported in grade level equivalents which employ tenths as the basic unit of measure; a child whose test score is equivalent to a fifth grader beginning the

60 DESCRIPTIVE STATISTICS

first month of school is assigned a 5.0 grade equivalent, a child whose score is equivalent to a fifth grader after six months of school is assigned a 5.6 grade equivalent, and so on. In this case our interval size is .1 and a score of 5.6 is thought of as representing a range from 5.55 to 5.65.

The following example illustrates the computation of the median in such a case:

GRADE EQUIVALENT	f	cf	GRADE EQUIVALENT	f	cf
7.9	2	50	5.9	0	33
7.8	0	48	5.8	2	33
7.7	1	48	5.7	0	31
7.6	0	47	5.6	0	31
7.5	3	47	5.5	5	31
7.4	1	44	5.4	6	26
7.3	0	43	5.3	2	20
7.2	2	43	5.2	2	18
7.1	0	41	5.1	3	16
7.0	1	41	5.0	4	13
6.9	0	40	4.9	0	9
6.8	0	40	4.8	0	9
6.7	2	40	4.7	0	9
6.6	0	38	4.6	0	9
6.5	1	38	4.5	3	9
6.4	0	37	4.4	0	6
6.3	0	37	4.3	2	6
6.2	3	37	4.2	0	4
6.1	0	34	4.1	0	4
6.0	1	34	4.0	2	4
			3.9	1	2
			3.8	0	1
			3.7	1	1

The median of the grade equivalents is calculated as

$$\text{Mdn} = L + \left(\frac{N/2 - cf_b}{f_w}\right)(i)$$

$$= 5.35 + \left(\frac{50/2 - 20}{6}\right)(.1)$$

$$= 5.35 + \left(\frac{5}{6}\right)(.1)$$

$$= 5.35 + (.83)(.1)$$

$$= 5.433$$

Notice that since $i = .1$ the lower limit of the interval containing the median is equal to $5.4 - .05 = 5.35$. In this case $(\frac{1}{2})(i)$ is equal to .05, not .5 as when $i = 1$.

The Median as a Position Average We see that the median is a position average. That is, it is determined by placing the scores in rank order and finding the middle score. The size of the measurements themselves do not affect the median. The highest score is placed in the highest position, but how large it actually is does not enter into the computation of the median. As we will see, this is an advantage when there are scores of extreme size in the distribution that would distort a calculated average. Note that in the following two distributions, the median is exactly the same.

Distribution A: 3, 5, 7, 9, 10, 11, 12

Distribution B: 3, 5, 7, 9, 10, 19, 25

The median is not influenced by the two deviant scores in Distribution B. Because the median is not sensitive to the presence of extreme scores, it may be considered as the most "typical" score in a distribution. For example, in the distribution: 20, 21, 23, 24, 25, 45, 66, the median is 24 and is the most typical score because it is closest to the aggregate of the other scores. The median is the score value that is closest to all of the scores contained in a distribution.

The median has a useful property which serves as the basis for its being the typical score in a distribution. This property is that the sum of the *absolute* deviations of the observations from the median is less than from any other point in the distribution. An absolute deviation is defined as the difference between two numbers without regard to the sign, or the direction, of the difference. This property is expressed by the following: $\Sigma |X - \text{mdn}| \le \Sigma |X - \text{any other point}|$. The symbol Σ, the upper case Greek letter sigma, is used to denote "sum of." The above expression reads: "The sum of the absolute differences between the scores and the median is less than or equal to the sum of their absolute differences from any other point." For example, assume the following observations: 4, 5, 6, 7, 9; the median is 6. The absolute deviations of the observations from the median are:

SCORES	ABSOLUTE DEVIATION FROM MEDIAN
9	3
7	1
6	0
5	1
4	2

The sum of the absolute deviations is 7. The sum of the absolute deviations from any other number would be larger than 7. Just try subtracting 4, 5, 6.2, or any of the other numbers and you will find that the sum of the absolute deviations will be larger.

Because the sum of the absolute deviations about the median is a minimum, it is often the best choice for a single point to represent the central tendency of a set of observations. It will be the point which is closest to the other scores in the distribution and as such is the "typical" score. Thus, if one wants an index of the most typical score, the median is preferred over the other measures of central tendency. For example, in the distribution considered earlier: 20, 21, 23, 24, 25, 45, 66, the median 24 was characterized as the most typical score. The sum of the absolute deviations of the scores from the median 24 is 72; the sum of the absolute deviations of the scores from the mean, which is 32, would be 94. Hence the median is closest to the other scores.

The median is not algebraically defined and thus may not be used in further calculations. For example, medians from separate groups may not be added to find the median of the combined group. Assume we know that the median income of professors at State University is $15,000, that of associate professors is $13,000, assistant professors $10,500, and instructors $8,500. There is no way that we could use these median figures to arrive at the median salary of all faculty members at State University. In order to do this, we would have to know the original distributions for each group, combine the distributions, and calculate a new median. This proves to be a disadvantage in some situations, where medians for separate groups are available but not the complete distributions.

Since the median is based on scores in rank order, it is by definition an ordinal statistic. The median may be determined for interval or ratio data, but the interval or ratio characteristics of the data are not involved in the computation of the median.

Exercises

Find the median of the following set of scores:
1. 3, 8, 7, 5, 4
2. 2, 7, 3, 5, 6, 4
3. 1, 2, 4, 3, 6, 3, 7

Solutions
1. 5
2. 4.5
3. 3.25

Mean

The best known and the most reliable measure of central tendency is the mean. The mean is the arithmetic average of a group of scores. It is found by adding all the scores in the distribution and then dividing the sum of the scores by N, the total number of scores:

$$\text{Mean} = \frac{\text{Sum of scores}}{N}$$

While the definition of the mean is the same, the symbol used for the mean depends on whether one is dealing with a parameter or a statistic. The symbol used for the population mean (parameter) is the lower-case Greek letter μ, mu (pronounced mew). A subscript may be added to identify the particular population involved; for example, μ_X would indicate the mean of a population of X scores, μ_Y would identify a population of Y's, and so on. A sample mean (statistic) is represented by the symbol $\overline{X}$ (pronounced X bar). If there are several sample means, each is identified by the symbol of the variable with the bar over it. For example, $\overline{Y}$ is the sample mean of the Y distribution and $\overline{Z}$ is the sample mean of the Z distribution.

Thus, the formula for the mean of a *population* of X scores is:

$$\mu_X = \frac{\Sigma X}{N} \quad \text{where:} \quad \begin{aligned} \mu_X &= \text{mean of a population of } X \text{ values} \\ X &= \text{score} \\ \Sigma X &= \text{the sum of the } X \text{ scores} \\ N &= \text{number of cases in the population} \end{aligned} \quad \text{Formula (3.2)}$$

The formula for the mean of a *sample* is:

$$\overline{X} = \frac{\Sigma X}{n} \quad \text{where:} \quad \begin{aligned} \overline{X} &= \text{mean of sample} \\ n &= \text{number of cases in sample} \\ &\text{(the other symbols are the same)} \end{aligned}$$

The computational procedure is exactly the same whether one is calculating a parameter or a statistic. The capital Greek letter Σ (sigma) means "the sum of"; operationally ΣX tells one to "sum the X's" (scores). The formula for either reads: The mean is equal to the sum of all the scores divided by the number of scores. Only the symbol differs.

Let us apply Formula (3.2) to determine the mean of the following set of scores: 15, 20, 21, 24, 25, 18, 22, 30, 25, 20.

$$\Sigma X = 15 + 20 + 21 + 24 + 25 + 18 + 22 + 30 + 25 + 20 = 220$$

If this set of data is a population, the formula for the mean is:

$$\mu_x = \frac{\Sigma X}{N} \qquad \mu_x = \frac{220}{10} = 22$$

if the data represent a sample, the formula for the mean is:

$$\bar{X} = \frac{\Sigma X}{n} \qquad \bar{X} = \frac{220}{10} = 22$$

In either case, the mean value is exactly the same.

If the definition, the computational procedure, and the numerical value of the mean are the same, whether the data are a sample or a population, you may question the need for the different symbolism. Let it suffice at this point to say that the distinction will be crucial later in statistical inference where we will use sample means to reach conclusions about population means. Note, in using Formula (3.2), that it is not necessary to put the scores in any order before they are summed.[2]

If the scores are arranged in a frequency distribution where X values may occur more than once, then each score must be multiplied by its frequency (f) and these products are then summed. This gives ΣfX, which is divided by the total number of scores. The formula for the mean then becomes:

POPULATION $\qquad\qquad$ SAMPLE

$$\mu = \frac{\Sigma fX}{N} \quad \text{or} \quad \bar{X} = \frac{\Sigma fX}{n} \qquad \text{Formula (3.3)}$$

Let us apply this formula to determine the mean of the sample frequency distribution shown in Table 3.3.

[2] The commutative principle for the addition of integers states that the order of the integers can be changed and the sum will be the same. In general: $a + b = b + a$.

TABLE 3.3 Computation of the Mean of a Frequency Distribution of Scores

SCORES X	FREQUENCY f	SCORE × FREQUENCY fX
10	3	30
9	2	18
8	4	32
7	4	28
6	3	18
5	4	20
4	4	16
3	2	6
2	3	6
	$N = 29$	$\Sigma fX = 174$

$$\mu = \frac{174}{29} = 6$$

Exercises

1. What is the mean of the following set of scores:

    1, 4, 4, 5, 5, 6, 7, 8

2. Compute the mean of the following distribution:

X	f
10	5
9	3
8	8
7	4
6	3
5	1

Solutions

1. 5 (40/8 = 5)
2. 8 ($\Sigma fX = 192$, $N = 24$; 192/24 = 8)

The mean as an arithmetic average is dependent on the numerical value of each observation in a distribution. If the value of a single score is changed either up or down, the value of the mean is changed in the same direction. The mean's sensitivity to all the scores is usually an advantage but under certain circumstances can be a disadvantage. Students are well aware of the effect of one or two very high test scores on the class average. If one wants extreme scores to influence the index of central tendency, the mean is the preferred index. If one does not want the index to be influenced by extreme scores, the median or the mode is preferred.

Perhaps the mean's sensitivity to extreme scores can best be illustrated by use of an example. Assume that we want to know the average salary earned by the members of the School Board. The salaries for the seven members are:

$100,000
25,000
20,000
17,500
16,000
15,000
12,500
$\Sigma = \$206,000$

The mean is 206,000/7 = $29,428+. The extreme salary of $100,000 greatly affects the value of the mean. The mean is at a value ($29,428) which is higher

66 DESCRIPTIVE STATISTICS

than all of the salaries except the highest one. The sensitivity of the mean to all the scores would negate its use in this situation. The median of $17,500 is much more representative of the salaries of these board members and would be the preferred measure of central tendency.

One of the most important properties of the mean is that it is the point in a distribution of scores about which the summed deviations are equal to zero. When the mean is subtracted from a score, the difference is called a *deviation score*. Those scores above the mean will have positive deviations while the scores below the mean have negative deviations. The sum of the deviations will be zero, that is, the sum of the negative deviations exactly equals the sum of the positive deviations. Thus, the mean may be considered the exact balance point in a distribution. Let us illustrate this with the following example: 5, 4, 3, 2, 1 (see Figure 3.3). The negative deviation scores sum to -3;

Deviations	-2	$+$	$-1 = -3$		$+1$	$+$	$+2 = +3$
	1	2		3	4		5

↑
Mean

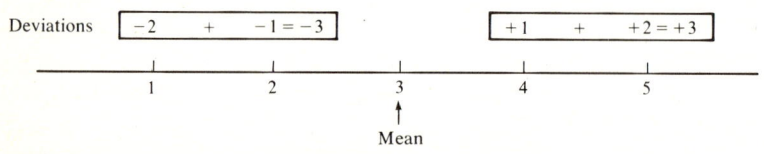

Figure 3.3 The Mean as Balance Point of a Distribution

the positive deviation scores sum to $+3$. The mean is the balance point about which the algebraic sum of the deviations is always zero. This characteristic may be expressed by the equation $\Sigma(X - \mu) = 0$, which states that the sum of the deviations of scores from the mean is equal to zero. This will be true in any distribution of scores.

This characteristic of the mean is analogous to a situation when children balance a teeter-totter. The lighter child sits farther from the fulcrum and the heavier child sits nearer it. In the same manner the mean is a fulcrum point which is influenced by both the "weight" (number of scores) and "distance on the teeter board" (value of scores). This is illustrated in Figure 3.4 where

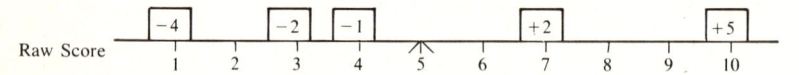

Raw Score	-4		-2	-1			$+2$			$+5$
	1	2	3	4	5	6	7	8	9	10

Figure 3.4 The Mean as the Fulcrum of a Distribution

there are three scores (1, 3, and 4) below the mean but near to it and two scores (7 and 10) above the mean but further from it. The negative deviation scores sum to -7; the positive deviation scores sum to $+7$. The mean is the center of gravity or fulcrum of the teeter-totter.

Since the mean is determined by the value of every score, it is the preferred measure of central tendency in any situation where the concern is with *quantity* rather than with *typicalness*. For example, a conglomerate corporation contemplating buying a factory and taking over its operation would be interested in the mean salary of the workers in the factory, since the mean multiplied by the number of workers would indicate the *total amount* of money required to pay all the workers. A sociologist studying the factory's community would probably be more interested in the median since the median indicates the pay of the *typical* worker.

Both mathematical theory and experience have shown that sample means are more *stable* than sample medians or sample modes. That is, if we randomly draw a number of samples from a population we can expect the means of our samples to be more similar to one another than are the medians or the modes. Therefore, a sample mean can be considered to be a good estimate of the corresponding population mean. The representativeness of sample means will be dealt with more fully in Chapter 9.

More mathematical operations can be performed with the mean than with the median or the mode. For example, if we know the mean of a group of observations of size N, we can determine the sum of the observations. Since $\mu_X = \Sigma X/N$, or $\bar{X} = \Sigma X/n$, then $n\bar{X} = \Sigma X$ and $N\mu = \Sigma X$. Due to this relationship between the product of the number of observations and the mean and the sum of the values of the observations, it is possible to find the combined mean for several sets or groups of observations. We only need to know the number of observations in each subgroup and the mean of the subgroup. The combined mean is found by multiplying each subgroup mean by the number in that subgroup, summing these products, and dividing this sum by N, the total number of observations in all groups. For example, assume three groups of observations with n's of 3, 2, 4, where the means are 4, 6, and 3, respectively. If $\Sigma X = n\bar{X}$, then the combined sum of all nine scores would $(3 \times 4) + (2 \times 6) + (4 \times 3) = 36$. Therefore, the mean of the three groups combined is $36/9 = 4$.

The above procedure can be stated in a general form that can be applied to any number of groups. Let k = the number of groups, $n_1 + n_2 + n_3 + \cdots + n_k$ represent the number of observations in the groups, and $\bar{X}_1 + \bar{X}_2 + \bar{X}_3 + \cdots + \bar{X}_k$ represent the means of the samples. Then the mean of the combined group will be equal to:

$$\bar{X} = \frac{n_1\bar{X}_1 + n_2\bar{X}_2 + n_3\bar{X}_3 + \cdots + n_k\bar{X}_k}{n_1 + n_2 + n_3 + \cdots + n_k}$$ Formula (3.4)

This formula tells us to multiply the number of observations in each group by the group mean, sum these products over all the subgroups, and divide by the total number of observations in all k groups.

If n is the same for all the k groups, the process of finding the combined mean is simplified. In this case, the mean of the combined groups is equal to the mean of the subgroup means. For example, assume that the means for four subgroups of ten students each are found to be 10, 13, 15, 18. The mean for the combined group of 40 students would be: $(10 + 13 + 15 + 18)/4 = 14$ (that is, we simply determine the mean of the means).

It is not possible to compute the medians or modes of combined groups as easily. For both statistics we must generate a new frequency distribution for the combined groups in order to locate the new median or mode.

It is interesting to note the effect on the mean of adding a constant to each score in the distribution or of multiplying each score by a constant. When a constant is added to each score in a distribution, the mean of the new distribution is equal to the mean of the original scores plus the value of the constant added. In equation form:

$$\overline{X}_{X+C} = \overline{X}_X + C$$ where $\overline{X}_X$ = mean of original distribution of scores

C = constant added

$\overline{X}_{X+C}$ = mean of the new distribution where a constant has been added to each score

Consider the distribution: 2, 4, 7, 8, 9 with a mean of 6. If we add a constant 2 to each score, then the mean of the new distribution (4, 6, 9, 10, 11) will be equal to the original mean, 6, plus the constant, 2, = 8.

When each score in a distribution of scores is multiplied by a constant, the mean of the new distribution is equal to the mean of the original score multiplied by the constant. In equation form: $\overline{X}_{XC} = \overline{X} \cdot C$. Again consider the distribution: 2, 4, 7, 8, 9 with a mean of 6. If we multiply each score in the distribution by the constant 2, then the mean of the new distribution (4, 8, 14, 16, 18) is equal to the original mean, 6, multiplied by 2, = 12.

Another important characteristic of the mean is that the sum of squares of deviations about the mean is less than the sum of squares of deviations about any other value. This property is expressed by the following: $\Sigma (X - \overline{X})^2 \leq \Sigma (X - A)^2$, where $A \neq \overline{X}$, which reads: the sum of the squared deviations from the mean is equal to or less than the sum of the squared deviations about any other score A (where A does not equal the mean). Let us illustrate this characteristic of the mean in the following example:

X	$(X - \overline{X})^2$	$(X - 4)^2$
4	4	0
5	1	1
6	0	4
7	1	9
8	4	16
$\Sigma X = 30$	$\Sigma (X - \overline{X})^2 = 10$	$\Sigma (X - 4)^2 = 30$
$\overline{X} = 6$		

$(X-5)^2$	$(X-7)^2$	$(X-8)^2$
1	9	16
0	4	9
1	1	4
4	0	1
9	1	0
$\Sigma(X-5)^2 = 15$	$\Sigma(X-7)^2 = 15$	$\Sigma(X-8)^2 = 30$

The sum of the squares of deviations about the mean of 6 is 10. Note that the sum of the squares of the deviations about any of the other values is always greater than 10.

Thus, the mean can be further described as the central value about which the sum of the squared deviations is a minimum. The mean is a measure of central tendency from a *least squares* point of view.

This concept of least squares has importance in statistics in the fitting of lines and curves, as we will see later.

Because the mean is amenable to numerous mathematical operations, it is widely used in further statistical calculations and in statistical inference. Most procedures in inferential statistics call for the mean rather than the other measures of central tendency.

Exercises

1. The mean test score on a statistics quiz for a class of 20 students is 60. The mean for a second class of 30 students is 45. What is the mean of the two groups combined?
2. Consider the following set of scores: 10, 9, 9, 8, 7, 6, 5, 8, 4, 4
   (a) Calculate the mean.
   (b) If 5 were added to each score in the distribution, what would be the value of the mean of the new distribution?
   (c) If each score in the distribution were multiplied by 2, what would be the value of the mean of the new distribution?
3. If the mean income for a group of 20 teachers is $13,000, what is the total income of this group?
4. Three groups of ten students each had means on a quiz of 79, 80, and 84, respectively. Determine the mean of the groups combined.

Solutions

1. 51 $\left[\dfrac{(20 \times 60) + (30 \times 45)}{50}\right]$
2. (a) 7 $\left(\dfrac{70}{10} = 7\right)$
   (b) 12 (7 + 5) The new mean is equal to the original mean plus the constant.

70 DESCRIPTIVE STATISTICS

> (c) 14 (7 × 2) The new mean is equal to the original mean multiplied by the constant.
>
> 3. $260,000 $\left(\bar{X} = \frac{\Sigma X}{n}, \quad n\bar{X} = \Sigma X, \quad 20 \times \$13{,}000 = \$260{,}000\right)$
>
> 4. 81 $\left[\frac{(79 + 80 + 84)}{3}\right]$

How To Mislead with Measures of Central Tendency

The chairman of the teachers' union in a small school complains that the average salary there is only $8,000. The president of the Taxpayers' League complains that the average salary is a whopping $10,500. The president of the PTA reports that the average is $9,000. The salaries are as follows: Who is lying?

x	f
$19,000	1
11,000	1
10,500	1
9,000	1
8,000	3

Actually, no one is lying; each is simply reporting the measure of central tendency that serves his purpose best. Three new teachers are at the starting salary, $8,000, which is the most frequently occurring score or mode. One teacher has been in the system for 40 years and has benefited from salary increments year after year so when we add the $19,000 salary in with the others we have $73,500 ÷ 7 or $10,500 for the mean. The median, $9,000, gives us the figure halfway through the distribution, or the typical salary.

It suits the union chairman's purpose to report the lowest figure, but no one can deny that more teachers receive $8,000 than any other figure. It suits the Taxpayers' League to report the highest figure but it is certainly true that it will take $10,500 × 7 of taxpayers' money to pay these seven teachers. The PTA president is thinking in terms of the typical teacher's pay.

Comparison of the Mean, Median, and Mode

The mode is a crude and unstable measure of central tendency that is not often used to describe a distribution. One generally prefers either the median or the mean because of their greater sensitivity and stability. One of the most important differences between the median and the mean is that the median is a rank or position statistic, unaffected by the numerical size of the scores while the mean is sensitive to the size of all scores including extreme scores in the distribution.

When a frequency distribution is represented graphically, for example as a polygon, an interesting comparison can be made of the three measures of central tendency. The mode is the score point with the greatest frequency, the point on the horizontal axis which corresponds to the tallest point of the curve. The median is the score point which bisects the total area, the point on the horizontal axis where an ordinate would divide the area under the curve into two equal parts. Half of the total area would fall to the left and half to the right of an ordinate drawn at the median. The mean is the score point on the horizontal axis which corresponds to the point of balance or fulcrum of the distribution. If the frequency distribution is unimodal and perfectly symmetrical, the mean, median, and mode will all fall at exactly the same point. However, this is not likely to happen in practice with distributions of empirical data.

In a bimodal symmetrical distribution, the mean and median will be equal but the modes will fall at other points. Figure 3.5 illustrates this relationship between the mean, median, and mode in symmetrical distributions.

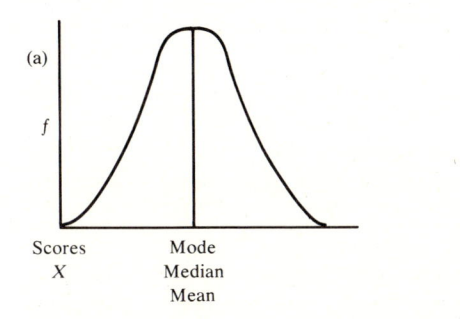

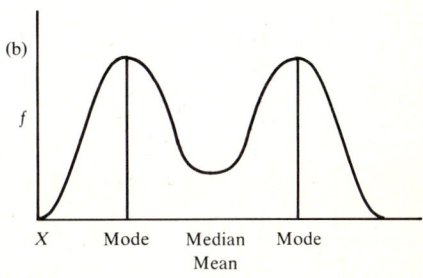

Figure 3.5 Mean, Median, and Mode in Two Symmetrical Distributions

Curve (a) in Figure 3.5 is unimodal and Curve (b) is bimodal.

If a distribution is skewed, that is, if scores are concentrated more at one end or the other, the curve will not be symmetrical and the three measures will not be equal. They will fall at different points in the distribution. Figure 3.6 illustrates the relationship among the three measures in skewed distributions.

72 DESCRIPTIVE STATISTICS

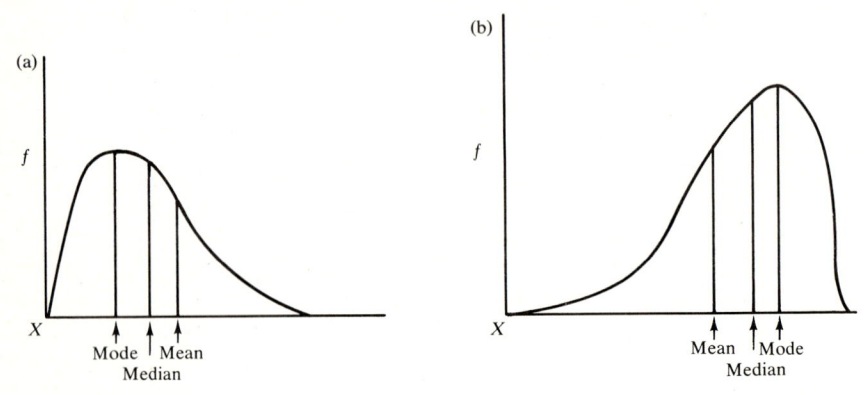

Figure 3.6 Relationship of the Mode, Median, and Mean in Positively and Negatively Skewed Curves

In the figure we can see that in the positively skewed distribution (a) the hump is on the left; this indicates that the mode corresponds to a low numerical value. The tail extends to the right, so the mean, which is sensitive to each score value, will be pulled in the direction of the extreme scores and will have a high value. The median, the middle score, is least affected by the hump and tail and hence will have a value between the other two measures. It will be lower than the mean but higher than the mode.

In the reverse situation, that is, in a negatively skewed distribution (b), the mean will have a lower numerical value than the median because the extremely low scores will pull the mean to the left. A hump usually occurs to the right to give the mode a high numerical value and again the median will be in the middle. This can be seen in Figure 3.6(b). It will be helpful to remember that the mean is always pulled in the direction of the skewed side of the distribution, a fact which illustrates quite succinctly the influence of extreme scores on the mean.

If we compute the values of the measures of central tendency for the frequency distribution shown in Table 3.4, we will have a different method of comparing them. Inspection of the distribution indicates that the bulk of the scores are of low value with a few high scores present; hence the distribution would be expected to be slightly skewed to the right, or positively skewed.

In this positively skewed distribution, the mean has the highest value, the mode has the lowest value, and the median is between the two (see Figure 3.7). In a positively skewed distribution the mean is *always higher* than the median. In such distributions the mode is usually lower than the median but this is not always the case. In negatively skewed distributions the mean is *always lower* than the median. The mode in a negatively skewed distribution is usually but not always higher than the median.

TABLE 3.4 Computation of Mean, Median, and Mode in a Positively Skewed Distribution

X	f	fX	cf
20	3	60	34
18	1	18	31
17	1	17	30
16	2	32	29
14	4	56	27
12	5	60	23
11	5	55	18
10	6	60	13
9	4	36	7
7	3	21	3
	34	415	

where:

$$\bar{X} = \frac{\Sigma fX}{n} = \frac{415}{34} = 12.21$$

$$\text{Mdn} = L + \frac{(N/2) - f_b}{f_w} i$$

$$= 10.5 + \frac{(34/2) - 13}{5} 1$$

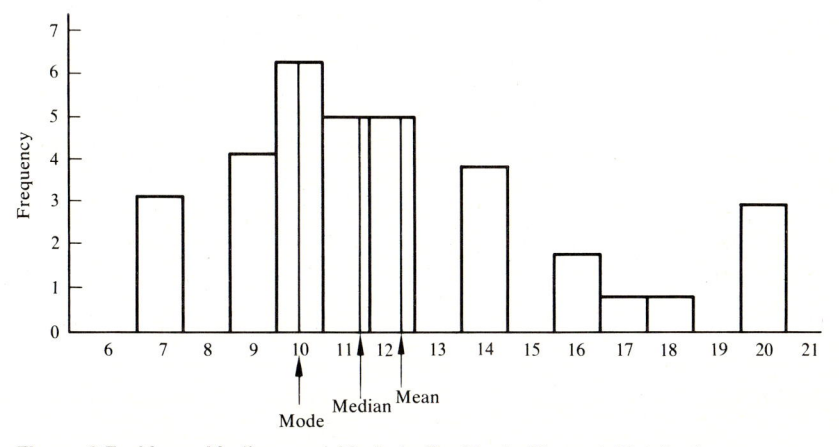

$$= 10.5 + .8(1)$$

$$= 11.3 \qquad \text{Mode} = 10$$

Figure 3.7 Mean, Median, and Mode in Positively Skewed Distribution

With knowledge of the mean and median we can determine the skewness of a distribution. For example, the median per capita income in the 1970 U.S. Census was $3,137. The mean per capita income was $4,138. From these figures we can determine that incomes in the United States were positively skewed and that individual incomes extended much farther above the median than they extended below the median.

The measures of central tendency may also be compared on the basis of their *stability*. The mean is the most stable or reliable measure of central tendency. If one were to draw numerous samples from a population, the means would show less fluctuation from sample to sample than the medians, and the medians less than the modes. Thus, if one is using a sample as a basis for inferring something about the population, which is a frequent procedure in statistics, the mean will provide the most reliable estimate of the corresponding population parameter.

Exercises

1. On a moderately easy classroom test where there are only a few low scores, one might expect the _____ (mean or median) to be higher in value.
2. Consider the following frequency polygon. Place the three measures of central tendency in order beginning with the one lowest in value.

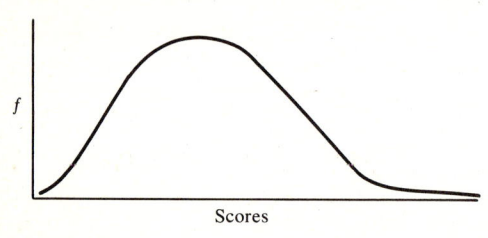

Solutions

1. The median would be higher in value. The curve is negatively skewed and the mean is pulled to the left (low values).
2. mode, median, and mean (The curve is positively skewed and the mean is pulled to the right. The mode is lowest with the median in between.)

Selecting a Measure of Central Tendency

There are several factors to consider when selecting a measure of central tendency. The first consideration is the type of scale represented by the data. If one has a nominal scale, then the mode is the only legitimate statistic to use. Recall that the mode is determined only by frequency of occurrence and not by the order of the variables or their numerical values. For example, suppose that a college population is divided into three groups on the basis of place of residence: 15 percent have apartments in town, 60 percent live in university housing, and 25 percent live in sororities or fraternities. We might report that the "average" student at the college lives in university housing. In this case, we are using the mode because the data are nominal.

If we were talking about the "average" salary of faculty members at a university, we would most likely use the median. That is, we would place all the salaries in order (ordinal scale) and then determine the middle value. The median would be preferred over the mean because the salaries of a few highly paid professors would distort the mean to a disproportionate extent and it would not be the most typical. The median is an ordinal statistic and is used when data are in the form of an ordinal scale. Of course, a mode can also be reported for ordinal data.

If one has an interval or ratio scale, then the mean is the recommended measure of central tendency, although the median or mode may be reported for these types of scales. For example, if we were reporting the average score made on a classroom test which is assumed to represent an interval scale, the mean would generally be used.

A second consideration in choosing a measure of central tendency is the purpose for which the measure is being found. If we want the value of every single observation to contribute to the average, then the mean is the appropriate measure to use. The median is preferred when one does not want extreme scores at one end or the other to influence the average or when one is concerned with "typical" values rather than with the value of every single case. If a community wanted to know the average taxable value of all its real estate, then the mean would be used since every type of real estate would be taken into consideration. However, if a community wanted to know the cost of the average family dwelling, the median would probably give a more accurate and typical picture of the typical residence.

In general, the mean is best for situations where *quantity* is the basic consideration, and the median is best where *typicalness* is the basic consideration.

If the purpose of the statistic is to provide a measure that can be used in further statistical calculations and for inferential purposes, then the mean is the best measure. The mean is amenable to the mathematical manipulation called for in more advanced statistical procedures. The median and mode are described as "terminal statistics" as they are not used in more advanced statistical calculations. Furthermore, the mean has greater sampling stability than the other measures. This is an important advantage when making inferences from samples to populations because the mean of a sample provides a better estimate of the population parameter than do the other measures of central tendency.

Because of their characteristics we find that the mean is the most widely used measure, the median is next, and the mode is used the least.

Table 3.5 summarizes the characteristics and uses of the measures of central tendency.

76 DESCRIPTIVE STATISTICS

TABLE 3.5 Measures of Central Tendency

CHARACTERISTICS	USE WHEN
MEAN	
1. An interval statistic	1. An interval interpretation is appropriate
2. A calculated average	2. One wants the value of each score to be represented
3. Value determined by every case in the distribution	3. The question is quantity rather than typicalness
4. Affected by extreme values	4. Further statistical computation is anticipated
5. Sum of deviations about the mean is zero	
6. Amenable to numerous mathematical operations	
7. Most stable measure	
8. Provides best estimate of population parameter	
9. Most widely used	
10. Represents average quantity	
11. It can be used with ratio or interval data	
MEDIAN	
1. An ordinal statistic	1. An ordinal interpretation is appropriate
2. A rank or position average	2. The middle score is desired
3. Value determined by scores near the middle of the distribution	3. One wants the typical value rather than an average determined by the value of each case
4. Insensitive to extreme values	4. When there are extreme scores which would distort a calculated average
5. Closest to the aggregate of all cases if the direction of the difference is ignored	
6. Amenable to only a few mathematical operations	
7. Less stable than mean, more stable than mode	
8. Better estimate of population parameter than mode but poorer than mean	
9. Less widely used than mean	
10. Represents typical score	
11. Can be used with ordinal, interval, or ratio data	
MODE	
1. A nominal statistic	1. A nominal interpretation is appropriate
2. An inspection average	2. One wants to know the most frequently occurring score
3. The most frequently occurring value	

Table 3.5 (Continued)

CHARACTERISTICS	USE WHEN
4. Usually occurs near the center of a distribution 5. Some distributions have more than one mode 6. Cannot be manipulated mathematically 7. Least stable 8. Poor estimate of population parameter 9. Rarely used 10. Most "popular" score 11. Can be used with nominal, ordinal, interval, or ratio data	3. A quick approximation of central tendency is desired

Exercises

Criticize the use of the measure of central tendency indicated in the following examples:

1. A math teacher knew the medians of each of her five math classes on the problem-solving section of a standardized test. They were 65, 68, 72, 79, and 80. She averaged these medians and reported that the median score for all her students on the test was 72.8.

2. A vacation Bible School director wanted to determine the average daily contribution. That day five children each contributed 5¢, fifteen contributed 10¢ each, fourteen contributed 15¢ each, and sixteen contributed 25¢ each. The director also found a $10 bill contributed by an adult visitor that day. She computed the mean contribution to be 35¢.

3. A reporter attending a school board meeting heard that the mean salary for all teachers in that city was $12,000. He wrote in his article, "Half of our teachers receive more than $12,000 per year."

4. A teacher found that the median grade equivalent for her fourth grade class in the five basic subjects was: reading 4.9, spelling 4.7, math 4.5, science 4.3, and social studies 4.7. She added these medians, divided by five, and reported that the average grade equivalent on all subjects was 4.6.

Solutions

1. Medians of separate groups cannot be added in order to find the median for the combined groups. The teacher would have to take all five distributions, put them together in one big distribution, and then compute the median.

78 DESCRIPTIVE STATISTICS

2. The mean should not be used because the $10 is an extreme score which influences the mean to a disproportionate extent. The median would be more typical and should be used in this case.
3. $12,000 is the average salary, the mean. It is not the point above which and below which one-half of the salaries would fall.
4. Medians cannot be added to find the average median.

FLOWCHART 3

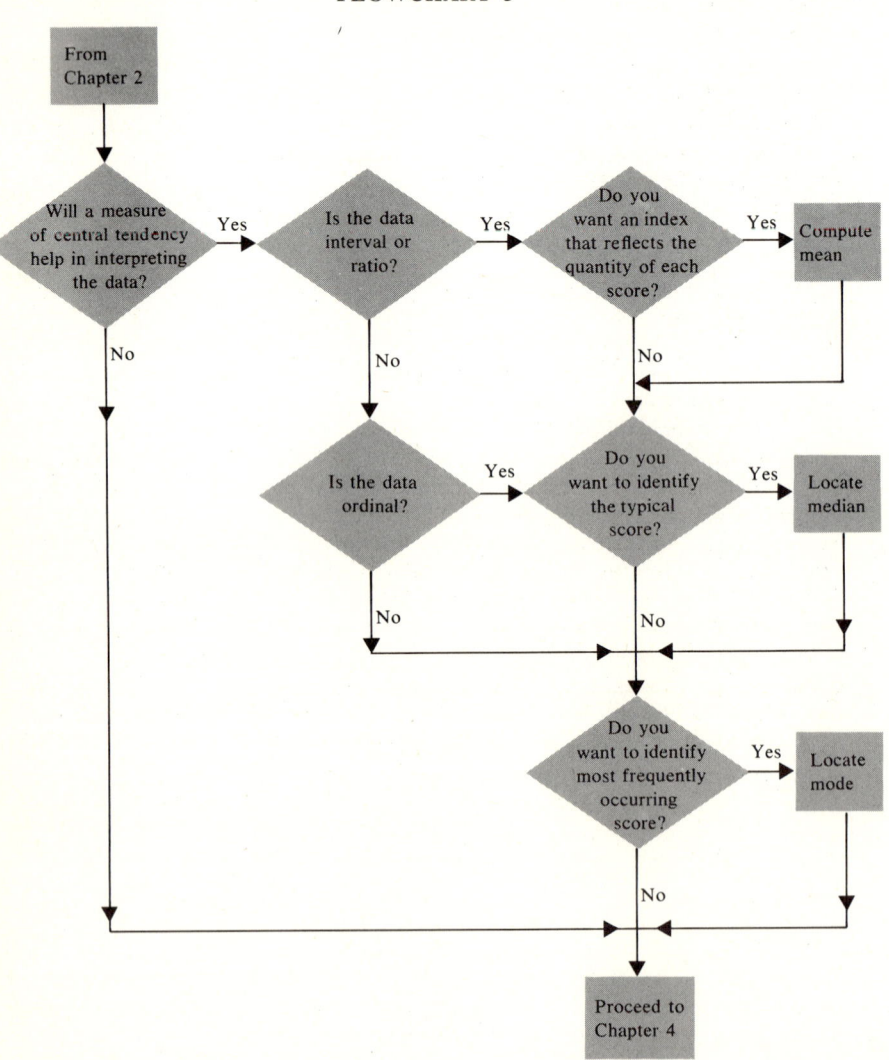

From Chapter 2

Will a measure of central tendency help in interpreting the data? — Yes → Is the data interval or ratio? — Yes → Do you want an index that reflects the quantity of each score? — Yes → Compute mean

No ↓ (from central tendency)

No ↓ (from interval/ratio) → Is the data ordinal? — Yes → Do you want to identify the typical score? — Yes → Locate median

No ↓ (from ordinal)

No ↓ (from typical score)

Do you want to identify most frequently occurring score? — Yes → Locate mode

No ↓

Proceed to Chapter 4

SUMMARY

Measures of central tendency are useful indices for summarizing a whole set of measures. There are measures of central tendency appropriate for each scale of measurement and they each yield a somewhat different type of information. The mode is a nominal statistic—it indicates that score which occurs most frequently. It is the least stable and least useful measure of central tendency. The median is an ordinal statistic equal to that point which divides the distribution in half. It is not sensitive to the magnitude of the scores but only to their rank and frequency. Thus, it is unaffected by extreme scores. The mean, an interval statistic, is generally the most stable and most widely used measure of central tendency. It takes into account every score in the distribution and can be used in computation for more sophisticated statistical analyses. It is equal to the sum of the scores divided by the number of scores.

Which measure of central tendency should be used is a function of the information that one wants to communicate with the statistic and of the particular circumstance of the score distribution. As we have seen, a measure of central tendency can be chosen to confuse as well as to clarify the information yielded by a distribution of measures.

In a positively skewed distribution the mean is higher than the median. In a negatively skewed distribution the mean is lower than the median.

EXERCISES

1. With the following distribution of scores from a spelling quiz, draw a histogram and indicate on the graph the mode, median, and mean.

SCORE	FREQUENCY
8	3
7	3
6	5
5	8
4	6
3	4
2	2
1	1
0	1

2. Given the following distribution: 15, 14, 14, 13, 11, 10, 10, 10, 8, 5
   (a) Calculate the mean.
   (b) Determine the value of the median.
   (c) Determine the value of the mode.
   (d) If 4 were added to each score in the distribution, what would be the mean of the new distribution?

80 DESCRIPTIVE STATISTICS

   (e) If each score in the distribution were multiplied by 4, what would be the value of the mean in the new distribution?
3. Determine the median of each of the following distributions.
   (a) 5, 9, 16, 25, 30
   (b) 4, 8, 17, 18, 30, 45
   (c) 10, 10, 12, 12, 15
4. Calculate the mean for each of the following subgroups.
   (a) Using Formula 3.4, determine the overall mean of the combined groups.

   Group 1:   5, 4, 3, 4
   Group 2:   6, 4, 2, 2, 1
   Group 3:   5, 3, 2, 1, 1, 0

   (b) What would be the value of the median of the combined groups?
5. Given four groups each with $N = 25$ whose means are 40, 72, 73, and 75, respectively, determine the value of the mean of the combined groups.
6. From the following data determine the mean of the combined groups.

   Group 1: $n = 10, \bar{X}_1 = 50$
   Group 2: $n = 8, \bar{X}_2 = 35$

7. Briefly explain the relationship between the skewness of a distribution of scores and the resulting values of the mean, median, and mode.
8. Match the following correctly.
   1. Mode   _____ ordinal
   2. Mean   _____ nominal
   3. Median   _____ interval
9. Match the following correctly.
   1. Mean   _____ the middle score
   2. Median   _____ the arithmetic average
   3. Mode   _____ the most frequently occurring score
10. If a person wants to choose a measure of central tendency that would indicate the typical score and yet be relatively unaffected by extremely high or low scores, he would probably choose the:
    (a) mean
    (b) median
    (c) mode
11. Go back to the distribution of scores in Problem 1 and add to it a score of 50. Recalculate the mean, median, and mode. What is the effect of adding this one score to the distribution?
12. To calculate the _____ you would have to first put the scores in rank order.
    (a) mode
    (b) mean
    (c) median
13. For each of the following indicate mode, median, or mean:
    (a) The score made by the typical examinee is the _____.
    (b) The measure of central tendency that is directly influenced by the value of every score is the _____.

(c) The measure that is based on rank or position is the _____.
(d) The measure that varies *most* from sample to sample within a population is the _____.
(e) The measure that varies *least* from sample to sample within a population is the _____.
(f) In a negatively skewed distribution one expects the _____ to be the lowest.
(g) A line drawn straight up from the _____ will divide the area of the frequency polygon exactly in half.

ANSWERS

1. mean = 4.73
   median = 4.81
   mode = 5
2. (a) mean = 11
   (b) median = 10.5
   (c) mode = 10
   (d) mean = 15
   (e) mean = 44
3. (a) 16
   (b) 17.5
   (c) 11.75
4. Group 1: $\overline{X} = 4$ (a) combined mean = 2.87
   Group 2: $\overline{X} = 3$
   Group 3: $\overline{X} = 2$ (b) median = 2.75
5. 65
6. 43.33
7. The three measures are not equal in a skewed distribution. The mean is pulled in the direction of the skewed side. Thus, in a positively skewed distribution, the mean has the highest value, the median is in between, and the mode is lowest in value. In a negatively skewed distribution, the mean is lowest in value, the median is in between, and the mode is highest in value.
8. median
   mode
   mean
9. median
   mean
   mode
10. median
11. mean becomes 6.06
    median becomes 4.875
    mode remains 5
12. median
13. (a)—median, (b)—mean, (c)—median, (d)—mode, (e)—mean, (f)—mean, (g)—median

CHAPTER FOUR
MEASURES OF VARIABILITY

OBJECTIVES

In this chapter, we consider the second group of measures used to describe distributions: measures of variability, which indicate how much spread or dispersion there is in the scores. After studying this chapter, the student should be able to:

1. Know the purpose of using measures of variability and name the commonly used measures.
2. Define range and determine the range for a distribution.
3. State the advantages and disadvantages of the range as a measure of variability.
4. Define quartile deviation and explain how it provides an index of variability.
5. Compute the quartile deviation for a distribution.
6. State the advantages and disadvantages of the quartile deviation as a measure of variability.
7. Define variance and compute the variance of a distribution.
8. State the advantages and disadvantages of the variance as a measure of variability.
9. Define standard deviation.
10. Compute the standard deviation using the definition formula and the computational formula.
11. Interpret standard deviation as an index of the variability of a distribution.

12. State the advantages and disadvantages of standard deviation as a measure of variability.
13. Know when each of the measures is the appropriate index of variability to use.

Introduction

We have seen in Chapter 3 the usefulness of measures of central tendency for providing a concise index of the average value of a set of scores. There is more to be known about sets of scores, however, than this one characteristic. Another very important attribute is the variability or spread of the scores. Variability is a universal characteristic of any set of scores with which the teacher, psychologist, or researcher might have to deal. For example, measures of achievement, intelligence, personality, or other characteristics may be expected to show variability in any sample of individuals. The measures of central tendency tell us nothing about the variability or spread of the scores in a distribution. In fact, distributions may have the same mean yet differ in the extent of variation of the scores around that measure of central tendency. It would be important for the teacher to know whether the range of intelligence test scores in the class is from 85–135 or from 95–110, since teaching methods and materials used would differ even though the mean intelligence score might be the same in both situations. The spread of the scores makes a difference. Figure 4.1 shows

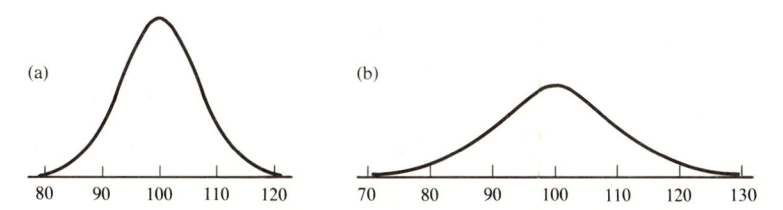

Figure 4.1 Two IQ Distributions with the Same Mean but Differing in Variability: (a) Small Variability; (b) Large Variability

two distributions of intelligence test scores that have the same mean yet differ in that one shows greater variability. In distributions (b) the scores are spread out to a greater extent than they are in distribution (a), where the scores are clustered closer to the mean.

We see that in order to adequately describe a distribution of scores, we need, in addition to a measure of central tendency, a measure of the variability of the scores. Information concerning variability may be as important or more important than information concerning central tendency.

84 DESCRIPTIVE STATISTICS

Consider the following weekly spelling test scores for three girls during a six-week period:

KAY	LOUISE	MARTHA
19	16	20
20	17	12
9	16	19
17	16	13
16	15	20
15	16	12
$\Sigma X = 96$	$\Sigma X = 96$	$\Sigma X = 96$
$\bar{X} = 16$	$\bar{X} = 16$	$\bar{X} = 16$

Each girl has a mean score of 16, yet it would be misleading to describe their spelling performances as being identical. They differ in the consistency of their performance. We see that Louise's scores are less variable, less spread out, than the scores of the other two girls.

The measures of variability provide a needed index of the extent of variation among the scores in a distribution. The range, quartile deviation, variance, and standard deviation are all measures used for this purpose. These measures of variability along with the measures of central tendency make up two types of descriptive statistics which are indispensable in describing distributions of scores.

The Range

The simplest measure of variability is the *range*. The range is defined as the difference between the largest and the smallest scores in a distribution of scores. Using a formula, the range is $R = X_h - X_l$, where X_h represents the highest score and X_l is the lowest score. In other words, the range is the distance between the highest score and the lowest score.

Range = $X_h - X_l$ where: X_h = highest score in distribution

X_l = lowest score in distribution   Formula (4.1)

In our example Kay's scores have a range of eleven (20 − 9), Louise's scores have a range of two (17 − 15), and Martha's scores have a range of eight (20 − 12).

Some statistics books define the range as the difference between the highest and the lowest scores plus 1; that is, range = $(X_h - X_l) + 1$. This definition considers the lower limits of the lowest score and upper limits of the highest score, in which case the range is one larger than that obtained by simply subtracting lowest score from highest score. For example, Kay's scores above would show a range of twelve (20.5 − 8.5) according to this interpretation. We will use the simple difference formula above, but the student should be aware that this other interpretation does exist and may appear in discussions of the range in other statistics books.

The range is a rough indication of the variation in a set of scores; it tells us the spread of scores from high to low. It is a very quick and simple statistic to compute and this ease of computation is the main advantage of the range. Unfortunately its disadvantages far outweigh that advantage and restrict its use to that of a quick estimate of variability.

Since the range takes into consideration only the two extreme scores it may not be truly descriptive of the distribution, especially when frequent or large gaps occur. The range does not reflect the variability of the scores between the two extremes. If we look again at Kay and Martha's scores we note that although Kay's scores have the greatest range (11), her scores were reasonably consistent with the exception of one very low score. The other five scores only have a range of five. Can we really say there is more variation in Kay's scores than there is in Martha's scores even though her range (8) is less than the range of Kay's scores?

The range is not considered a stable measure of variability because its value can fluctuate greatly with a change in just a single score—either the highest or the lowest.

Use of the range as an index of variability is restricted to an interval scale, where the *differences* between the numbers have an empirical interpretation. Only with an interval scale can one meaningfully speak of largest and smallest scores and the difference between them. The numbers used in a nominal scale are used only for identification of categories and in an ordinal scale only to represent the order of the objects under consideration. The difference between the numbers in nominal and ordinal scales has no empirical meaning; hence the range is not an appropriate statistic to use with these scales.

Exercises

1. The IQ scores for two classes are given below:

    Class A: 102, 101, 100, 99, 98
    Class B: 150, 130, 100, 70, 50

    (a) Compute the mean IQ for each class. Would this measure of central tendency be a sufficient basis for making decisions about instructional methods, materials, and so on?
    (b) What additional information would the teacher need?
2. Find the range for each of the sets of scores below:
    (a) 6, 4, 4, 9, 11, 7, 10, 8
    (b) 7, 8, 8, 10, 9, 5, 7, 6
3. Compute the range of the IQ scores for the two classes in Exercise 1.

> Solutions
>
> 1. (a) The mean is 100 in both classes. No, the teacher should know more than just the mean score. If she considered the mean only, she might conclude that the two classes are similar in ability and attempt to teach them in the same way.
> (b) She would need information about the variability of the scores—how spread out they are in each class. Inspection of the scores shows that Class B is highly variable in IQ, while Class A is homogeneous.
> 2. (a) 7
> (b) 5
> 3. Range in Class A is 4; range in Class B is 100.

Quartile Deviation

A second index of variability is the *quartile deviation*. It is a more stable measure than the range, because it is based on the spread through the center of the distribution of the total group rather than using only the two extreme scores. The quartile deviation is defined as half the distance between first and third *quartiles* in a distribution. Quartiles are points on the score scale that divide the total number of observations or scores in a distribution into four equal groups. The first quartile, Q_1, is the score point that sets off the lowest 25 percent of the groups (it is also referred to as the 25th percentile). The middle quartile, Q_2, is the median score point (the 50th percentile). The third quartile, Q_3, is the score point that sets off the upper 25 percent of the group of scores (it corresponds to the 75th percentile). The interval from Q_1 to Q_3 contains the middle 50 percent of the cases in a distribution and is called the *interquartile range*. To calculate the interquartile range, one first determines the score points of the first and third quartiles (Q_1 and Q_3) and then finds the difference between these two.

The next step is to divide the interquartile range by two, which gives the average distance from the median to each of the quartiles. This is called the *semi-interquartile range* or quartile deviation (QD).

To illustrate let us consider a simple example of eight scores, as follows: 6, 8, 9, 11, 12, 13, 14, 17. The score point 8.5 separates the lowest fourth of the scores from the rest and is identified as Q_1. The score point 13.5 separates the highest fourth from the rest and is identified as Q_3. The difference between these two points $13.5 - 8.5 = 5$ is the interquartile range. Half this, $5 \div 2 = 2.5$, is the semi-interquartile range.

Distributions differ, of course, in the size of the semi-interquartile range. When the scores in a distribution are spread out, this means that the average distance from the median to the quartiles is greater and this spread will be reflected in a greater QD or semi-interquartile range. Conversely, if the scores are more closely concentrated around the median, the QD will be smaller because the average distance from the median to the quartiles is less. Figure 4.2 shows two distributions which differ in the size of the QD.

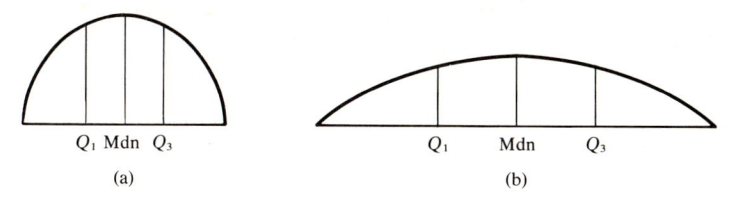

Q_1 Mdn Q_3
(a)

Q_1 Mdn Q_3
(b)

Figure 4.2 Distributions with Different Quartile Deviations

Distribution (b) in Figure 4.2 has greater spread and hence a greater quartile deviation than distribution (a).

The procedure for calculating Q_1 and Q_3 is similar to the procedure for calculating the median (Q_2). For Q_1 the formula is:

$$Q_1 = L + \left(\frac{N/4 - cf_b}{f_w}\right) i$$

where: Q_1 = the first quartile

N = the number of cases in the distribution

$N/4$ = 25 percent of total $N (\frac{1}{4}N)$

L = the lower limit of the interval within which the first quartile lies

cf_b = the cumulative frequency below the interval containing the first quartile

f_w = the frequency of cases within the interval containing the first quartile

i = the interval size   Formula (4.2)

The formula for the third quartile is:

$$Q_3 = L + \left(\frac{\frac{3}{4}N - cf_b}{f_w}\right) i$$ Formula (4.3)

Let us work an example using this formula.

X	f	cf
8	2	20
7	5	18
6	4	13
5	3	9
4	2	6
3	4	4

In the formula we must begin with the ratio $N/4$ first. Since there are 20 cases, then Q_1 is the point below which $20/4 = 5$ cases fall. Counting up from the bottom, we find that four cases fall below the interval 4, the lower limit of which is 3.5. We need one more case from interval 4 to make 5. There are two cases in that interval. Thus we will go $\frac{1}{2}$ of the way into the interval 4. Thus $Q_1 = 3.5 + (\frac{1}{2})1 = 4$. Or Q_1 may be calculated with Formula (4.2) as follows:

$$Q_1 = 3.5 + \left(\frac{20/4 - 4}{2}\right)1$$
$$= 3.5 + \tfrac{1}{2}$$
$$= 4$$

Calculating the third quartile with Formula (4.3) we have:

$$Q_3 = 6.5 + \left(\frac{3 \cdot 20/4 - 13}{5}\right)1$$
$$= 6.5 + \tfrac{2}{5}$$
$$= 6.9$$

Once we have the values of Q_3 and Q_1 we divide the difference between them by 2 to obtain the quartile deviation:

$$QD = \frac{Q_3 - Q_1}{2} \qquad \text{Formula (4.4)}$$

In our example:

$$QD = \frac{6.9 - 4}{2}$$
$$= 1.45$$

It can be seen that the computation of the quartile deviation is more time-consuming than finding the simple range, but the result is a considerably more stable statistic. The quartile deviation is less influenced by chance differences in samples derived from a common population.

Since the quartile deviation, like the median, is insensitive to extreme scores, it is the preferred measure in situations where one wants an index that does not reflect the extreme scores. For instance, assume one is comparing the spread of annual family incomes in two communities. In Bluecollarton the first quartile is $4,500 and the third quartile is $9,500. The quartile deviation is $2,500. In Racetrack Heights the first quartile is $3,600 and the third quartile is $19,600, which gives a quartile deviation of $6,500. There is much greater variation in income in Racetrack Heights than in Bluecollarton. Assume that a family with a $500,000 per year income should move into Bluecollarton. This would increase the range of incomes there, but the quartile deviation would be affected little, if any. Since the addition of one rich family would probably have little impact on the social structure of Bluecollarton, the quartile deviation would be a more realistic index of the variability in family incomes than the range. Since quartile deviation is an index reflecting the spread of scores throughout the *middle* part of a distribution, it is preferred whenever extreme scores may distort the data. Thus the circumstances which require the use of the quartile deviation as a measure of variability are the same as those which call for the median as the measure of central tendency. The quartile deviation and the median belong to the same statistical family, since both are based on quartile points and both are insensitive to extreme scores.

One point about the semi-interquartile range or QD needs clarification. Although data are arranged in rank order when finding quartiles, the Q_1 and Q_3 points are computed on the basis of an interval scale; that is, on the distribution of raw scores. Q_1 is found by locating the point in the distribution setting off the lower 25 percent of the cases but it is reported as a *score value*, which implies an interval scale.

Exercises

1. Compute the range and QD for the following distribution:

X	f
8	3
7	4
6	3
5	3
4	2
3	1

2. Given the following information about a distribution of scores: $Q_1 = 28$ and $Q_3 = 78$:
   (a) What percent of the scores in the distribution would fall below score 78?
   (b) Twenty-five percent of the scores would fall below which score?
   (c) Calculate the QD for this distribution.
   (d) If this were a symmetrical distribution, 50 percent of the scores would fall below which score?
3. Assume that the largest score in a distribution was increased by 25 points; what effect would this score have on (a) the range; (b) the quartile deviation?
4. If a symmetrical distribution of 36 scores has a median of 65 and a QD of 7:
   (a) What score values would set off the middle half of the scores in the distribution?
   (b) About how many scores would you expect to have below 58?

Solutions

1. Range = 5 (8 − 3)
   QD = 1.21
   $Q_1 = 4.5 + (\frac{1}{3})1 = 4.83$
   $Q_3 = 6.5 + (\frac{3}{4})1 = 7.25$
   $$QD = \frac{7.25 - 4.83}{2} = \frac{2.42}{2} = 1.21$$

2. (a) 75 percent
   (b) 28
   (c) 25 $\left(\frac{78-28}{2}\right)$
   (d) 53 (the median) (78 − 25)

3. (a) increase the range
   (b) no effect on the QD

4. (a) 58 and 72
   (b) 9 (25 percent of 36)

Variance

The disadvantage of the quartile deviation as a measure of variability is that it does not take into consideration the value of each of the raw scores in the distribution. It identifies only the distance that is half the range between the first and third quartiles. To arrive at a more reliable indicator of the variability or spread in a distribution, one should consider the value of each individual score and determine the amount by which each varies from the mean of the distribution. You recall that the mean was identified as the most stable measure of central tendency. If one can derive an index based on how each score deviates from the mean score, one will have a more stable index of the variability in a distribution. This procedure is followed when computing the measures of variability called *variance* and *standard deviation*. Let us consider variance first.

The crux of this approach to variability is to consider the extent to which each individual score in a distribution deviates from the mean of that distribution. That is, one subtracts the mean from each score to determine the deviation of the score from the mean. This is expressed as $X - \mu = x$, where X = score, μ = mean, and x = deviation score. In this text, a small x always represents a deviation score. Scores above the mean will have positive x values and scores below the mean will have negative x values.

$$X - \mu = x \quad \text{where: } X = \text{the individual score}$$
$$\mu = \text{the mean of the distribution}$$

x = the deviation of the score from the mean

Formula (4.5)

After determining how much each score deviates from the mean, we need to add all of these deviations in order to arrive at a measure of the average deviation. A problem exists, however, when we attempt to sum the deviations of the scores from the mean. We find that the sum of those scores above the mean (positive deviation scores) exactly balances the sum of scores below the mean (negative deviation scores). The sum of all the deviation scores is always equal to zero. This is true by definition since the mean is the point around which the positive and negative deviation scores exactly balance. We can illustrate this with our example.

KAY		LOUISE	
SCORE	DEVIATION	SCORE	DEVIATION
X	$X - \mu$	X	$X - \mu$
19	+3	16	0
20	+4	17	+1
9	−7	16	0
17	+1	16	0
16	0	15	−1
15	−1	16	0
	$\Sigma x = 0$		$\Sigma x = 0$

MARTHA	
SCORE	DEVIATION
X	$X - \mu$
20	+4
12	−4
19	+3
13	−3
20	+4
12	−4
	$\Sigma x = 0$

The deviation scores provide a good basis for measuring the spread of scores in a distribution. However, we cannot use the sum of these deviations in order to get an index of spread because this sum in any distribution will be zero. To overcome this problem we square all the deviation scores. When the negative scores are squared, they become positive in sign (a minus times a minus is a plus), so all the squared scores will be positive scores. Then these squared deviation scores are added to give a measure called the sum of the squared deviation scores or simply the sum of squares (Σx^2). Thus, $\Sigma x^2 = \Sigma (X - \mu)^2$. The following shows the computation of the sum of the squares in our example.

92 DESCRIPTIVE STATISTICS

	KAY		LOUISE
DEVIATION SCORE	SQUARED DEVIATION	DEVIATION SCORE	SQUARED DEVIATION
x	x^2	x	x^2
+3	9	0	0
+4	16	+1	1
−7	49	0	0
+1	1	0	0
0	0	−1	1
−1	1	0	0
	$\Sigma x^2 = 76$		$\Sigma x^2 = 2$

MARTHA	
DEVIATION SCORE	SQUARED DEVIATION
x	x^2
+4	16
−4	16
+3	9
−3	9
+4	16
−4	16
	$\Sigma x^2 = 82$

Variance is a measure of variability which is derived from the deviation of scores from their mean. It is defined as the mean of the squared deviation scores. Thus, to calculate variance we sum the squared deviation scores and then divide by the number of cases. The population variance is symbolized by the lower-case Greek letter sigma (σ) raised to the second power, that is, σ^2. Thus, σ^2 represents the mean of a *population* of deviation scores squared (population variance).

$$\sigma^2 = \frac{\Sigma(X-\mu)^2}{N} = \frac{\Sigma x^2}{N}$$

where: σ^2 = the variance
X = the individual score
μ = the mean of the distribution
N = the number of scores in the distribution
Σx^2 = the sum of the squared deviations

Formula (4.6)[1]
Variance
(definition formula)

---

[1] Many statistics have both a definition formula and one or more computational formulas. The definition formula is the verbal definition of the statistic in formula form. Computational formulas simplify the arithmetic typically involved in calculating a statistic.

Formula (4.6) indicates that to calculate variance one first subtracts the mean from each score to obtain a series of deviation scores, x. Each of the deviation scores is squared and then summed. The sum of the squared deviations is divided by the number of cases.

Let us apply Formula (4.6) to calculate the variance of the spelling scores for the three girls.

$$\text{Kay:} \quad \frac{\Sigma x^2}{N} = \frac{76}{6} = 12.67$$

$$\text{Louise:} \quad \frac{\Sigma x^2}{N} = \frac{2}{6} = .33$$

$$\text{Martha:} \quad \frac{\Sigma x^2}{N} = \frac{82}{6} = 13.67$$

Comparing the variances we see that Louise's scores are much less diverse than those of the other two girls and that Martha's have the greatest variance. Although Kay's scores have the greatest range, when we consider the deviation from the mean for each girl's scores, Martha's scores vary the most. This illustrates the greater sensitivity of variance as an index of total dispersion, or spread, compared to the range which only indicates the dispersion of two scores, the highest and the lowest.

This measure of variation makes use of the average squared difference between the scores in a distribution and the mean of all the scores. It is an index of the spread or dispersion of the observations. If there are large differences between scores and the mean score this will be reflected in a large variance. If the differences between scores and the mean are small, then the variance will be small.

Variance is more widely used for inferential statistics than for descriptive statistics. Its usefulness for descriptive purposes is limited because in the process of squaring the deviation scores we depart from our original unit of measurement. If we measure height and have a distribution of measurement in feet and then square the deviation scores from the mean, we will have square feet and not the original unit of feet. In our example the variance of Martha's scores is 13.67 square words. The units in variance are often outlandish units such as square words, square hours, or square psychologists!

For example, we might find by survey that in Upper Uphoria the mean number of psychologists per elementary school is .68 psychologists and that the variance is .39 square psychologists. As silly as these results seem at face value, they do indeed yield useful information. If we find in Lower Melancolia that the mean number of psychologists per elementary school is equal to .28 and the variance is equal to .15 square psychologists per building, then we would know that Upper Uphoria has more psychologists per elementary school than Lower Melancolia and that the number of

94 DESCRIPTIVE STATISTICS

psychologists per elementary school varies more in Upper Uphoria than in Lower Melancolia.

Exercises

1. The following are retention scores of children who studied a list of nonsense syllables: 8, 6, 5, 4, 4, 3. Compute the variance of these retention scores.
2. What is the variance of the following set of scores: 10, 9, 10, 7, 5, 8, 6, 5, 4, 6?
3. If the variance of a sample of 50 scores is calculated to be 9.5, what is the sum of the squares of the deviations about the mean?

Solutions

1. 2.67   ($\Sigma x^2 = 16$, $N = 6$, $\sigma^2 = 16/6$)
2. 4.2   ($\bar{X} = 7$, $\Sigma x^2 = 42$, $\sigma^2 = 42/10$)
3. 475   (50×9.5)

Standard Deviation

This problem of departing from one's original unit of measurement can be remedied easily, however, by taking the square root of the variance. This takes us back to the same unit as the original measure and provides a useful descriptive statistic known as the *standard deviation*. The standard deviation is defined as the square root of the mean of the squared deviation scores, which is equivalent to saying that the standard deviation is the square root of the variance. The symbol for standard deviation is σ, which is the square root of σ^2, the variance. The definition for the standard deviation is expressed in the following formula:

$$\sigma = \sqrt{\frac{\Sigma x^2}{N}}$$

where: σ = standard deviation

Σx^2 = the sum of the squared deviation scores

N = the number of scores in the distribution

Formula (4.7)
Standard Deviation
(definition formula)

The steps to be followed in the computation of standard deviation through the use of the definition formula are:

1. Find the deviation of each score from the mean of all the scores.

2. Square each of these deviation scores.
3. Add the squared deviation scores together and divide the sum by N. This gives the mean of the squared deviation scores (variance).
4. Extract the square root of the variance.

The standard deviation of Martha's scores using Formula (4.7) is thus:

$$\sigma = \sqrt{\frac{\Sigma x^2}{N}} = \sqrt{\frac{82}{6}} = \sqrt{13.6} = 3.67$$

Exercise

1. Find the standard deviation of the following set of scores using the definition formula or the deviation score method: 5, 4, 3, 2, 1.

Solution

1. 1.414 ($\sqrt{2}$)

Computation of Standard Deviation from Raw Scores

The formula $\sigma = \sqrt{\Sigma x^2/N}$ is a direct translation of the verbal definition into mathematical symbols. When the mean is a whole number, this definition can be conveniently used for computation. When the mean is not a whole number the definition formula becomes unwieldy. Furthermore, in some cases its use may introduce inaccuracy due to rounding errors. Table 4.1 shows the calculation of the standard deviation where the mean is not a whole number. Although the raw scores were small and easy to handle whole numbers, the mean was a mixed number so that our computation required the squaring of numbers such as 1.2 and −2.8. Imagine how difficult the process would be with a mean such as $2\frac{6}{7}$ or 3.169841.

Fortunately it is possible to compute the standard deviation directly from raw scores without calculating all the deviation scores and squaring them. It can be shown algebraically that the sum of the squared deviation scores is equal to the sum of the squared raw scores minus the sum of the raw scores squared divided by N. In formula form this is expressed as $\Sigma x^2 = \Sigma X^2 - [(\Sigma X)^2/N]$. The first value, ΣX^2, requires us to square each raw score first and then sum those squares; the second value, $(\Sigma X)^2$, requires us to sum the raw scores first and then square that sum. Let us calculate the sum of squares, Σx^2, by using the raw scores from Table 4.1.

TABLE 4.1 Calculation of the Standard Deviation Using the Definition Formula

X	x	x^2
6	2.2	4.84
5	1.2	1.44
4	.2	.04
3	−.8	.64
1	−2.8	7.84
$\Sigma X = 19$	$\Sigma x = 0$	$\Sigma x^2 = 14.80$

$$\mu = 3.8$$

$$\sigma = \sqrt{\frac{\Sigma x^2}{N}}$$

$$= \sqrt{\frac{14.80}{5}}$$

$$= \sqrt{2.96}$$

$$= 1.72$$

In our example:

X	X^2
6	36
5	25
4	16
3	9
1	1
$\Sigma X = 19$	$\Sigma X^2 = 87$

$$\Sigma x^2 = \Sigma X^2 - \frac{(\Sigma X)^2}{N}$$

$$= 87 - \frac{(19)^2}{5}$$

$$= 87 - \frac{361}{5}$$

$$= 87 - 72.2$$

$$= 14.8$$

The resulting sum of squares, 14.8, is identical to the sum of squares we obtained with the definition formula as shown in Table 4.1 since the two formulas are mathematically equivalent. Putting this calculation of sum of squares into the numerator of Formula (4.7) we have

$$\sigma = \sqrt{\frac{\Sigma X^2 - \frac{(\Sigma X)^2}{N}}{N}}$$

where: $\Sigma X^2 =$ the sum of the squared raw scores
$(\Sigma X)^2 =$ the sum of the raw scores, squared
$N =$ the number of scores in the distribution
$\sigma =$ the standard deviation

Formula (4.8)
Standard Deviation
(raw score formula)

Measures of Variability 97

Using a computational formula such as Formula (4.8) is usually the most convenient procedure for calculating the standard deviation. Table 4.2 shows the calculation of the standard deviation using the computational formula.

TABLE 4.2 Calculation of the Standard Deviation by the Raw Score Method

X	X^2	
5	25	Computing standard deviation with the
5	25	computational formula we have:
5	25	
4	16	
3	9	$\sigma = \sqrt{\dfrac{\Sigma X^2 - \dfrac{(\Sigma X)^2}{N}}{N}}$
2	4	
1	1	
$\Sigma X = 25$	$\Sigma X^2 = 105$	$= \sqrt{\dfrac{105 - \dfrac{(25)^2}{7}}{7}}$
$\mu_x = \dfrac{25}{7} = 3.571429$		$= \sqrt{\dfrac{105 - 89.285714}{7}}$
		$= \sqrt{\dfrac{15.714286}{7}}$
		$= \sqrt{2.244898}$
		$= 1.498298$

This method of computation avoids the extra steps of determining and squaring deviation scores since it can be done using raw score data only.

To illustrate that the computational formula is not only equivalent but also easier to use, let us compute the standard deviation of the above example using the definitional formula: $\sigma = \sqrt{\Sigma x^2 / N}$.

X	x	x^2	
5	1.428571	2.040815	
5	1.428571	2.040815	$\sigma = \sqrt{\dfrac{\Sigma x^2}{N}}$
5	1.428571	2.040815	
4	.428571	.183673	
3	.571429	.326531	$= \sqrt{\dfrac{15.714285}{7}}$
2	1.571429	2.469389	
1	2.571429	6.612247	$= \sqrt{2.244898}$
$\Sigma X = 25$		$\Sigma X^2 = 15.714285$	$= 1.498298.$
$\mu_x = 3.571429$			

98 DESCRIPTIVE STATISTICS

Although we arrive at the same answer using either formula, computation is usually must less cumbersome and complicated when the raw score formula is used.

The raw score formula for Σx^2 can also be incorporated into the formula for variance:

$$\sigma^2 = \frac{\Sigma X^2 - \frac{(\Sigma X)^2}{N}}{N}$$

Formula (4.9)
Variance
(raw score formula)

Let us review the steps involved in the calculation of standard deviation directly from the raw scores:

1. Count the number of scores, N.
2. Find the sum of the raw scores, ΣX, then square this value. This gives $(\Sigma X)^2$.
3. Square each of the raw scores, then find the sum of these squared values; this yields ΣX^2.
4. Substitute ΣX^2, $(\Sigma X)^2$, and N in the formula.
5. $(\Sigma X)^2$ is to be divided by N and the quotient subtracted from ΣX^2.
6. Divide the quantity found in Step 5 by N. This yields the variance.
7. Extract the square root of the quantity found in Step 6. This gives standard deviation.

Exercises

1. Find the standard deviation of the following group of scores using the raw score formula: 5, 4, 3, 2, 1.
2. Find the standard deviation of the following set of scores using both the raw score method and the deviation score method: 8, 3, 5, 6, 1, 1.

Solutions

1.

X	X^2
5	25
4	16
3	9
2	4
1	1
15	55

$$\sigma = \sqrt{\frac{\Sigma X^2 - \frac{(\Sigma X)^2}{N}}{N}}$$

$$= \sqrt{\frac{55 - \frac{(15)^2}{5}}{5}}$$

$$= \sqrt{\frac{55 - 45}{5}}$$

$$= \sqrt{\frac{10}{5}} \text{ or } \sqrt{2}$$

$$= 1.414$$

2. $\sigma = 2.58$

Deviation method: $\Sigma x^2 = 40$; $\sigma = \sqrt{\dfrac{40}{6}}$; $\sigma = \sqrt{6.66}$; $\sigma = 2.58$

Raw score method: $\Sigma X^2 = 136$, $\Sigma X = 24$

$$\sigma = \sqrt{\dfrac{136 - \dfrac{(24)^2}{6}}{6}}$$

$$= \sqrt{\dfrac{40}{6}}$$

$$= \sqrt{6.66}$$

$$= 2.58$$

Computing Standard Deviation from a Frequency Distribution

Unless N (the number of cases) is very small, data are usually arranged in a frequency distribution. The steps for calculating the standard deviation are the same except that each raw score, X, and raw score squared, X^2, must be multiplied by its frequency, f.

$$\sigma = \sqrt{\dfrac{\Sigma fX^2 - \dfrac{(\Sigma fX)^2}{N}}{N}}$$

where: ΣfX^2 = the sum of the products found by squaring each raw score and multiplying it by the frequency with which it occurs

$(\Sigma fX)^2$ = the quantity found by squaring the sum of the products obtained when each score is multiplied by its frequency

N = the total number of scores in the distribution; found by summing the frequencies ($\Sigma f = N$)

σ = the standard deviation

Formula (4.10)
Standard Deviation
(of a frequency distribution)

Table 4.3 shows the computation of the standard deviation of a frequency distribution using Formula (4.10).

100 DESCRIPTIVE STATISTICS

TABLE 4.3 Calculating of the Standard Deviation from a Frequency Distribution

X	f	fX	fX^2
9	1	9	81
8	2	16	128
7	1	7	49
6	2	12	72
5	2	10	50
4	1	4	16
2	1	2	4
$N = 10$		$\Sigma fX = 60$	$\Sigma fX^2 = 400$

$$\sigma = \sqrt{\frac{\Sigma fX^2 - \frac{(\Sigma fX)^2}{N}}{N}}$$

$$= \sqrt{\frac{400 - \frac{(60)^2}{10}}{10}}$$

$$= \sqrt{\frac{400 - 360}{10}}$$

$$= \sqrt{4}$$

$$= 2$$

Exercise

1. Calculate the standard deviation of the following distribution:

X	f
15	1
11	1
9	2
7	2
5	1
3	2
1	1

Solution

1. $\sigma = 4 \quad (\Sigma fX^2 = 650, \Sigma fX = 70)$

$$\sigma = \sqrt{\frac{650 - \frac{(70)^2}{10}}{10}} = \sqrt{\frac{650 - 490}{10}} = \sqrt{16} = 4$$

The Variance and the Standard Deviation of Samples

The formulas that we have considered so far [Formulas (4.6) and (4.7)] for variance and standard deviation are to be used for calculating the variance and standard deviation of a *population*. They apply only when one wants a measure of variability in data that represent the total population.

When the data represent a sample, the formulas and the symbols for variance and standard deviation are slightly different. Let us compare the formulas and the symbolism used for the population and sample variance and standard deviation.

POPULATION (parameters)		SAMPLE (statistics)
$\sigma^2 = \dfrac{\Sigma x^2}{N}$	Variance	$s^2 = \dfrac{\Sigma x^2}{n-1}$
$\sigma = \sqrt{\dfrac{\Sigma x^2}{N}}$	Standard Deviation	$s = \sqrt{\dfrac{\Sigma x^2}{n-1}}$

Note that the symbol used for the sample variance is s^2 (a lower-case s raised to the second power) and the symbol for sample standard deviation is s. The corresponding symbols for the population parameters are σ^2 and σ. The main difference is in the formulas, however. When calculating sample variance, the sum of squares (Σx^2) is divided by $n - 1$ instead of N as is the case with the population variance. The sum of squares is found in exactly the same way and has the same symbol for both population and sample. Once the sum of squares is calculated, one must determine whether the data are a population or a sample. If it is a population, the sum of squares is divided by N; if a sample, it is divided by $n - 1$.[2] If the standard deviation is needed, then the square root of the appropriate variance is taken.

You can see that slightly different values will be obtained for variance depending on whether the data are considered a population or a sample. For example, in computing the variance of Kay's spelling scores earlier, we considered the six scores as a population and $\sigma^2 = 12.67$. However, if we treat those six scores as sample data, we get $s^2 = 15.2$. This represents a

---
[2] *The computational formula for variance [Formula (4.9)] is also modified.*

Population variance: $\quad \sigma^2 = \dfrac{\Sigma X^2 - \dfrac{(\Sigma X)^2}{N}}{N}$

Sample variance: $\quad s^2 = \dfrac{\Sigma X^2 - \dfrac{(\Sigma X)^2}{n}}{n-1}$

102 DESCRIPTIVE STATISTICS

rather marked difference in value. However, this large difference is due to the small number of cases involved. As n gets larger, the difference between the variance computed as a parameter and the variance computed as a statistic decreases.

You may question the need for a different formula for the sample and population variances. The reason for this is because the formula for the sample variance gives a better estimate of the population variance than if one used the regular population formula. Most often we must work with samples rather than with a population, and we usually want to use the sample data to estimate the unknown population values.

It can be demonstrated mathematically that the variance of a sample computed by Formula (4.6) tends to underestimate the variance of the population from which the sample was drawn. It can also be demonstrated mathematically that if the value $n - 1$ is substituted for N in the formula, the result will be a better estimate of the population variance. Thus, the formula for the sample variance which is used to give the best estimate of the population variance is shown in:

$$s^2 = \frac{\Sigma x^2}{n - 1}$$

where: s^2 = variance of a sample

Σx^2 = sum of the squared deviation scores = $\Sigma (X - \bar{X})^2$

$n - 1$ = number in the sample minus one

Formula (4.11)
Sample Variance

Then the formula for the sample standard deviation is shown in:

$$s = \sqrt{\frac{\Sigma x^2}{n - 1}}$$

where: s = standard deviation of a sample

Σx^2 = sum of the squared deviation scores

$n - 1$ = number in the sample minus one

Formula (4.12)
Sample Standard Deviation

To summarize, when computing variance and standard deviation, one first finds the sum of squares. Then if the data represent a *population*, the sum of squares is divided by N to give σ^2, the population variance. If the data represent a *sample*, which is usually the case, the sum of squares is divided by $n - 1$ to give s^2, the sample variance. This sample variance provides

Exercises

1. Listed below are the number of errors made by the *population* of left-handed boys at Central High School on a test of manual dexterity. Calculate the standard deviation of the error scores for this population: 10, 9, 8, 7, 7, 6, 5, 4, 3, 1.
2. Assume that the ten boys above are to be considered a *sample* of the population of left-handed boys in that school system. Calculate the standard deviation of this sample to be used as an estimate of the standard deviation of the population.
3. The Σx^2 for a sample of 30 cases was found to be 420. What would be the best estimated value of the population variance?

Solutions

1. $\sigma = 2.65$ $\left(\sigma = \sqrt{\dfrac{\Sigma x^2}{N}} = \sqrt{\dfrac{70}{10}} = \sqrt{7} = 2.65\right)$

2. $s = 2.79$ $\left(s = \sqrt{\dfrac{\Sigma x^2}{n-1}} = \sqrt{\dfrac{70}{9}} = \sqrt{7.777} = 2.79\right)$

3. $\left(s^2 = \dfrac{\Sigma x^2}{n-1}\right.$

   $\left.s^2 = \dfrac{420}{29} = 14.48\right)$

the best estimate of the unknown population variance. Then to find the standard deviation, one takes the square root of the variance.

Interpretation of Standard Deviation

Standard deviation is an indicator of the spread, scatter, or variability that characterizes a distribution of scores or other measures. The standard deviation increases in proportion to the spread of the scores. If a distribution has cases that are widely variable or spread out from the mean, then the squared deviation scores are large and as a result the variance and standard deviation are large. On the other hand, if all the scores are clustered closely around the mean, the squared deviation scores are small and the variance and standard deviation are also small. In the distributions below we can see the relationship between the spread of scores and standard deviation.

By visual inspection of the scores in Table 4.4 we see that those in Group A are more closely concentrated around the mean than those in Group B. The standard deviation provides a concise index of the spread in each, which can be used for comparison purposes. Computation shows that the deviation scores in Group A are small, the squared deviation scores are smaller, and

DESCRIPTIVE STATISTICS

TABLE 4.4

GROUP A			GROUP B		
X	x	x^2	X	x	x^2
16	3	9	20	11	121
15	2	4	15	6	36
14	1	1	12	3	9
13	0	0	10	1	1
13	0	0	10	1	1
13	0	0	9	0	0
12	−1	1	6	−3	9
12	−1	1	4	−5	25
11	−2	4	3	−6	36
11	−2	4	1	−8	64

$\mu = 13 \quad \Sigma x^2 = 24 \qquad\qquad \mu = 9 \quad \Sigma x^2 = 302$

$$\sigma = \sqrt{\frac{\Sigma x^2}{N}} = \sqrt{\frac{24}{10}} = 1.55 \qquad \sigma = \sqrt{\frac{\Sigma x^2}{N}} = \sqrt{\frac{302}{10}} = 5.49$$

Exercise

1. Below are data for two sections of seventh grade math students:

Section	N	Mean IQ	SD of IQ Scores
1	25	103.2	9.1
2	24	103.6	12.5

(a) How would you interpret these data?
(b) If you were given a choice, which section would you prefer to teach and why?

Solution

1. (a) The two sections are practically identical in their average IQ; however, Section 2 shows greater variability in IQ scores than does Section 1. The scores in Section 2 are more spread out from the mean score.

(b) Your preference (many teachers would prefer Section 1 because they show less variability in IQ and it would be easier to select materials, methods, and so on).

the variance and standard deviations are smaller than in Group B. A comparison of the standard deviations for the two distributions (1.55 and 5.49) shows that the spread of the scores in Group B is almost four times greater than in Group A. It might also be noted that while Group B has the greater variability it has a smaller measure of central tendency (mean), which emphasizes again the fact that measures of central tendency tell us nothing about variability, and vice versa.

One might question when a standard deviation is "large" or "small." We use the labels "large" or "small" to describe the standard deviation only in a relative sense. That is, a distribution has a large or small standard deviation as compared with the standard deviation of scores in another distribution. For example, we might say that scores on Test X have a smaller standard deviation than scores on Test Y. A fifth grade with a high standard deviation in reading achievement would be more heterogeneous or more variable in their reading performance than a fifth grade with a low standard deviation. The latter would be described as being more homogeneous in their performance. Different teaching techniques might be required for the two groups.

Advantages of Standard Deviation

The standard deviation is the most reliable measure of variability of a sample, since it takes into account the numerical value of each score. It provides a more stable estimate of the population variability than any of the other measures of variability. That is, there would be less difference between the standard deviations of samples randomly selected from a population than there would be for other measures of variability in the same circumstances. Thus, the standard deviation of a random sample taken from a population can be considered a relatively good estimate of the variability in the population as a whole.

Since the standard deviation can be manipulated mathematically, it is used in further statistical computations. It is widely used in statistical inference when one wishes to draw conclusions about population variability.

A Comparison of the Measures of Variability

The measures of variability differ as to their relative stabilities. The standard deviation is the most stable, the range is the least stable, and the quartile deviation is less stable than the standard deviation but more stable than the range. Since the range takes into consideration only the two extreme scores it is unlikely to be descriptive of the entire distribution, especially when frequent or large gaps occur. Furthermore, since the range is based on only the highest and lowest scores it is likely that the value obtained will vary considerably from sample to sample in a population. The presence in a sample of just one extremely high or low score can alter the range drastically. For this reason, the range is not regarded as a very stable statistic. Because of this lack of stability, a sample range, especially from a small sample, may not be a very good estimate of the range that might exist in the population.

Another disadvantage of the range is that it is dependent upon the size of the sample. Generally, the larger the sample, the larger the range that will be observed because the chances of drawing an extreme score are increased. Because of the influence of sample size on range, it is not useful to directly compare ranges that have been computed for samples of varying size.

In most cases there is a need for a measure of variability that reflects more of the scores in a distribution than just the two extreme scores. The semi-interquartile range or quartile deviation indicates the distance between two percentile points in the middle of the distribution. It is not influenced by extremely high or low scores and shows less variation from sample to sample in a population. This means that quartile deviation is more stable than the range as a measure of variability. It still does not take into consideration the value of *each* of the raw scores in the distribution, which constitutes a disadvantage. When the median is the appropriate index of central tendency of a distribution, QD is usually used as the measure of variability; for instance, when the distribution is highly skewed.

The standard deviation is the most reliable measure of the variability of a sample, since it takes into account the numerical value of each score. It provides a more stable estimate of the variability of a population than any of the other measures of variability since there will be less fluctuation in the standard deviation from one sample to another than for other measures of variability. The standard deviation of a random sample taken from a population can be considered a good estimate of the variability in the population as a whole. It is often the preferred index of variability in inferential statistics where further mathematical manipulations must be performed on the data. It should be used with caution with highly skewed distributions because the extreme scores would distort its value. Since deviation scores are squared in the computation of standard deviation, the extreme or widely deviant scores when squared would carry disproportionate weight and would increase the value of the standard deviation. Its use is recommended in those circumstances where the mean is considered the appropriate measure of central tendency. It requires at least interval data.

> Exercise
>
> 1. A teacher found that the range of test scores in Section 1 was from 32 to 89; in Section 2 the scores ranged from 65 to 88. The teacher concluded that individual differences in math achievement were much more pronounced in Section 1 than in Section 2. Criticize this conclusion.
>
> Solution
>
> 1. One would not want to reach such a conclusion on the basis of the range alone. The score of 32 in Section 1 could be the one extreme score and the rest could be clustered more closely together as in Section 2. She should calculate the standard deviation or quartile deviation for each section and compare these before reaching such a conclusion.

SUMMARY

Measures of variability are necessary for adequately describing quantitative distributions. The three measures most frequently used for this purpose are: range, quartile deviation, and standard deviation. The range is the distance between the highest and lowest scores in a distribution. Quartile deviation is equal to half the distance between the lower and upper quartiles in a distribution. It yields a measure of one-half of the range of the scores in the middle 50 percent of the distribution. Variance is an interval scale measure and is the mean square of deviations from the mean. However, variance is in squared units of the original measure. The standard deviation is the most common measure of variability and is the square root of the mean of the squared deviation scores. It belongs to the interval scale of measurement. Each of the measures of variability yields unique information. An index of variability should be chosen specifically because it yields the necessary information and because it is the most stable, accurate measure for the conditions at hand.

INDEX	ADVANTAGES	DISADVANTAGES
Range	Easy to compute.	Unstable—a change in a single score can alter its value to a great extent.
Quartile Deviation	More stable than the range because it reflects a larger portion of the scores. Not influenced by extreme scores.	Does not reflect the value of every score. Cannot be mathematically manipulated in further computations.

108 DESCRIPTIVE STATISTICS

INDEX	ADVANTAGES	DISADVANTAGES
Standard Deviation	Most reliable measure of variability; takes into account the value of every score. Can be mathematically manipulated in further computations.	Distorted by extreme scores.

EXERCISES

1. The following problems are based on the scores on a math quiz for ten students (1, 5, 3, 7, 6, 4, 8, 2, 5, 9):
   (a) What is the value of N in the above example?
   (b) What is the mean for this set of scores?
   (c) What is the standard deviation?
   (d) What is the quartile deviation?
   (e) What is the range of this distribution?
   (f) What does the mean tell you about the spread of the scores?

2. If bowler A has an average score of 170 with a standard deviation of 15 and bowler B has a mean of 175 and a standard deviation of 25 what do you know about their bowling performances?

3. The scores below represent the vocabulary test scores from a seventh grade class of 20 pupils. Calculate the range, standard deviation, and quartile deviation and discuss the benefits and disadvantages of each as a measure of the variability of these scores.

X	f	fX	X^2	fX^2
16	1	16	256	256
15	0	0	225	0
14	0	0	196	0
13	0	0	169	0
12	2	24	144	288
11	0	0	121	0
10	2	20	100	200
9	1	9	81	81
8	1	8	64	64
7	1	7	49	49
6	4	24	36	144
5	2	10	25	50
4	1	4	16	16
3	1	3	9	9
2	4	8	4	16

FLOWCHART 4

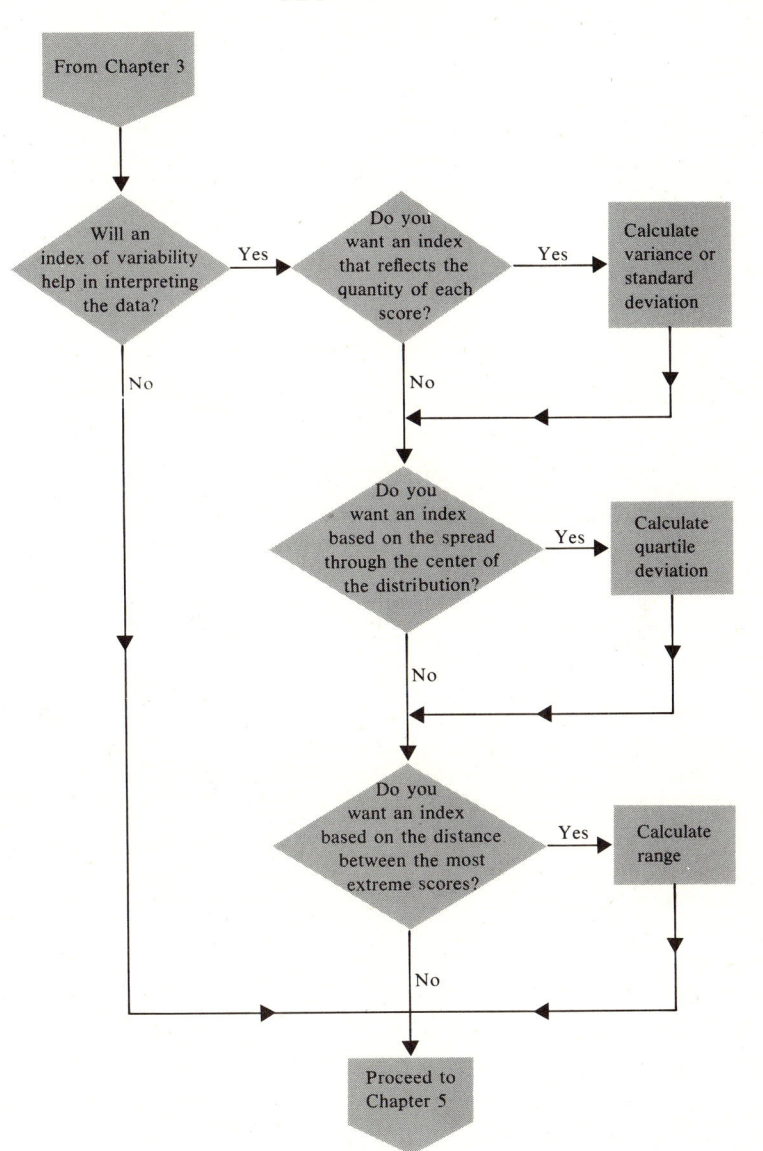

```
From Chapter 3
      │
      ▼
 ┌──────────────┐
 │ Will an      │         ┌──────────────┐        ┌──────────────┐
 │ index of     │  Yes    │ Do you want  │  Yes   │ Calculate    │
 │ variability  │────────▶│ an index that│───────▶│ variance or  │
 │ help in      │         │ reflects the │        │ standard     │
 │ interpreting │         │ quantity of  │        │ deviation    │
 │ the data?    │         │ each score?  │        └──────┬───────┘
 └──────┬───────┘         └──────┬───────┘               │
        │ No                     │ No                    │
        │                        ◀───────────────────────┘
        │                        ▼
        │                 ┌──────────────┐        ┌──────────────┐
        │                 │ Do you want  │        │              │
        │                 │ an index     │  Yes   │ Calculate    │
        │                 │ based on the │───────▶│ quartile     │
        │                 │ spread thru  │        │ deviation    │
        │                 │ the center of│        └──────┬───────┘
        │                 │ distribution?│               │
        │                 └──────┬───────┘               │
        │                        │ No                    │
        │                        ◀───────────────────────┘
        │                        ▼
        │                 ┌──────────────┐        ┌──────────────┐
        │                 │ Do you want  │        │              │
        │                 │ an index     │  Yes   │ Calculate    │
        │                 │ based on the │───────▶│ range        │
        │                 │ distance btwn│        │              │
        │                 │ most extreme │        └──────┬───────┘
        │                 │ scores?      │               │
        │                 └──────┬───────┘               │
        │                        │ No                    │
        └───────────────────────┐▼◀──────────────────────┘
                         ┌──────────────┐
                         │ Proceed to   │
                         │ Chapter 5    │
                         └──────────────┘
```

109

110 DESCRIPTIVE STATISTICS

4. Why is the standard deviation described as the most stable of the three measures of variability?
5. What percent of scores lie between Q_1 and Q_3?
6. Students who have equal means from a series of math exams have performed almost identically.

    True   False

7. Distributions that have large means also have large standard deviations.

    True   False

8. The calculation of the range involves more scores from a distribution than any other measure of variability.

    True   False

9. The range represents the distance in raw score units between the highest and lowest scores.

    True   False

10. Distributions with small standard deviations always have small ranges.

    True   False

11. The problem with variance as a measure of variability is that its units are smaller than those of the original measuring scale.

    True   False

12. The standard deviation is the most important and widely used of all the measures of variability.

    True   False

13. If one wanted to minimize the effect of an extreme score, he would choose the (quartile deviation, standard deviation) as the index of variability.
14. An instructor found that a test item was ambiguous and decided to add 5 points to each test score. What effect would this have on the mean and standard deviation of the scores?

ANSWERS

1. (a) 10
   (b) 5
   (c) 2.449
   (d) $Q_1 = 3, Q_3 = 7$ $QD = \dfrac{Q_3 - Q_1}{2} = \dfrac{7 - 3}{2} = \dfrac{4}{2} = 2$
   (e) 8
   (f) mean tells nothing about variability
2. Bowler A is more consistent.

3. Range = 16 − 2 = 14

$$\sigma = \sqrt{\frac{1173 - \frac{(133)^2}{20}}{20}} = \sqrt{14.4275} = 3.798$$

$$QD = \frac{Q_3 - Q_1}{2} = \frac{9.5 - 3.5}{2} = \frac{6}{2} = 3$$

4. It takes into account the numerical value of each score. Also, there would be less variation among the standard deviations of samples randomly selected from a population than there would be for other measures of variability.
5. 50 percent
6. False
7. False
8. False
9. True
10. True
11. False
12. True
13. Quartile deviation
14. The mean would be increased by 5; no effect on the standard deviation. The deviations from the mean would be the same in both distributions.

CHAPTER FIVE
INTERPRETATION OF INDIVIDUAL SCORES

In this chapter, we consider methods of expressing raw scores so that they can be interpreted meaningfully. Essentially these methods involve the use of measures which indicate the relative position of scores within a given group.

OBJECTIVES

After studying this chapter, the student should be able to:

1. Define percentile rank and state the purpose of using this measure.
2. Calculate the percentile rank of scores in a frequency distribution.
3. State the advantages and disadvantages of the percentile rank as an indicator of relative position.
4. Explain the precautions that must be followed in the interpretation of percentile ranks.
5. Define the term percentile.
6. Distinguish between percentile and percentile rank.
7. Determine the score value which lies at any given percentile in a distribution.
8. Define the terms quartile and decile and interpret symbols such as D_1, D_5, Q_1, Q_3, and so on.
9. Define standard score and write the formula for the standard z-score.
10. Apply the z-score formula and express the raw scores from a distribution as z-scores.

11. List the characteristics of a z-score distribution.
12. State the advantages of standard scores as indicators of relative position.
13. Convert z-scores to other standard scores with different means and standard deviations, such as Z-scores or stanines.
14. Relate the indicators of relative position to the appropriate scale of measurement.

A raw score by itself has little meaning. What does it mean if you are told that you have a score of 19 on a statistics test? Even if you remember that the test had 25 items and convert the 19 to 76 percent correct you have no information about the relative worth of your score. If the test was very difficult, 19 may have been the highest score in your class, but it could be the lowest score on an easy test.

An informative way to interpret an individual score is to show where it is located in a distribution of scores. Is your score of 19 one of the highest scores in your statistics class or one of the lowest? Is it above the mean or below the mean? How many students in the class had scores lower than 19? Various indices have been developed for indicating the relative position of a score in a distribution. The *percentile rank* (an ordinal measure) gives the relative rank of a score, and the various forms of *standard scores* (interval measures) locate scores in terms of the mean and standard deviation of the distribution. In this chapter we also introduce percentiles which, although they are not indices of location like percentile ranks and standard scores, do mark off a distribution in a way that helps us interpret individual scores.

Percentile Rank

A percentile rank gives us the relative rank of a score within a distribution, based on a scale of 100. A score's percentile rank tells what *percent* of the scores in a distribution fall below that score. If your score has a percentile rank of 6 you know that 6 percent of the scores in the total distribution fall below yours. A percentile rank of 95 means that 95 percent of the scores fall below yours.

Since a percentile rank locates a score in terms of its rank in the distribution it is an ordinal statistic. Rank alone has little meaning unless we know the number in a distribution. A score that ranks eighth in a distribution of 200 is a relatively high score. A score that ranks eighth in a distribution of ten is a relatively low score. Through the use of a scale of 100 the percentile rank provides an index of position that is independent of the number in a distribution.

Percentile ranks have a universal meaning which raw scores do not have. Any percentile rank near zero always means a score that is low in its group,

114 DESCRIPTIVE STATISTICS

any percentile rank near 50 always means near the group average, any score near 100 always means high within the group. A percentile rank of 96 in a class statistics test tells you that you are among the best in your group. A percentile rank of 96 in a spittoon marksmanship contest tells you exactly the same thing. Of course one must always remember that the percentile rank gives one's position within a group and is not an absolute value. A young lady with a percentile rank of only 4 for physical beauty within a group of Miss America contestants is still an attractive woman. A percentile rank of 92 for honesty within a prison population does not necessarily indicate an honest man. It is very important to know the nature of the group with whom one is being compared.

It is all too easy to confuse percentile rank with percent correct. A student with 15 items correct out of 20 items has 75 percent of the items correct but we do not know his percentile rank until we know the scores of the other students in his group. If 15 correct is a high score within his group he will have a high percentile rank. If most of the group had more than 15 correct he will have a low percentile rank.

Exercises

1. A student has a percentile rank of 15 on a standardized test. This means that he surpassed _____ percent of the comparison group.
2. What is your percentile rank if you have 48 correct on a 50 item test?

Solutions

1. 15 percent
2. Impossible to determine without knowing the scores of others who took the test

---

Calculation of Percentile Rank

The first step in the calculation of percentile rank is to arrange the scores in a frequency distribution as shown in Table 5.1. Now say we want to find the percentile rank of the score of 38. From the cumulative frequency column we see that 15 scores fall below 37.5, the lower real limit of the interval containing the score of interest, 38. These 15 scores are 60 percent of the 25 scores. From the frequency column we see that 8 scores (32 percent of the distribution) fall above 38.5, the upper real limit of the interval containing the score 38. The remaining 2 scores (8 percent) are within the interval 37.5–38.5. To summarize, the people with a score of 38 have scored higher than 60 percent of the group, lower than 32 percent of the group, and 8 percent of the scores are tied at 38.

It is not very convenient to state that scores of 38 have a percentile rank of

TABLE 5.1 Statistics Test Scores for a Class of 25

SCORE x	FREQUENCY f	CUMULATIVE FREQUENCY cf
43	1	25
42	2	24
41	1	22
40	3	21
39	1	18
38	2	17
37	3	15
36	2	12
35	2	10
34	1	8
33	2	7
32	1	5
31	2	4
30	0	2
29	1	2
28	0	1
27	1	1

60–68, so we arbitrarily assign the midpoint of this range to the scores in question to get a percentile rank of 64. This means that 64 percent of the scores in that distribution are lower than the score point 38. The score point 38 divides the real limits of the interval 37.5–38.5 in half so we think of half of the scores within that interval as falling below that point and half of the scores falling above that point.

Let us derive a percentile rank for a score of 32 in Table 5.1. Four scores, 16 percent, are lower than 32; 20 scores, 80 percent, are above 32. The one score of 32 represents 4 percent of the total with a range of 16–20 percent. The midpoint of this range is 18 percent, which is the percentile rank assigned to the score value of 32 in this distribution. Eighteen percent of the entire distribution of scores fall below the score point 32.

The formula for calculating percentile rank is:

$$PR = \frac{cf_b + (f_w/2)}{N}(100)$$

where: PR = percentile rank

cf_b = the cumulative frequency below the interval containing the score of interest

f_w = the number of scores within the interval containing the score of interest

N = the total number of scores in the distribution

Formula (5.1)

116 DESCRIPTIVE STATISTICS

Using the formula we calculate the percentile rank for a score of 43 from Table 5.1 as follows:

$$PR = \frac{cf_b + (f_w/2)}{N}(100)$$

$$= \frac{24 + \frac{1}{2}}{25}(100)$$

$$= \frac{24.5}{25}(100)$$

$$= 98$$

Note that although 43 is the highest score in the distribution its percentile rank is 98 not 100. A percentile rank indicates the rank of a score within a distribution. Since each score is a part of the distribution no score can be higher than all the distribution. A percentile rank of 99.9 is possible but a percentile rank of 100 is not. In our example the person with a score of 43 has scored higher than the other 24 people in the distribution. This score occupies the highest 4 percent of the total distribution, that is, 96–100 percent. So his percentile rank will fall at 98 percent, the midpoint of this range. In the same way the person with the score of 27 occupies the bottom 4 percent of the distribution, 0–4 percent. Thus his percentile rank is 2. A percentile rank indicates the relative rank of a score *within* a distribution; that is, it only considers scores in terms of their relationship to other scores within the comparison group.

Exercise

From Table 5.1 calculate the percentile rank for a score of 29; a score of 35; a score of 40.

Solution

$$\frac{1+\frac{1}{2}}{25}(100) = 6, \quad \frac{8+\frac{2}{2}}{25}(100) = 36, \quad \frac{18+\frac{3}{2}}{25}(100) = 78$$

Using Percentile Ranks

Percentile ranks transform scores into a form that is easily interpreted and readily understandable. They provide one with a convenient frame of reference for interpreting and comparing scores. A cautionary reminder is in order, however. Since percentile ranks are relative position measures, it is extremely important to know with whom an individual is being ranked. A college entrance exam score which has a percentile rank of 75 among applicants at a state university might have a percentile rank of only 35 at a select private college. A percentile rank tells us the position of an individual only *within* a specific population. If a percentile rank is to be meaningful, one must know the nature of the comparison group. Another point is that raw scores are usually not evenly distributed throughout a distribution. For this reason percentile ranks do not represent equal units along a scale of performance. The percentile rank for a score is based on the number of scores that particular score exceeds. If there are many other scores near a particular score a slight increase in raw score value will move that score above several other raw scores and thus increase the percentile rank dramatically.

If there are very few scores near a particular score, a considerable change in raw score will be necessary to produce a change in position relative to other scores and thus a change in percentile rank. For example, a statistics professor has recorded the weights of the students in his class and used them to illustrate the computation of percentile rank. The result is the polygon shown in Figure 5.1. A 160-lb student and a 210-lb student both resolve to lose weight and actually lose 10 lb each. The 10-lb loss moves the 160-lb student from a percentile rank of 75 to a percentile rank of 50. The same weight loss only changes the heavier student's percentile rank from 95 to 90.

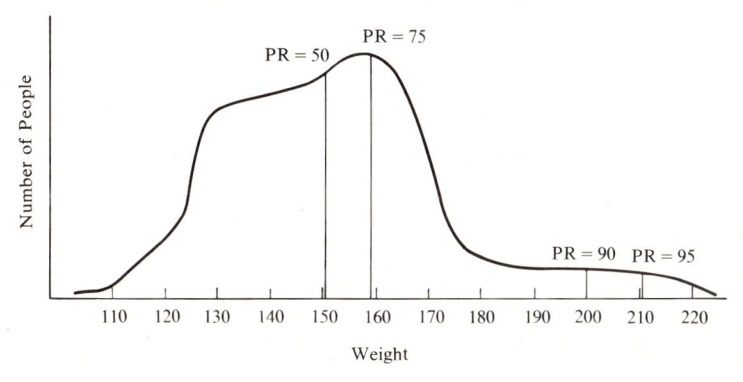

Figure 5.1 Weights of Students in Statistics Class

Generally one finds a concentration of cases near the middle of a distribution, then a tapering off of cases at either end. In such distributions, minor differences in raw scores will appear as major differences in percentile ranks among those scores which are near the center of the distribution, where a large number of the scores typically are located. At the extreme ends of the distribution, where there are few scores, major differences in raw score will have only minor effects on percentile rank. We will look at the phenomenon more closely when we consider the normal curve in Chapter Six.

Percentiles

Percentiles, like percentile ranks, are ordinal measures. With percentiles one determines what score point will divide the distribution so that the predetermined percent of scores fall below that point. We have already studied the median, the point which divides a distribution in half. The median is also termed the 50th percentile, since 50 percent of scores fall below it. We have also considered the first and third quartiles which are in fact the 25th and 75th percentiles.

In Table 5.1, we found that score 38 had a percentile rank of 64. Conversely, we could say that the 64th percentile in this distribution is 38. Percentiles are points on the *raw* score scale below which given percentages of the cases in the distribution fall. Thus, percentiles are points on the score scale which correspond to given percentile ranks. For example, the 80th percentile is the point on the score scale that has exactly 80 percent of the cases below it; the percentile rank of this point is 80, but the particular raw score value of this point is referred to as the 80th percentile. On an IQ test, a score of 100 usually corresponds to the 50th percentile. Thus, the percentile rank of score 100 is 50 and the 50th percentile is 100.

It is important to see the distinction between percentile rank and percentile. Again, the *percentile rank* of a given score is the percentage of scores in the entire distribution which are below it. As mentioned earlier, the percentile rank is a point on the original score scale corresponding to the particular percentile. On the other hand, the xth *percentile* of a given distribution is the point on the raw score scale below which x percent of the scores fall. If 90 percent of the individuals taking a test score less than 75, then 75 is the 90th percentile point. The percentile rank of score 75 is 90. In calculating percentile rank, one starts with the raw score X and determines the percent of cases falling below it. In the cases of percentiles, one determines the location of the point on the score scale below which x percent of the cases fall.

Percentiles are symbolized by the letter P_x, with x denoting the particular percentile. Thus, the 90th percentile is written P_{90}.

Percentiles are used for decision making when part of a population is to be selected because of its position within the total. A college may only admit students who score above the 40th percentile on an admissions test, that is, those in the top 60 percent. A school may give additional screening tests to the bottom 10 percent on a reading test (those scoring below the 10th percentile) in selecting students for a remedial reading program. The top 5 percent (those above the 95th percentile) in a senior class may be considered for scholarships.

Exercises

1. If the percentile rank of raw score 65 in a distribution is 73, the 73rd percentile would correspond to raw score _____.
2. Forty percent of the scores in a distribution lie below the raw score that is at the _____ percentile.
3. A student's raw score of 88 on the math achievement test falls at the 90th percentile. What percentile rank would raw score 88 represent?
4. In a population of elementary school children, 84 percent fall below an IQ of 115. Mary receives an IQ of 115 on a test of intelligence.
    (a) Her percentile rank on this test is _____.
    (b) Her score of 115 falls at the _____ percentile.
5. In a population of elementary school children, 50 percent fall below an IQ of 100.
    (a) On an intelligence test, the 50th percentile would correspond to _____ IQ score.
    (b) The percentile rank of this IQ score is _____.

Solutions

1. 65
2. 40th percentile
3. 90
4. (a) 84   (b) 84th
5. (a) 100   (b) 50

Computation of Percentiles

In previous chapters we computed three specific percentiles: the median which is the 50th percentile, or P_{50}, and the first and third quartiles which are the 25th and 75th percentiles, P_{25} and P_{75}. Let us review by computing the median for the data in Table 5.2. We want to locate the score point below which 50 percent of the cases will fall. This point will be the 50th percentile, or the median as it is commonly called.

The first step is to determine *how many* scores represent 50 percent of the total number of scores. If we use the data from Table 5.2 to illustrate the calculation, 50 percent of the total number in this case is 62. Our first step is to look in the cumulative frequency column. We find that the score of 10 has a *cf* of 56 and the score of 11 has a *cf* of 69. The point with a *cf* of 62 must therefore be above the upper real limit of the score of 10 and will be found within the real limits of the score of 11 (10.5–11.5). The 56 scores which fall below the upper real limit of the score of 10 are six scores short of our necessary 62. This tells us that the 62nd score is the sixth score up from the lower real limit of the score of 11. Since in this case there are 13 scores

120 DESCRIPTIVE STATISTICS

TABLE 5.2 Frequency Distribution of 124 Test Scores

X	f	cf	X	f	cf
20	1	124	9	10	45
19	2	123	8	9	35
18	3	121	7	8	26
17	6	118	6	4	18
16	7	112	5	6	14
15	6	105	4	3	8
14	8	99	3	2	5
13	10	91	2	1	3
12	12	81	1	2	2
11	13	69			
10	11	56			

within this interval the point we are seeking is six-thirteenths or .4615 above the lower real limit of the interval. Rounding this to the nearest hundredth and adding it to the lower real limit of the interval, 10.5, we have 10.96 as our median or 50th percentile.

Now let us use the same process to find the 95th percentile in the same distribution. Since $N = 124$, the 95th percentile will be the point below which 95 percent of 124 (.95 · 124) or 117.8 of the cases fall. Referring to Table 5.2 we find that the score point having a cumulative frequency of 117.8 occurs in the interval with the real limits of 16.5–17.5. The cumulative frequency at the lower real limit (16.5) is 112, which is 5.8 short of our needed 117.8. There are six cases within the interval. Therefore the point with 117.8 cases beneath it, the 95th percentile, is located 5.8/6 or .967 of the way through that interval. Adding .967 to 16.5 we arrive at 17.47 after rounding. The 95th percentile is 17.47.

Figure 5.2 illustrates the six scores between the score points 16.5 and 17.5 divided so that 5.8 of the six fall below the 95th percentile and .2 fall above it.

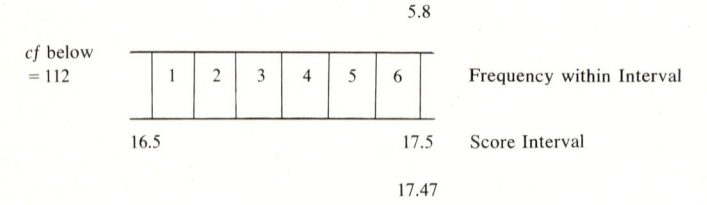

Figure 5.2 Graphic Illustration of Interpolation within an Interval to Locate Score Corresponding to a Given Percentile Point

Percentile points can be calculated with Formula (5.2).

$$P_x = L + \left(\frac{pN - cf_b}{f_w}\right) i$$

where: P_x = score at any given percentile

L = the lower real limit of the interval containing the given percentile (P_x)

pN = the desired percentage of N, the total frequency

cf_b = the total frequency in all intervals *below* the interval containing the desired percentile (the cf below L)

f_w = the frequency of cases *within* the interval containing the desired score

i = interval size

Formula (5.2)
Percentile

Note that the formula for the median (50th percentile) in Chapter 3 and the formulas for the 25th and 75th percentiles in Chapter 4 are adaptations of this formula for specific percentile points. Formula (5.2) is the general formula which can be used for any value of P_x.

Let us illustrate the use of the formula by again calculating P_{95}, this time with the formula. Substituting in the formula, pN, $(.95)(124) = 117.8$; $L = 16.5$; $cf_b = 112$; and $f_w = 6$. The interval size is 1, the difference between the upper and lower real limits (17.5–16.5) of the interval containing P_x. (The most common interval size is 1 but other interval sizes occur if the unit of measure is other than 1 or if the data have been grouped.)

Substituting our figures into the formula we have:

$$P_{95} = 16.5 + \left(\frac{(.95)(124) - 112}{6}\right) 1$$

$$= 16.5 + \left(\frac{5.8}{6}\right) 1$$

$$= 16.5 + .97$$

$$= 17.47$$

Let us find various percentile points for the distribution shown in Table 5.3. Since grade equivalents are expressed in tenths of a year the interval size for this data is .1.

To find P_{40} in this distribution we substitute into the formula $P_x = P_{40}$, $pN = (.40)(24) = 9.6$; $L = 2.45$ (halfway between 2.4 and 2.5), $cf_b = 8$, $f_w = 2$,

122 DESCRIPTIVE STATISTICS

TABLE 5.3 Arithmetic Grade Equivalents for a Class of 24 Second Graders

x	f	cf
4.1	1	24
4.0	0	23
3.9	0	23
3.8	0	23
3.7	0	23
3.6	1	23
3.5	0	22
3.4	1	22
3.3	1	21
3.2	2	20
3.1	0	18
3.0	0	18
2.9	1	18
2.8	2	17
2.7	3	15
2.6	2	12
2.5	2	10
2.4	2	8
2.3	0	6
2.2	3	6
2.1	1	3
2.0	1	2
1.9	0	1
1.8	1	1

and $i = .1$. The computation becomes:

$$P_{40} = 2.45 + \left(\frac{(.40)(24) - 8)}{2}\right).1$$

$$= 2.45 + \left(\frac{1.6}{2}\right).1$$

$$= 2.45 + .08$$

$$= 2.53$$

Sometimes the percentile point of interest falls at the upper real limit of an interval and it is not necessary to apply the formula. For example, if we want the 50th percentile for the data in Table 5.3 we find that 50 percent of scores fall below the upper real limit of the score 2.6. Therefore that upper real limit (2.65) is the median and no further computation is required.

If the percentile point of interest falls at the upper real limit of an interval and if the interval above it has a zero frequency, we have a problem inasmuch as more than one point fits the definition of that percentile. For example, to establish the 75th percentile for the data in Table 5.3 we look for the point that has 18 scores below it. The upper real limit of the score 2.9 has a cumulative frequency of 18, so the upper real limit of that score, 2.95, fits the definition of the 75th percentile. However, the upper real limits of the next two intervals, 3.05 and 3.15, also fit the definition of the 75th percentile as they, too, have a cumulative frequency of 18. By convention, whenever more than one point fits the definition of a percentile that percentile is recorded as halfway between the lowest and highest points which fit its definition. In our example, the 75th percentile is recorded as (3.15 + 2.95)/2 = 3.05.

Exercises

1. Using the data in Table 5.3 find the 95th percentile.
2. What is the 25th percentile in this distribution?

Solutions

1.
$$P_{95} = 3.55 + \left(\frac{(.95)(24) - 22}{1}\right).1$$
$$= 3.55 + .08 = 3.63$$

2. By convention halfway between 2.25 and 2.35 or 2.30.

Quartiles and Deciles

Some percentiles are more widely used than others and have been given special names. These are the quartiles and deciles.

The 25th, 50th, and 75th percentiles are referred to as quartiles. These three quartiles divide a frequency distribution into four equal parts. The first or lower quartile, denoted by the symbol Q_1, is the point on the score scale below which 25 percent (or one-quarter) of the cases fall; thus Q_1 corresponds to P_{25}. The second quartile (Q_2) is the point in the distribution below which 50 percent of the cases lie; hence Q_2 corresponds to the 50th percentile, or median. The third quartile (Q_3) is the point below which 75 percent or three-quarters of the cases in a distribution fall; thus it corresponds to P_{75}.

You will sometimes encounter the use of the term quartile to describe a range rather than a point. It is important to remember that quartiles are *points* on a distribution and not fourths of the distribution. It is correct to say, then, that an individual who scores below Q_1 is in the bottom fourth of the class but incorrect to describe his score as being in the lowest quartile.

124 DESCRIPTIVE STATISTICS

Decile points are also often used to mark off a distribution. As the name indicates, deciles divide a distribution into tenths. Thus there are 9 deciles, 1 through 9, which divide a distribution into ten equal parts. D_1 is the first decile; below D_1 lie the bottom 10 percent of the group. Hence D_1 is equivalent to P_{10}. The second decile, or D_2, is the point in the distribution below which 20 percent of the cases fall. D_5 corresponds to P_{50} or Q_2. Like quartiles, deciles are points in a distribution, not segments. Thus an individual may score at the 9th decile or above the 9th decile, but not *in* the 9th decile.

Chase[1] has suggested that since quartiles and deciles are frequently incorrectly used to describe segments of a distribution we should adopt new terms that can be used for that purpose. He proposes the words *decoid* and *quartoid* for that purpose. Scores below the 25th percentile would be in the first quartoid, those between P_{25} and P_{50} in the second quartoid and so on. In the same manner a distribution would be divided into ten decoids.

Exercises

1. Complete the following:

    (a) Q_1 = _____ percentile
    (b) D_4 = _____ percentile
    (c) P_{60} = _____ decile
    (d) P_{75} = _____ quartile

2. In the following distribution, determine the score values which fall at:
    (a) P_{60} (b) Q_1 (c) D_4

X	f
10	1
9	2
8	4
7	5
6	4
5	3
4	1

Solutions

1. (a) 25th (b) 40th (c) 6th (D_6) (d) 3rd (Q_3)
2. (a) P_{60} = 7.3 (6.5 + 4/5(1) = 7.3)
    (b) Q_1 = 5.75 (5.5 + 1/4(1) = 5.75)
    (c) D_4 = 6.5

---

[1] Clinton I. Chase, *Elementary Statistical Procedures* (New York: McGraw-Hill, 1967).

Standard Scores

We have seen how the percentile rank may be used to make raw scores more meaningful by indicating the relative position of a score in a distribution. We noted, however, that the percentile rank has some disadvantages. Percentile ranks are ordinal. They only tell us how a person *ranks* relative to other persons in the group. They do not indicate how much better one person's score is than another's. Furthermore, it was pointed out that differences in percentile ranks are usually not proportional to differences in raw scores. A small change in percentile rank near the extremes of the distribution, as for example, from the 96th to the 98th percentile, would typically represent a large difference in raw scores, whereas a change from the 60th to the 62nd percentile would typically represent a minor difference.

Standard scores locate an individual in terms of how much above or below the mean his score falls as measured in standard deviation units. Standard scores are interval measures which serve a purpose similar to that of the ordinal percentile ranks; that is, they both enable us to locate an individual within a group.

Standard scores have two characteristics which percentile ranks lack: (1) they are a direct transformation of raw scores and therefore reflect the magnitude of a score; (2) since they are interval measures they can be used in a wide variety of mathematical computations. Standard scores indicate how many standard deviations above or below the mean a score is located.

The z-Score

The z-score is the basic standard score. It is defined as the deviation of a raw score from the mean ($X - \mu_x$ or x) divided by the standard deviation (σ). It is symbolized by a small letter z. The formula for a standard score is:

$$z = \frac{X - \mu_x}{\sigma} = \frac{x}{\sigma}$$ where: X = raw score

μ_x = the mean of the distribution

σ = the standard deviation of the distribution

x = deviation of score from the mean ($X - \mu_x$)

z-Score   Formula (5.3)

To convert a raw score to a z-score, one subtracts the mean from the raw score and divides by the standard deviation. If the raw score is above the mean, z will be a positive value; if the raw score is below the mean, z will be negative. Thus the mean is the reference point and the standard deviation is the basic unit for measuring distance from that point. A score is reported in terms of its distance from the mean in standard deviation units. The z-scores represent abstract numbers rather than the particular units of the original scores (pounds, inches, reading achievement, IQ scores). When one divides a deviation score by the standard deviation, one converts from the original score units into standard deviation units, which become the basic unit for measuring distance. Thus z-scores have a universal meaning. A z-score of +2.0 means two standard deviations above the mean whether we are talking about IQ scores, rifle marksmanship scores, size of cabbages, ages of kings, or any other variable.

Suppose you receive a score of 95 on a test of manual dexterity where the mean score for the group is 75 and the standard deviation is 10. When your raw score is converted to a z-score, $z = (95 - 75)/10 = 2$, you can see that you have scored exactly two standard deviations above the mean. The z of 2 indicates that your performance is relatively high in this distribution. On the other hand, if you had a raw score of 60 on this same test, your z-score would be $(60 - 75)/10 = -15/10 = -1.5$. A z of -1.5 indicates that you are below the mean by one and one-half standard deviations. Thus you are in the lower end of the distribution and your performance would be considered as relatively poor.

When all of the scores in a distribution are expressed as z-scores, one has, by a simple mathematical procedure, transformed the original scores into a new distribution which will in fact have a new mean and a new standard deviation. The z-score transformation produces a distribution of z-scores which has a mean of 0 and a standard deviation of 1. The distribution has been standardized; that is, it now has a "standard" mean and a "standard" standard deviation and the scores are appropriately called standard scores. The resulting z-scores have standard meaning, so that when one knows the z-score one immediately knows the relative position of the original score in terms of the mean and the standard deviation of the distribution from which it came.

TABLE 5.4 Raw Scores, Deviation Scores, and z-Scores

CASE	X	x	z
A	14	6	1.5
B	10	2	.5
C	8	0	.0
D	6	−2	−.5
E	2	−6	−1.5
Sum	40	0	.0
Mean	8	0	.0
SD	4	4	1.0

A z-score transformation and the resulting new distribution is shown in Table 5.4. The raw scores (X) have a mean of 8 and a standard deviation of 4. The next column shows the deviation scores (x) found by subtracting the mean from each score; note that the sum of the deviation scores and hence the mean are 0 but the standard deviation of the deviation scores is still 4. The z-scores shown in the last column are found by dividing the x by the standard deviation. The z-scores have a sum and mean of 0 but a standard deviation of 1.00. One may also see the relative positions of the individuals in the group. Case A is 1.5 standard deviations above the mean, while Case E is 1.5 standard deviations below the mean.

Exercise

Given a distribution with a mean of 50 and a standard deviation of 8 what are the z-scores for individuals with a raw score of 56, a raw score of 50, and a raw score of 38?

Solution

$$\frac{56-50}{8} = .75; \qquad \frac{50-50}{8} = 0; \qquad \frac{38-50}{8} = -1.5$$

Characteristics of z-Scores

When all of the scores in *any* distribution have been converted to z-scores the transformed distribution has a mean of 0 and a standard deviation of 1. This is inevitable given the way z-scores are defined.

If one graphs a distribution of z-scores this graph will have exactly the same shape as the original raw score distribution. Transforming raw scores into standard z-scores does not change the shape of the original distribution. If the original distribution is skewed, then the z-score distribution derived from it will be skewed. If the original distribution is normally distributed, then the standard score distribution derived from it will be normally distributed.

Unlike percentile ranks, differences in z-scores reflect the differences in the original raw scores; that is, z-score differences are proportional to the differences in raw scores. Any ratio between two raw score differences will be unchanged by the z-score transformation. For example, let us compare the scores of three students whose scores on a test are $X_1 = 60$, $X_2 = 75$, and $X_3 = 85$, respectively, where the mean is 70 and the standard deviation is 10. These scores convert to z-scores of $z_1 = -1$, $z_2 = .5$, and $z_3 = 1.5$. Note then that the difference between X_1 and X_2 is 1.5 times as great as the difference between X_2 and X_3. This is still true when we compare the z-score differences.

128 DESCRIPTIVE STATISTICS

$$\begin{array}{cc} X & z \\ \dfrac{60-75}{75-85} = & \dfrac{-1-.5}{.5-1.5} \\ \dfrac{-15}{-10} = & \dfrac{-1.5}{-1.0} \\ 1.5 & = 1.5 \end{array}$$

Thus the relative distance between the original score measures is not changed by the transformation. Since this is true, it is possible to use the z-scores to indicate how much better one score is than another. If one individual has a z-score of .5 and another has a z-score of 2.5, one knows that the latter score is much higher than the former and, furthermore, one knows how much higher. The latter score is higher by two standard deviations.

The useful characteristics of z-scores that have been described are due to the fact that z-scores represent *linear* transformations of the original raw scores. In linear transformations (as contrasted with transformations to percentile ranks), the shape of the distribution of standard scores is exactly the same as that of the original scores. Furthermore, linear transformations preserve the relative distance relationships that existed in the original scores. A linear transformation is defined as a procedure in which the equation for changing from one kind of score to another can be graphed as a straight line. To illustrate a linear transformation, let us plot the data from Table 5.4 as a graph. Figure 5.3 shows the transformation from raw scores to z-scores.

In Figure 5.3 note that the transformation line for the z-scores is a straight line, that is, it is a linear transformation. The graph can be used to get z-scores for any X scores. For example, assume we want the z for an X of 14. It is only necessary to go vertically from score 14 to the transformation line and horizontally to the z-axis in order to read the corresponding z-score (1.5).

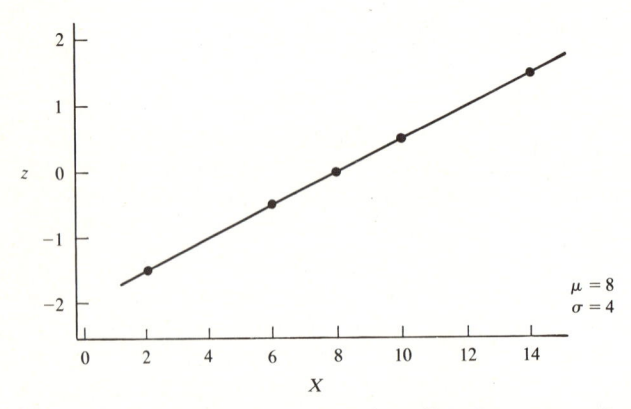

Figure 5.3 Linear Transformation from Raw Scores to z-Scores

It can also be seen that the relative distance between the z's is the same as it was in the original distribution. Consider two scores 14 and 10: The distance of the score 14 above the mean is three times the distance of score 10 [$14 - 8 = 6$; $10 - 8 = 2$]. This same relationship exists in the z-scores where the z of 1.5 (score 14) is three times the distance above the mean as the z of .5 is (score 10).

Because it represents a linear transformation a z-score distribution is an interval statistic because z-scores not only indicate the relative position of individuals but also how much better one score is than another in terms of standard deviations. Percentile ranks, as you will recall, permitted a rank ordering of the individuals in a group, but one could not say that the difference between P_{94} and P_{96} was the same as the difference between P_{54} and P_{56}. However, a word of caution is in order. A z-score of 0 is not an absolute zero but rather a point on a line. For this reason, although one can say that the difference between z of .5 and z of 1 is equal to the difference between z of 1.5 and z of 2, one cannot say that a z of 3 represents twice as much ability on a test as a z of 1.5.

In summary, the characteristics of z-scores are:

1. The unit of measurement of z-scores is the standard deviation.
2. Raw scores above the mean of the original distribution will have positive z-values; scores below the mean will have negative z-values.
3. The mean of a z-score distribution is zero.
4. The standard deviation of a z-score distribution is one.
5. A z-score distribution has the same shape as the original X distribution of scores.
6. A z-score distribution is an interval scale.

Exercises

1. If a student received a z-score of 0, one would know that this student's raw score was ——————— (below, above, or equal to) the mean.

2. Below are listed the scores made by five students from a class of 30 on math and spelling tests. The math test had a mean of 30 and a standard deviation of 3. The spelling test had a mean of 45 and a standard deviation of 5. For each, tell whether the student did better on the math or the spelling test or did equally well on both:

	MATH	SPELLING
John	36	35
Peter	33	55
Mike	27	40
Chris	33	45
Beth	36	55

3. Mary Smith, who has an IQ of 102, has applied to two colleges. At Western State College, the mean IQ of the freshmen class is 115 with a standard deviation of 13. At Eastern State College, the mean IQ of the freshmen class is 115 with a standard deviation of 6. At which college would she have the higher ranking compared to the other freshmen?

4. A student was told that he had made a z-score of .95 on a class test where the mean was 82 and the standard deviation was 6. What raw score did this student obtain?

Solutions

1. equal to

2.

	z-SCORES	
	MATH	SPELLING
John—math	2	−2
Peter—spelling	1	2
Mike—equally well on both	−1	−1
Chris—math	1	0
Beth—equally well on both	+2	+2

3. She would rank higher at Western State ($z = -1$). Her z-score at Eastern would be −2.16.

4. 87.7 or 88

Use of z-Scores Since the mean and standard deviation of any standard distribution are zero and one, the z-score transformation equates the mean and variance of all distributions. As a result one is able to compare the performance of an individual on different measures, provided that such comparisons of z-score standings are made with respect to the same reference group. Let us assume that two examinations have been given to a class. The means and standard deviations for the two examinations are shown below along with the scores made by two pupils in the group.

	SCIENCE	z_s	MATH	z_m
Mean	66		55	
Standard deviation	8		10	
John's raw score	74	1.0	65	1.0
Mary's raw score	70	.5	70	1.5

It would be difficult to compare the raw scores, but converting them to z-scores permits a comparison. We see that John has performed equally well on both tests; that is, in relation to the performance of the other individuals in the class, a score of 65 in math places John in the same relative position as a score of 74 on the science test because both scores are 1 standard deviation above the mean ($z = 1.0$). On the basis of the raw scores, we might have thought that his performance in science was higher than his performance in math. Mary, who had a raw score of 70 on both tests, actually had a higher relative position in math than she had in science. In math, her z-score is 1.5, $(70-55)/10$, and in science her z-score is .5, $(70-66)/8$. Mary's relative standing was lower in science, although this is not apparent in the original raw scores. It can also be seen that John was superior to Mary in science, but Mary was superior in math. Such comparisons can only be made when the z-scores are calculated for the same group of subjects.

Another use of z-scores is one that is particularly helpful to teachers. At the end of a grading period, teachers often add the raw scores from various tests that have been given during that period in order to arrive at a grade for each student. For example, a teacher may add all the spelling test scores that the child has earned during a nine-week period and arrive at a composite or an average test score upon which the grade would be based. Statistically, this practice of arriving at a composite by adding individual test scores is questionable unless all those tests have the same standard deviation, which is highly unlikely. This is because the effective weight of each component test is determined by its standard deviation. For example, if one test had a standard deviation of 10 and the other had a standard deviation of 5, then the former test would have two times as much effect in determining the individual's standing on the composite score. If each test is to have the same weight in the composite score, each must have the same standard deviation. A good way to arrange this is to convert each of the test scores to a standard score where the means and standard deviation would be equivalent. A composite could then be calculated using the several z-scores and one would know that all the tests were being weighted equally. However, if differential weighting is desired on some of the tests, this can be done when adding the z-scores for the composite. For instance, if one wishes the final test to count twice as much as any of the other tests, then it is only necessary to multiply the student's z-scores from the final by 2 and enter that into the composite.

A disadvantage of z-scores is that they are not so easily interpreted to the layman as are percentile ranks. A z-score is not easily understood unless one is familiar with the concepts of the mean and standard deviation; for this reason most test manuals use percentile ranks to report test scores.

Many regard the negative values and the decimal values that z-scores sometimes have as being another disadvantage of a z-score scale. They see the use of minus signs and decimals as awkward and as subject to error. Other standard scores are available which eliminate decimal points and minus signs.

Other Standard Scores

A z-score distribution may be transformed to a new distribution where there are no decimal points and negative values. The decimal point is eliminated by multiplying each z-score by some convenient constant and the minus sign is eliminated by adding another constant to each z-score. As a result the mean and the standard deviation of the new distribution will be other than 0 and 1, respectively. The new mean and standard deviation will be determined by the constants chosen for the multiplication and addition. The choice of constants is arbitrary but certain constants enjoy wide use.

The Z-Score One of the most popular and convenient transformations is to convert the z-scores to a distribution having a mean of 50 and a standard deviation of 10. This transformation is called the Z-score. (Some texts call it the *T*-score.) One converts to the desired standard deviation of 10 by multiplying each z-score by 10 and then adding the constant 50 to each converted z-score to obtain a mean of 50. The formula for Z then becomes:

$$Z = 10(z) + 50$$

or

$$Z = 10\left(\frac{X - \mu}{\sigma}\right) + 50$$

Formula (5.4)
Z-score

The formula indicates that one first multiplies the z-scores by 10 and then adds 50. For example, if an individual has a z of 1.62, then his Z-score is $Z = 10(1.62) + 50 = 66.2 = 66.$[2] If the z-score is $-.5$, then $Z = 10(-.5) + 50 = (-5) + 50 = 45$.

Thus, interpretations are made using the mean of 50 as the origin and the standard deviation of 10 as the basic unit for measuring distance from the mean. A Z of 60 would thus be interpreted as indicating a score that is 1 standard deviation above the mean and a Z of 75 would be 2.5 standard deviations above the mean. A Z of 40 would be 1 standard deviation below the mean. The Z-scores will all have a positive value and will usually range from 20 to 80.

[2] *Z-scores are rounded to the nearest whole number.*

Exercises

1. Express the following raw scores as Z-scores: 45, 30, 50, 25, where the mean is 40 and the standard deviation is 5.
2. Mike received the same raw score (82) on a social studies and a math test. The mean for the social studies test was 86 and the standard deviation was 4. The mean for the math test was 78 and the standard

deviation was 2. Express Mike's raw score of 82 as a Z-score and compare his performance in the two subjects.
3. If one scored exactly at the mean on a test, he would have a Z-score of _____.

Solutions
1. 45 = Z of 60   50 = Z of 70
   30 = Z of 30   25 = Z of 20
2. Mike ranked higher in math. (Z of 70 in math; 40 in social studies)
3. 50

Other z-Score Conversions Other conversions of the z-score are often used. For example, many intelligence tests are transformed to a scale with a mean of 100 and a standard deviation of 15. The formula is:

$$IQ = 15\left(\frac{X - \mu}{\sigma}\right) + 100 \quad \text{or} \quad IQ = 15(z) + 100.$$

An individual who scored exactly at the mean of the original raw score distribution would thus have a derived score on the intelligence test of 100. An individual who was 1 standard deviation above the mean would have an IQ of 115, while a person who scored 2 standard deviations below the mean would have an IQ of 70. The general formula for converting a z-score to a different standard score is:

Standard score = $z\sigma_s + \mu_s$,   where: z = z-score
σ_s = standard deviation of the standard score
μ_s = mean of the standard scores.

Formula (5.5)

For example, results of the verbal and quantitative subtests of the Graduate Record Examination are reported in standard score form with a mean of 500 and a standard deviation of 100. A score of 1.45 standard deviations above the mean is reported as $(1.45)(100) + 500 = 645$. A score .55 standard deviations below the mean is reported as $(-.55)(100) + 500 = 445$.

Exercises

1. What IQ's are represented by the following standard scores? (mean = 100, σ = 15)
    (a) $z = 2$ (b) $z = -1$ (c) $z = 1.5$ (d) $z = -.8$ (e) $z = 0$
2. A college student received a standard score of 750 on the Graduate Record Examination. How would you interpret this score?

> 3. Assume another test publisher had developed an aptitude test where the mean was set at 200 and the standard deviation at 50. What would a score of 150 mean on this new test?
>
> Solutions
>
> 1. (a) 130 (b) 85 (c) 122.5 (d) 88 (e) 100
> 2. This score is 2.5 standard deviations above the mean ($z = 2.5$). The student ranks very high in the comparison group.
> 3. 1 standard deviation below the mean

Stanines Stanines are a standard score system which provides a single-digit score scale running from 1 to 9. The mean is assigned a value of 5 and the standard deviation a value of 2. This system, originally used by the Air Force during World War II, is called *stanines* which means *sta*ndard scores with *nine* categories.

The formula for computing a stanine from a raw score is:

$$\text{Stanine} = 2\left(\frac{X_i - \mu}{\sigma_x}\right) + 5 = 2z + 5$$

The lowest stanine 1 represents a score that is 2 or more standard deviations below the mean and the highest stanine 9 represents a score that is 2 or more standard deviations above the mean. A stanine of 2 represents 1.5 standard deviations below the mean and a stanine of 8 represents a $z = 1.5$. Stanines are rounded to the nearest whole number. Therefore, a stanine of 8, for example, is assigned to z-scores ranging from 1.25 to 1.75.

SUMMARY

Percentile ranks and standard scores enable us to identify the position of an individual within a group in a manner which has the same meaning in any distribution. Percentile ranks are ordinal measures which indicate the percent of scores falling below a given score. Percentile ranks indicate a score's position in relation to other scores but do not reflect differences in raw scores. Therefore, differences in percentile rank in parts of the distribution where there are many scores represents less difference in performance than differences in percentile rank where there are few scores.

Standard scores are a direct transformation of raw scores. Differences in standard scores have the same meaning in any part of a distribution. Standard scores tell us how many standard deviations above or below the mean a score is located. The basic standard score is the z-score with a mean of 0 and a standard deviation of 1. Other standard scores such as Z-scores and stanines are conversions of the z-score. Percentiles divide a distribution

FLOWCHART 5

into parts. A percentile marks off a distribution by identifying the score point below which a given percentage of scores are located. The three quartiles, nine deciles, and the median are specific percentile points that are widely used. In locating percentiles one starts with a given percentage and locates the score point that has that percentage of the distribution below it. In computing percentile rank one starts with a given score point on the distribution and determines the percentage of the distribution below it.

EXERCISES

1. Using the data in Table 5.2, calculate the percentile rank of the following scores:
   (a) 9 (b) 15 (c) 19
2. Using the data in Table 5.1, calculate the scores corresponding to the following percentiles:
   (a) P_{30} (b) Q_3 (c) P_{90}
3. Using the data in Table 5.3, determine the score values corresponding to:
   (a) P_{20} (b) D_6
4. If Mary Smith has a reading achievement score at the 85th percentile, how would you interpret her score with respect to the total group?
5. What percentile corresponds to:
   (a) Q_1 (b) third quartile (c) D_1 (d) fifth decile
6. The mean score on a test is 40 and the standard deviation is 4. Express each of the following raw scores as a z-score:
   (a) 41 (b) 30 (c) 48 (d) 36 (e) 46
7. Give the T-score equivalent ($\bar{X} = 50$, $S = 10$) for each of the z-scores in the above problem.
8. Change each of the following z-scores to a scale that has a mean of 100 and a standard deviation of 15.
   (a) +1 (b) 0 (c) 2.5 (d) −3 (e) −1.65
9. A student has a raw score of 92 where the mean is 85 and the standard deviation 5. Express his raw score as a standard score on a scale that has a mean of 500 and a standard deviation of 100.
10. Below are listed the student's raw score and the mean and standard deviation for tests given in four subjects.

	RAW SCORE	CLASS MEAN	STANDARD DEVIATION
Arithmetic	75	84	15
Science	82	84	2
Spelling	86	80	3
English	98	90	8

Evaluate the student's performance in the various subjects (that is, in which did she perform the best, and so on).

11. Test 1 has a mean of 20 and a standard deviation of 4. Test 2 has a mean of 45

and a standard deviation of 5. Below are listed the scores obtained by three students on the two tests. For each student, tell whether the student scored higher on Test 1 than on Test 2, or better on Test 2 than on Test 1, or equally well on both.

	TEST 1	TEST 2
Bob	26	35
Sue	24	50
Amy	18	42

12. The median is equivalent to the _____ percentile, _____ quartile, and _____ decile.
13. Match the following:

MEASURE	TYPE OF SCALE
(a) Z-score	(1) Interval scale
(b) Percentile	(2) Ordinal scale
(c) Percentile rank	
(d) Stanine	

14. John tells you that he has received a music aptitude score of 45. How would you interpret this score?

ANSWERS

1. (a) 32 (b) 82 (c) 98
2. (a) 34 (b) 39.75 (c) 41.75
3. (a) 2.21 (b) 2.73
4. 85 percent of the scores were lower than her score, or she surpassed 85 percent of the group.
5. (a) P_{25} (b) P_{75} (c) P_{10} (d) P_{50}
6. (a) .25 (b) −2.5 (c) 2 (d) −1 (e) 1.5
7. (a) 52.5 (b) 25 (c) 70 (d) 40 (e) 65
8. (a) 115 (b) 100 (c) 137.5 (d) 55 (e) 75.25
9. 640
10. She was highest in Spelling ($z = 2$), then English ($z = 1$), then Arithmetic ($z = -.6$), and lowest in Science ($z = -1$).
11. Bob scored higher on Test 1.
    Sue did equally well on both.
    Amy scored higher on Test 1.
12. 50th percentile, 2nd quartile, and 5th decile
13. (a) 1 (b) 2 (c) 2 (d) 1
14. No interpretation could be made of this raw score of 45. One would need to know the mean and the standard deviation of the test or at least have a frequency distribution of the scores of the comparison group in order to determine percentile rank.

CHAPTER SIX
THE NORMAL CURVE

OBJECTIVES

After completing this chapter the reader should be able to:

1. Describe the general nature of normal curves.
2. Recognize situations where normal curves can be used in descriptive statistics.
3. Use the table of the standard normal curve to find percentile rank equivalents for z-scores.
4. Use the table of the standard normal curve to find the z-score or raw score equivalent for a percentile rank.
5. Use the table of the standard normal curve to determine the proportion of a distribution falling between two z-scores.
6. Find the interval on any normal curve that will include specified proportions of a distribution.
7. Distinguish between lepto-, meso-, and platykurtic curves.

The Concept of the Normal Curve

One of man's most interesting discoveries was the determination of a relationship between measurements of many types of natural phenomena and the mathematical laws of chance. The discovery was made by the German mathematician Gauss (1777–1855). Gauss studied the distribution of errors (discrepancies between repeated measures) made in astronomical observations and found that the distribution of these errors was approximated by a particular probability curve developed earlier by DeMoivre (1667–1754). DeMoivre's probability curve graphically described the mathematical distribution of chance happenings such as the flip of coins, the rolling of dice, and so on.

To illustrate DeMoivre's probability curve, imagine that you shake up and then toss five pennies into the air and let them land on the floor. You count the number of heads that appear and record this number as X. If you repeat this process a number of times and plot the frequency distribution of X, the results might appear as the histogram in Figure 6.1 with the number of heads on the abscissa, and the number of throws in which each number of heads

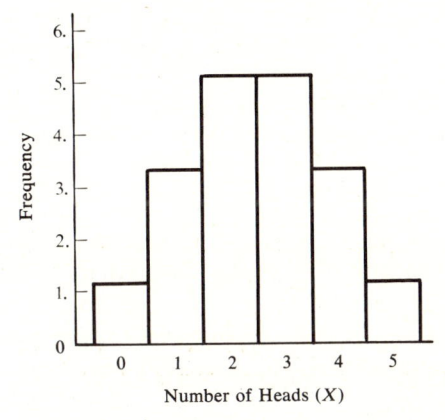

Figure 6.1 Histogram Showing the Number of Heads Appearing when 5 Coins are Tossed

appeared on the ordinate. By repeated tossing of the pennies and counting the number of times a given number of heads appeared, one could determine the chance, or probability, of getting a certain number of heads on a given toss. (However, the probability of this event could be determined more easily by applying the mathematical laws of chance which we will consider later.) Assume now that the number of coins thrown is increased to 100 and the process of throwing the coins is repeated, say, 500 times. If a histogram were plotted for these data and the ends of the bars of the histogram connected, the result would be a continuous-appearing curve. If this process of tossing the coins was continued an infinite number of times, the curve would appear as the curve shown in Figure 6.2. Most of the tosses would result in between 40 and 60 heads. Very few tosses would produce less than 35 or more than 65 heads.

The symmetrical bell-shaped type of curve shown in Figure 6.2 is known as a *normal curve*. The concept of normal curves is very basic in statistics. The reason for its importance results from the fact that the frequency distributions of many natural events have shapes similar to the shape of this curve. Not only the errors of measurement observed by Gauss but also many physical and psychological phenomena when shown in a frequency polygon resemble the normal curve. For example, the wing spans of adult

140 DESCRIPTIVE STATISTICS

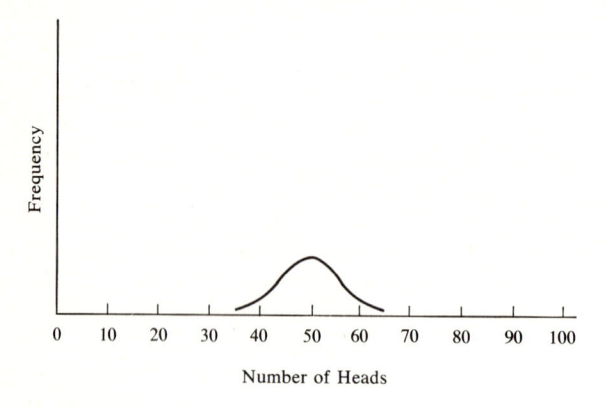

Figure 6.2 Curve Showing the Number of Heads Appearing on a Given Toss when 100 Coins are Tossed

Canada geese, the weights of American girls on their 10th birthday, the reading achievement scores of sixth graders, and so on, all tend to give bell-shaped frequency distributions similar to the normal curve. Galton (1889)[1] observed that the distribution of men's heights is closely approximated by a normal curve. To illustrate, imagine several thousand college men lined up on a football field according to height with the tallest at one end and the shortest at the other. Assume that those of the same height are standing behind one another to form lines. The view from the press box above the field would look something like the polygon in Figure 6.2. Near the middle where the men are of average height, the vertical lines would extend far back; that is, the greatest concentration of men would be at the middle. As one looked from the middle toward the two ends where the tallest and shortest men are, one could see that the lines would grow shorter until at the very extreme ends there may be only one man.

Similarly, the distributions of scores on many psychological tests, such as IQ tests and tests of school achievement, are approximated by the normal curve. Figure 6.3 shows a histogram of IQ scores with a normal curve superimposed on the columns of scores. Although the fit is not perfect, we see that the distribution of scores closely approximates the normal curve.

Figure 6.4 shows the distribution of scores on a personality test measuring submission-dominance tendencies in the personal relationships of a group of college women. Again, we see that this distribution of scores approximates the symmetrical, bell-shape of the normal curve.

Actually, the normal curve is a graph of a rather complex mathematical

---

[1] F. Galton, *Natural Inheritance* (London, 1889), cited by Helen Walker, *Studies in the History of Statistical Method* (Baltimore: Williams & Wilkins, 1929).

The Normal Curve   141

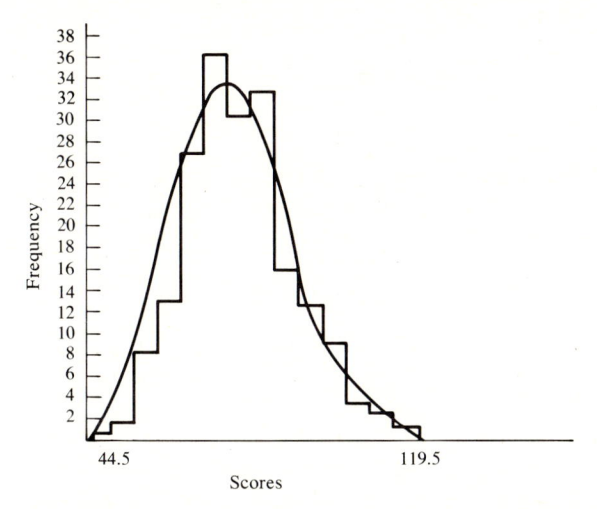

Figure 6.3 Distribution of Scores on a Test of Intelligence[2]

relationship. This relationship, when plotted, results in a graph which is unimodal and symmetrical. There is an equation for the normal curve, which relates the height of the curve to the X-score values. Formula (6.1) shows the equation which defines the normal curve:

$$Y = \frac{1}{\sigma\sqrt{2\pi}} e^{-\frac{(X-\mu)^2}{2\sigma^2}}$$

where: Y = height of the curve corresponding to particular values of X
X = a score value corresponding to a particular height
π = a constant equal to 3.1416
e = a constant equal to 2.7183
μ = the mean of the X variable
σ = the standard deviation of the X variable

Formula (6.1)
Normal Curve

Formula (6.1) can be used to construct *a* normal curve. We say "a" normal curve, because while Formula (6.1) imparts the characteristic bell-shape to the curve, the form of the curve depends on the mean and the standard deviation. Thus, Formula (6.1) defines not one but a whole family of curves, any one of which is determined by specifying the values of μ_x, the

[2]From Henry E. Garrett, *Statistics in Psychology and Education* (New York: McKay Co., 1958), p. 104.

142 DESCRIPTIVE STATISTICS

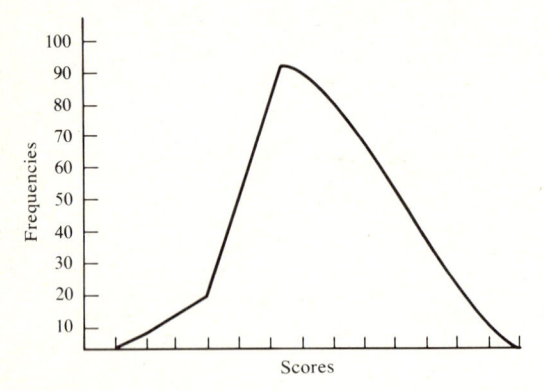

Figure 6.4 Distribution of Scores on a Personality Test[3]

population mean, and σ_X, the population standard deviation of the X distribution. That is, given the value of the X-score and the two population parameters, μ and σ, one can substitute in Formula (6.1) and solve for the corresponding value of Y, the height of the curve. If all the different values of X are substituted in the equation and the corresponding values of Y obtained, then the paired values of X and Y may be plotted graphically and the result will be a normal curve with mean μ and standard deviation σ. However, since the mathematical calculations are rather involved, the beginning student in statistics is not asked to master this formula. You will not have to use it in any computations. It is included only to inform you that there is a mathematical formula for constructing a normal curve. The important thing to remember is that this one normal curve equation may result in many normal curves. Given the values of the population parameters, the equation specifies the particular form that a normal curve assumes. Each time that μ and σ change, a different normal curve results. There is not one or *the* normal curve. The three curves in Figure 6.5 all fit the definition of the

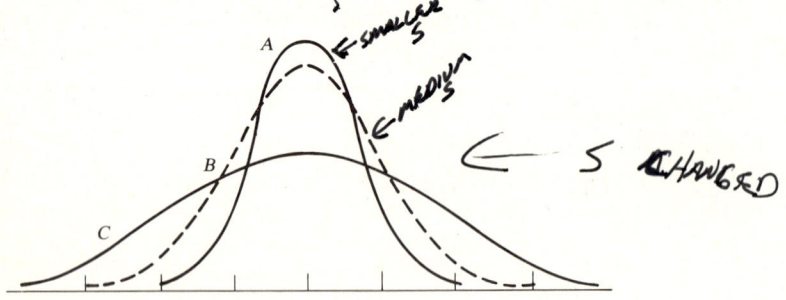

Figure 6.5 Three Normal Curves

[3] From Anne Anastasi, *Differential Psychology* (New York: Macmillan, 1958), p. 32.

normal curve, yet they are different in appearance. We also see that they differ in variability. Curve A is rather narrow and the ordinates would be relatively long; Curve C is rather wide and the ordinates would be relatively short. Curve B is the most widely used to illustrate normal curves.

The important properties of normal curves are:

1. A normal curve is symmetrical with its maximum height at the mean. It is often described as a bell-shaped curve. (Some say it looks more like a well-weathered manure pile.)
2. Given the way normal curves are defined, the mean, median, and mode fall at the same point.
3. The height of the curve decreases as one moves to the left and right of the point of maximum height (the mean, median, and mode). From this point the decrease in the height of the curve accelerates and the curve is convex, until one reaches 1 standard deviation above or below the mean. From these points the decrease decelerates and the curve becomes concave.
4. Although the height of the curve continues to decrease as one moves farther and farther from the mean it never actually reaches zero. Therefore, the theoretical range of the normal curve is from plus infinity ($+\infty$) to minus infinity ($-\infty$). (Actually, so little of the curve extends above $+3$ standard deviations or below -3 standard deviations that for many purposes these are the practical limits of the normal curve.) A problem always exists when illustrating the normal curve because the printed line cannot really show how very close the curve is to the baseline beyond ± 3 standard deviations.

When discussing the form of frequency polygons, the normal curve is frequently used as the basis for certain comparisons. For instance, curves are described as symmetrical (like the normal curve) or nonsymmetrical (skewed). Skewness was discussed in Chapter 2. Another characteristic of the form of curves is *kurtosis* (from the Greek word *kyrtos*, which means curved). If a curve has *short tails* in comparison to the normal curve, it is said to be *platykurtic* (*platy* means flat in Greek). A platykurtic curve has a broad, relatively flat appearance. If a curve has thinner tails than the normal curve, it is said to be *leptokurtic* (*lepto* means thin in Greek). A leptokurtic curve is relatively peaked in the middle. If the kurtosis is the same as in a normal curve, it is said to be *mesokurtic*. Curve (a) in Figure 6.6 is platykurtic; Curve (b) is leptokurtic; Curve (c) is mesokurtic.

The Standard Normal Curve

We have seen that two or more distributions may be normal and yet differ in terms of their means and standard deviations. It is possible, however, to

(a) platykurtic

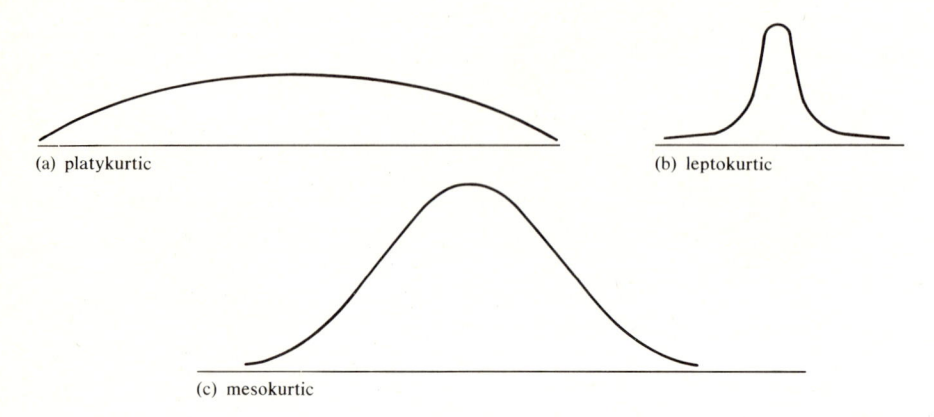

(b) leptokurtic

(c) mesokurtic

Figure 6.6 Kurtosis

transform any normal distribution into a distribution of standard scores (z-scores). As you may recall from Chapter 5, this is done simply by expressing each value of X in the population distribution as a deviation from the mean and then dividing this deviation by the standard deviation $[z = (X - \mu)/\sigma]$. When scores in a normal distribution are transformed into z-scores the result is a distribution of scores with a mean of 0 and a standard deviation of 1, which is called the *standard normal curve*. The base line of the standard normal curve is marked off in terms of z-scores rather than the raw scores (X's). The standard normal curve is very useful in statistics because instead of having to work with an infinite number of normal curves differing from one another in mean and standard deviation, one has to consider only the one normal distribution of z-scores where the mean μ is always 0 and the standard deviation σ is 1. In the standard normal distribution there is a fixed proportion of cases between a vertical line, called an ordinate, drawn at any one z-score point and an ordinate drawn at any other z-score point.

Exercises

1. Which of the following sets of data would most likely be represented by a normal curve?
    (a) the heights of professional basketball players
    (b) the heights of 10,000 army recruits
    (c) the intelligence test scores of 5,000 secondary school students
    (d) the intelligence test scores of 5,000 secondary school students who have applied to take the CEEB (College Entrance Examination Board) examination.
2. If 5,000 test scores are normally distributed, how many scores lie below the mean?
3. Although all normal curves are symmetrical in form, they may differ in their _____ and _____.

4. Which is more likely in a distribution: z of 4 or Z of 40?

Solutions

1. (b) and (c)
2. 2,500
3. location (mean) and variability (standard deviation)
4. Z of 40

Landmarks on the Standard Normal Curve

Most statistical procedures involving the normal curve involve the proportions of the curve, or areas, that occur in various sections of the curve. For that reason it is useful to consider certain "landmarks" on the curve. The most important of these is the mean. Since the normal curve is symmetrical, .5000 or 50 percent of a normal distribution is located on either side of the mean. The other important landmarks are z-scores of ±1.00, ±2.00, and ±3.00. These landmarks and the areas between them are shown in Figure 6.7. The figure indicates that there are 34.13 percent of all cases falling between the mean and a z-score of 1. That is, between the mean and 1 standard deviation above the mean, there are 34.13 percent of all the cases. Similarly, 34.13 percent of all cases fall between the mean and a z of −1.

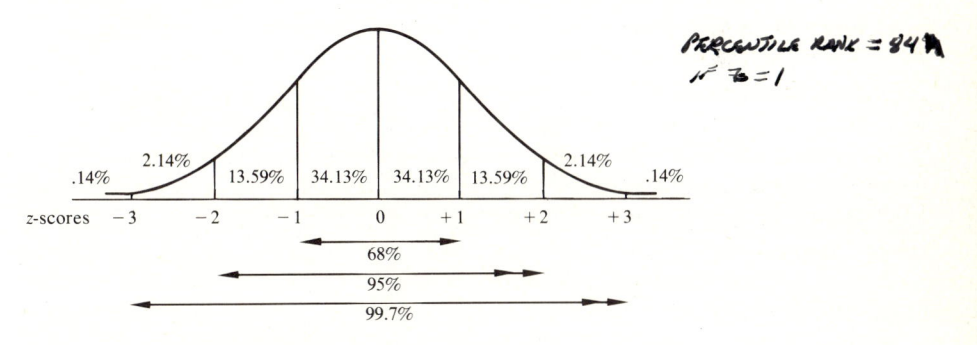

PERCENTILE RANK = 84
IF Z = 1

Figure 6.7 Proportions of the Normal Curve

Thus, the area between +1 z and −1 z includes 68.26 percent of all cases in a normal distribution. Another way of expressing this is that approximately two-thirds of the cases will fall between +1 and −1 standard deviations from the mean of a normal distribution. Between the mean and 2 standard deviations above the mean ($z = 2$) are found 47.72 percent of all cases in a normal distribution (34.13 percent + 13.49 percent). Similarly, 47.72 percent of the cases are found between the mean and 2 standard deviations below the mean ($z = -2$). Thus, 95.44 percent (47.72 percent + 47.72 percent) of the

146   DESCRIPTIVE STATISTICS

cases will fall between +2 and −2 standard deviation points in the normal curve. In a normal distribution, 99.72 percent of all cases will fall between +3 and −3 standard deviations, that is, 49.86 percent on each side of the mean (34.13 percent + 13.59 percent + 2.14 percent). Between the z-scores of 1 and 2 we find 13.59 percent of the cases. Approximately 2 (2.14) percent of the cases fall between the z-scores of 2 and 3 and only about .14 percent or 14 in 10,000 cases fall above a z-score of +3 and another 14 in 10,000 fall below a z-score of −3.

The Standard Normal Curve Table

Since distributions of standard scores (z-scores) always have a mean of 0 and a standard deviation of 1, it has been possible to develop just one normal curve table which can be used for all normal distributions expressed in standard score form. This eliminates the problems associated with the fact that normal distributions may have different means and standard deviations. The z-score is a sort of universal language because it has the same meaning regardless of the raw score values in the original distribution. The table of this standard normal curve is shown in Table A2 in the Appendix.

The standard normal curve table gives the proportion of the total area under the curve between the mean and any z-score. Since the normal curve is symmetrical, the areas from the mean are the same for both positive and negative z-scores; therefore, only positive z-scores are shown.

Before we continue discussing the normal curve table, perhaps a brief review of z-scores would be helpful. A z-score is defined as the deviation of a raw score from the mean divided by the standard deviation. The formula is $z = (X - \mu)/\sigma$. Since normal curve problems involve populations of data, we will use population symbols in the z-score formula. If a score of 75 is obtained from a distribution with a mean of 70 and a standard deviation of 10, the z-score is $(75 - 70)/10 = .5$. A z of .5 means that the score is .5 standard deviations above the mean. A score of 62 in the same distribution would give a z of $(62 - 70)/10 = -.8$, which means that this score is .8 standard deviations below the mean.

Let us examine the normal curve table in the Appendix. In Table A2, note that there are five columns which can be cross-referenced with one another. Column 1 contains z-score values and is used for both positive and negative z-scores. Column 2 indicates the *proportion* of the normal curve between the mean and the z-score value in the same row in column 1. Column 3 is the value in column 2 plus .5. This value is the proportion in the *larger* part of the curve. For example, in the row with $z = 1.50$ in column 1, we find .9332. This tells that 93.32% of the curve is below a z of +1.50 and 93.32% of the curve is above a z of -1.50. Column 4 gives the proportion in the *smaller* part of the curve. For example, column 4 indicates that 6.68% of the curve is above a z of $+1.50$ and 6.68% of the curve is below a z of -1.50. Column 4 is equal to 1.00 minus column 3. The areas in columns 2 through 4 are proportions and may be converted to percentages by multiplying them by 100. Column 5 gives the height of the curve at various z score points. We will not refer to column 5 until Chapter 9.

To illustrate the use of Table A2, let us consider a z-score of .85. First, it is necessary to locate a z of .85 in column 1 of the table. Then column 2 indicates that .3023 of the area of the curve falls between the mean and a z of .85. Column 3 indicates that .8023 of the area lies below $z = +.85$ and .8023 of the area lies above $z = -.85$ (.5000 + .3023). Column 4 indicates that .1977 of the area lies below $z = -.85$ and .1977 of the area lies above $z = +.85$. These areas are indicated in the normal curve shown in Figure 6.8.

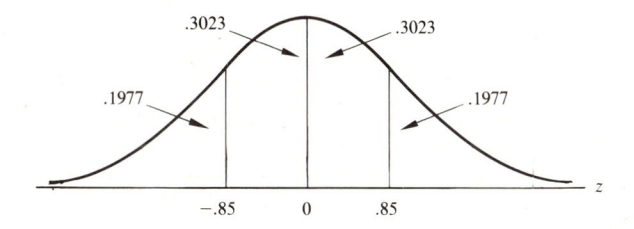

Figure 6.8 Area Under the Normal Curve Between the Mean and a z of $\pm.85$ and Beyond z of $\pm.85$

To check on your understanding of the normal curve table, try to find the proportion of the normal curve between the mean and a z of 2.95. It is .4984. Next, what is the proportion of the area of the normal curve lying between the mean and a z-score of -1.42? It is .4222. What proportion of the area would fall below a z of -1.42? Column 4 indicates that it would be .0778.

Exercises

1. Examine the normal curve in Figure 6.7:
   (a) What percent of cases fall below the mean?
   (b) What percent fall below a z of $+1$?
   (c) What percent fall below a z of -1?
   (d) What percent fall above a z of $+1$?
   (e) What percent fall more than 3 standard deviations above the mean?
   (f) What are the approximate limits of the middle two-thirds of the cases?

2. Assuming a normal distribution of scores, what percent of scores might be expected to fall between Z-scores of 50 and 70?

Solutions

1. (a) 50 percent
   (b) 84.13 percent
   (c) 15.87 percent
   (d) 15.87 percent
   (e) .14 percent
   (f) 1 standard deviation above and below the mean
2. 47.72 percent   (34.13 + 13.59) Z of 50 and 70 = z of 0 and +2.

Using the Normal Curve Table

We have said that many distributions of scores, such as standardized intelligence and achievement test scores, closely approximate a normal distribution. Thus, the normal curve is useful in both descriptive and inferential statistics. In descriptive statistics it helps us interpret z-scores in terms of percentile rank equivalents or vice versa and it helps us determine where to expect certain proportions of a normal distribution to be located. We will leave the discussion of the use of the normal curve for inferential statistics to later chapters. Let us now consider the types of problems in descriptive statistics that may be solved through the use of the normal curve. Of course, when the normal curve table is used, it is necessary to assume that the population data under consideration are normally distributed.

Finding Percentile Rank Equivalents for z-Scores

Many people find percentile ranks easier to understand than z-scores. For example, suppose a layman asks you: "What does a Binet IQ test score of 112 mean?" Knowing that the mean for Binet IQ scores is 100 you could explain that it is an above-average score. But how much above average? Knowing that the standard deviation for Binet scores is 16 you might explain that a score of 112 is $\frac{3}{4}$ a standard deviation above the mean but you will probably not have helped your questioner very much with this statement. As you become familiar with z-scores you develop a general picture of what various z-scores mean. Probably it would be more meaningful to the layman if you could give him the percentile rank for the score of 112. Since Binet IQ scores are approximately normally distributed we can use the normal curve table to find the percentile rank for him. Whenever a score distribution is known to be normal we can find the percentile rank of a score by using the normal curve table.

The Normal Curve 149

You recall that the percentile rank of a score is defined as the percentage of scores in a distribution of scores that fall *below* that particular score. Thus, in the normal curve the percentile rank would correspond to the area falling to the *left* of the given score, since this would indicate the portion or percentage lower in value. To use the normal curve table we convert the score to a z-score and consult the table to find the proportion of cases falling below that particular z-value.

The determination of the percentile rank for a positive z-score requires that we include all the area below the mean plus the area between the mean and our z-score. For positive z-scores one finds the proportion in column 4 (the area in larger portion). To determine the percentile rank for a negative z-score we take our value from column 3. When multiplied by 100, this proportion becomes the percentage falling below and hence the percentile rank of that z-score and of any raw score equivalent to that z-score.

In our example a Binet score of 112 has a z-score of .75. To determine the percentile rank of a z of .75 we want to find all the area or the percentage of scores from the *left* end of the curve up to a z of .75. Consulting the normal curve table (Table A2), we find in column 4 that the area below a z of +.75 is .7734. The percentile rank is thus, $.7734 \times 100 = 77.34$, which indicates that 77.34 percent of the cases in a normal distribution will fall below a z-score of .75. We round this to 77. We could tell our layman that a score of 112 on the Binet exceeds the scores of 77.34 percent of the population. Figure 6.9 illustrates this graphically.

To determine the percentile rank for a negative z-score we use Column 4. Since negative scores fall below the mean it is necessary to determine the area in the left end of the curve below the z-value. Column 4 in Table A2 gives the area in each of the extreme ends of the distribution. For negative z-score one multiplies the appropriate value in this column by 100 to get the percentile rank. For a z of $-.65$ Table A2 shows that .2578 of the area

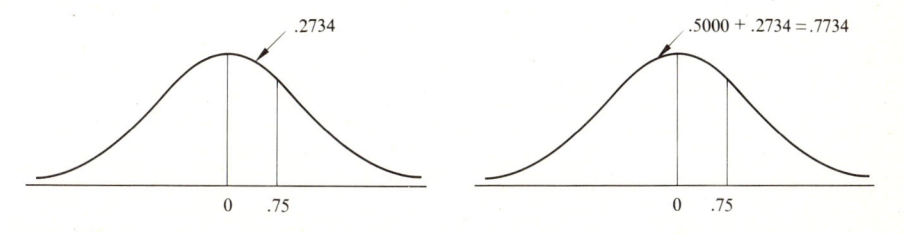

Figure 6.9 Proportion of the Area of Normal Curve Below a z-Score of .75

150   DESCRIPTIVE STATISTICS

falls below that point. (See Figure 6.10.) Therefore, the percentile rank is (.2578)(100) = 25.78. For $z = -.70$ the percentile rank is (.2420)(100) = 24.20. For a Binet score of 96 the z is $(96 - 100)/16 = -.25$. The percentile rank for $-.25$ is 40.13 which we round to 40.

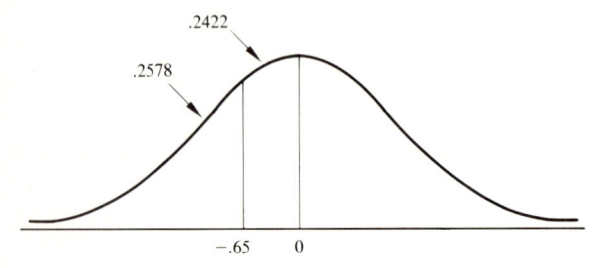

Figure 6.10 Proportion of the Area of Normal Curve Below a z-Score of $-.65$

When a distribution is known to be normal the percentile rank equivalents for z-scores are exact. There are many situations in which you would like to make a reasonable estimate of percentile ranks but you really do not know whether your distribution is normal or not. In such cases you can still use the normal curve table but you must remember that your results will be estimates based on an assumption of normality.

For example, you know that Professor X assigns A's to the top 25 percent of his class. You have a score of 44 on his last test which has a mean of 40 and a standard deviation of 10. You do not know anything about the shape of the distribution. Is it likely that your score has a percentile rank of 75 or greater so that you can look forward to receiving an A? Converting a score of 44 to a z-score we have a .40($z = (44 - 40)/10 = .4$). In Table A2 we find the value .6554 in column 3. *If* the distribution is normal, your percentile rank is 65. If the distribution is negatively skewed, you would expect your score to have a higher percentile rank than that. If the distribution is positively skewed, you would expect your percentile rank to be lower than that. A percentile rank of 65 is your estimate, assuming normality. Therefore, it is unlikely that your score falls in the A category.

To summarize, the steps in using the normal curve to find a percentile rank for a score are:

1. Convert the score to a z-score. Find this z-score in column 1, Table A2.
2. If the z-score is negative, locate the value in the same row in column 4.
3. If the z-score is positive, locate the value in the same row in column 3.
4. Multiply the result of step 2 or 3 by 100 and round to nearest whole number to find the percentile rank.
5. If the distribution is known to be normal, the percentile rank is exact. If the distribution shape is unknown, the result is approximate.

Exercises

1. In a normal distribution what is the percentile rank of a z-score of 1.75; a z-score of 0; a z-score of −.25?
2. In a normal distribution with a mean of 20 and a standard deviation of 2, what is the percentile rank for a score of 17.0; a score of 20.4?
3. What percentile corresponds to a score:
    (a) 1 standard deviation above the mean
    (b) 1.5 standard deviations above the mean
    (c) 2 standard deviations below the mean
    (d) 1.2 standard deviations below the mean
4. Assume a normal distribution of IQ scores with a mean of 100 and standard deviation of 15. Find the percentile rank of persons with IQ scores of:
    (a) 90   (b) 120   (c) 135
5. Given a score 1.05 standard deviations above the mean in a normal distribution, find the:
    (a) percent of cases between the mean and the score
    (b) percent of cases falling below the score
    (c) percent of cases falling above the score
    (d) percentile rank of the score
6. Given a score that is 2.45 standard deviations below the mean in a normal distribution, find the:
    (a) percent of cases between the mean and the score
    (b) percent of cases falling below the score
    (c) percent of cases falling above the score
    (d) percentile rank of the score
7. On a test the mean is 70 and the standard deviation 12. Assuming a normal distribution, find the z-score and the percentile rank for the following scores:
    (a) 92   (b) 52

Solutions

1. 96; 50; 40
2. 7; 58
3. (a) 84th percentile
   (b) 93rd percentile   (50 + 43.32)
   (c) 2nd percentile
   (d) 11.5th percentile
4. (a) 25   ($z = -.67$)
   (b) 90.82   ($z = 1.33$)
   (c) 99   ($z = 2.33$)
5. (a) 35.31   (b) 85.31   (c) 14.69   (d) 85.31
6. (a) 49.29   (b) .71   (c) 99.29   (d) .71
7. (a) $z = 1.83$;   PR = 96.64   (b) $z = -1.5$;   PR = 6.68

Finding the z-Score or Raw Score Equivalent for a Given Percentile Rank

In some situations, one has the percentile rank and wants to know the z-score or the raw score that would correspond to that percentile rank. We will again turn to the normal curve table but will use the reverse procedure from that followed in the previous section. For percentile ranks less than 50 one locates the number in column 4 closest to that needed and then finds the corresponding z-value from the first column of the table. When the percentile rank is below 50 the area in the smaller portion (column 4) will represent the percentile rank.

For example, we want to locate the score equivalent to a percentile rank of 20, the score that separates the bottom 20 percent. In column 4 of Table A2 we find .2005 is the value nearest .20. In the same row we find a z-score of .84. Approximately 20 percent of the normal curve falls below a z-score of $-.84$. Another way of expressing this is to say that the twentieth percentile point (P_{20}) in the normal curve is $z = -.84$. In the same way we find the 10th percentile point (P_{10}) is $z = -1.28$, the 30th percentile point (P_{30}) is $z = -.52$, and so on.

To find z-score equivalents for percentile ranks greater than 50, find the proportion nearest to that needed in column 3. This is because with a percentile rank greater than 50 the area in the greater portion will represent the percentile rank. For example, to find a z-score corresponding to a percentile rank of 75, find the value closest to .75 in column 3 which is .7486. The z-score in that row is .67. Therefore, a z-score of $+.67$ is the z-score equivalent to the percentile rank of 75. Remember that for percentile ranks below 50, the z-score values will be negative, while those above 50 will have positive values.

Once we have the z-score equivalent for any percentile rank we can go one step further and convert that z-score to a raw score if we know the mean and standard deviation for the distribution to which the score belongs. We have just determined that in a normal distribution a percentile rank of 75 has a z-score equivalent of $+.67$. Now let us take this a step further, using the Binet IQ scale as our example. In this scale the mean is 100 and the standard deviation is 16. To convert the z-score to a raw score, we substitute in the familiar z-score formula. We know z in this case and are looking for X, the raw score:

$$z = \frac{X - \mu_x}{\sigma_x}$$

$$.67 = \frac{X - 100}{16}$$

$$X - 100 = .67(.16)$$

$$X = 100 + 10.72$$
$$X = 110.72$$

First, multiply the z-score by the standard deviation: $(+.67)(16) = +10.72$. Then add this product to the mean: $10.72 + 100 = 110.72$. The 75th percentile point on the Binet scale is 110.72.

Suppose a college will only admit those above the 40th percentile on the College Board Examination. What score is necessary to qualify given that the examination scores are normally distributed with a mean of 500 and standard deviation of 100? Since our percentile rank is less than 50 we use column 4 in Table A2 and our z-score will be negative. The z-score in this case is $-.25$. Converting this into a College Board score we multiply by 100 (the standard deviation) and add 500 (the mean) to give us a score value of 475 $[(-.25)(100) + 500]$.

As was the case when we were converting in the other direction, percentile ranks converted to z-scores are exact when a distribution is known to be normally distributed and are reasonable estimates when the distribution is unknown but it is reasonable to assume normality. If you know you had a percentile rank of 89 on a test but have no other information about the test, you can determine that your z-score is 1.23 and therefore your best estimate is that you scored 1.23 standard deviations above the mean.

Exercises

1. In a normal distribution what is the z-score equivalent for a percentile rank of 28; 98?
2. In a normal distribution with a mean of 25 and a standard deviation of 5, what raw score would be associated with a percentile rank of 90; a PR of 14?
3. A student was told that he had a percentile rank of 87 on the Graduate Record Exam. The mean is 500 and the standard deviation 100. What was his: (a) z-score; (b) raw score on this examination?
4. If a student scored at the 25th percentile on the Graduate Record Exam., what raw score would this represent?
5. In a distribution of IQ scores (mean = 100, σ = 15), identify the: (a) 76th percentile; (b) 45th percentile.

Solutions

1. $-.58$; 2.05
2. 31.40; 19.60
3. (a) $z = 1.13$ (b) 613
4. 433
5. (a) 111   (b) 98

154 DESCRIPTIVE STATISTICS

Finding the Proportion of a Distribution Falling between Any Two z-Score Points

In Figure 6.7 we saw that 68.26 percent of the normal curve fell between +1 and −1 standard deviation from the mean, 95.44 percent between $z = +2.00$ and $z = -2.00$, and 99.72 percent between $z = +3.00$ and $z = -3.00$. We can use Table A2 to find the proportion of a normal distribution lying between any pair of z-scores.

If one of the z-scores is positive and the other is negative we find the proportion of the curve between them by adding values from column 2 in Table A2. What proportion of the curve lies between z of −1.5 and z of +.5? Column 2 indicates that the proportion between $z = -1.5$ and the mean is .4332, and from the mean to $z = +.5$ is .1915. Therefore, the proportion between $z = -1.5$ and $z = +.5$ is .4332 + .1915, or .6247. This means that 62.47 percent of the cases in a normal distribution will fall between these z-scores.

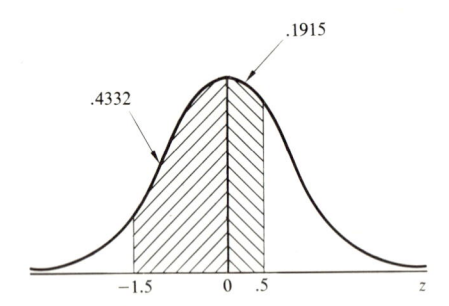

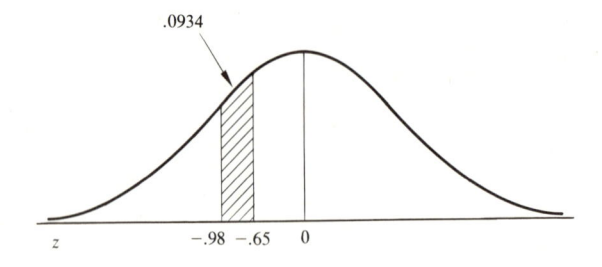

When we want the proportion of the normal curve falling between two z-scores with the same sign we subtract the area for the smaller z-score from the area for the larger z-score. For example, let us find the proportion of cases between a z of −.65 and a z of −.98 in a normal distribution. Column 2

in Table A2 indicates that the area between the mean and a z of .98 is .3365, while the area between the mean and a z of .65 is .2422. Thus, the area between $z = .65$ and $z = .98$ is found by subtracting the area for the smaller z-score from the area for the larger z-score; in this case, $.3365 - .2422 = .0943$. We would expect 9.43 percent of the cases in a normal distribution to fall between these z-score points. We can use raw scores as our input in this process if we first convert them to z-scores.

We can determine the number of cases instead of the percent if the situation warrants it. Let us assume that we have a normal distribution of 5,000 WISC intelligence test scores with a mean of 100 and a standard deviation of 15. Suppose we wanted to know how many of these 5,000 individuals could be expected to fall in the normal range of IQ scores, 90–110. The first step would be to convert these raw scores to z-score form so that we may consult the normal curve table:

$$\frac{90 - 100}{15} = -.67 \qquad \frac{110 - 100}{15} = .67$$

We now find the area between a z of $-.67$ and a z of $+.67$. The table indicates that in each case the area is .2486. The total area included between a z of $-.67$ and a z of .67 is thus $.2486 + .2486 = .4972$, or 49.72 percent of the cases. Therefore, $.4972 \times 5,000 = 2,486$ individuals out of the group of 5,000 could be expected to fall within the range of 90–110 on the intelligence test.

Exercises

1. What percent of a normal distribution falls between a z-score of -1.0 and $-.5$; between $-.5$ and $+.5$?
2. In a normal distribution of IQ scores where the mean is 100 and $\sigma = 15$, what percent of children fall between 85 and 120?
3. In a normal distribution where the mean is 70 and $\sigma = 5$, what percent of the area falls between 66 and 76?
4. In a normal distribution where the mean is 500 and $\sigma = 100$, what percent of the cases could be expected to fall between 625 and 725?

Solutions

1. 14.98 percent; 38.30 percent
2. 74.95 $(.3413 + .4082 \times 100)$
3. 67.30 $(.2881 + .3849 \times 100)$
4. 9.34 $(.4878 - .3944) \times 100$

Finding the Interval on the Curve that Will Include Some Specified Proportion of the Total Area

Sometimes one wants to determine the z interval (or raw score interval) that will include a given proportion of the cases. That is, one knows the proportion and is solving for the z or raw score interval. Suppose one wanted to determine the cut-off points in terms of z-scores that would include the middle .90 of the total area; that is, one wants to leave .05 in each end of the distribution. In other words, one wants to locate the middle 90 percent of the cases in the distribution.

One must find the z-score that would include .45 of the area above the mean and .45 below the mean. In column 2 the area values nearest to .45 are .4495 and .4505. Since our .45 is exactly halfway between these two values we will approximate and use a z-score halfway between the two z values in column 1; that is, 1.645. The limits of the area we are concerned with are, therefore, $z = \pm 1.645$. The remaining proportion, .10 or 10 percent, falls outside these limits.

If one wanted to apply this to a normal curve where the mean and standard deviation are known, one could report the limits in raw score form. For example, let us repeat the process of determining the limits of the middle 90 percent using WISC intelligence test scores (mean = 100 and standard deviation = 15), and continuing the calculation to determining raw scores:

$$z = \frac{X - \mu}{\sigma}$$

$$1.645 = \frac{X - 100}{15} \qquad -1.645 = \frac{X - 100}{15}$$

$$X - 100 = 1.645(15) \qquad X - 100 = -1.645(15)$$

$$X = 100 + 24.675 \qquad X = 100 - 24.675$$

$$X = 124.675 \qquad X = 75.325$$

This shows us that 90 percent of the cases in this normal distribution of WISC intelligence test scores would fall between raw scores of 75 and 125. If one wanted to provide special classes for the highest 5 percent and the lowest 5 percent and assign the remainder to regular classes, those with IQ scores between 75 and 125 would be assigned to regular classes.

One can follow the above procedure to locate any other specified proportion of a normal curve. If one knows the mean and standard deviation of the test scores, one can convert the z-scores to raw score cut-offs as illustrated above.

Exercises

1. How far above and below the mean in a normal distribution do we need to go to include the middle 30 percent of the distribution; the middle 80 percent?

2. On a classroom test, the mean was found to be 62 and the standard deviation 5. The instructor gave the top 10 percent of the group an A. What was the minimum score needed to get an A? Assume a normal distribution.
3. If the bottom 5 percent received a failing grade on the above test, what was the highest failing score?
4. A teacher wishes to identify the middle 50 percent of a group on a test where the distribution is known to be normal.
    (a) What z-score range would include the middle 50 percent?
    (b) If the mean is 60 and the standard deviation 9, what raw scores would set off the middle 50 percent?
5. Knowing that the mean IQ is 100 and $\sigma = 15$, what IQ score would set off the upper 20 percent of the elementary school population?
6. On a classroom test, the mean was 65 and the standard deviation 10. If the top 15 percent got an A and the next 20 percent got a B, find the highest and lowest scores to get a B grade.

Solutions

1. $z = \pm .39$; ± 1.28
2. 68   ($z = 1.28$)
3. 53.775 or 54   ($z = -1.645$)
4. (a) $-.67$ to $+.67$
   (b) 54 to 66
5. 112.6   (z of .84)
6. 68.9 or 69 is lowest and 75.4 or 75 is the highest   (z of .39 to z of 1.04)

Let us summarize the types of problems that can be solved using the normal curve:

1. One can approximate the percentile rank of any specified z-score and, if μ_x and σ_x are known, of any specified raw score.
2. One can approximate the z-score for a percentile rank.
3. One can find the proportion or percent of cases between any two z-score points and, if μ_x and σ_x are known, between any two raw scores.
4. One can locate the interval on the curve in either z-scores or raw scores that will include a given percent of the cases.

SUMMARY

The normal curve is a frequency polygon of a theoretical distribution which is unimodal and symmetrical with the mean, median, and mode at the same point. The height is greatest at this point and decreases on both sides of this point to form a bell-shaped curve. The theoretical ranges of the normal curve are from $-\infty$ to $+\infty$ but most of the area lies between $+3$ and -3

FLOWCHART 6

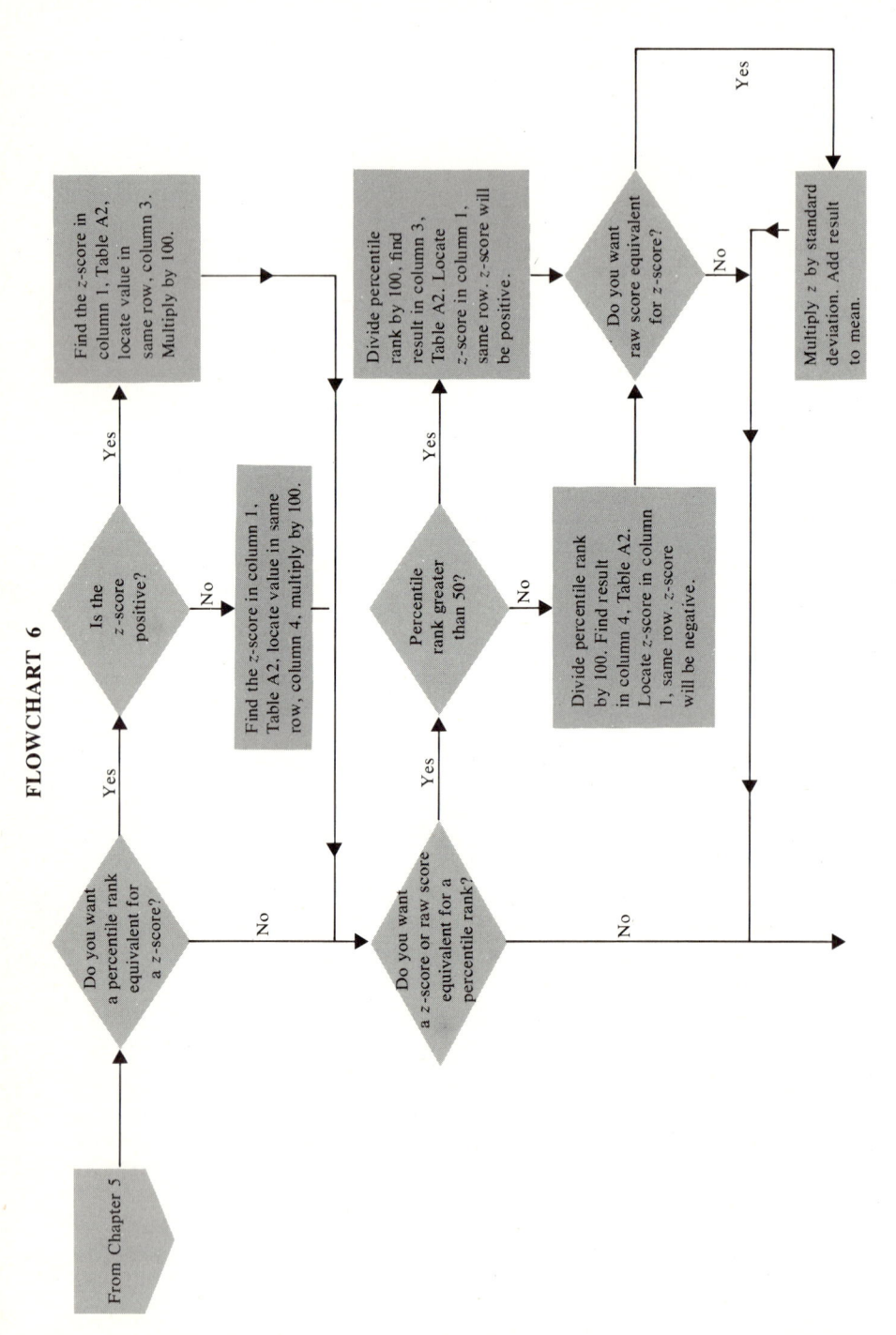

From Chapter 5 → **Do you want a percentile rank equivalent for a z-score?**

- **Yes** → **Is the z-score positive?**
  - **Yes** → Find the z-score in column 1, Table A2, locate value in same row, column 3. Multiply by 100.
  - **No** → Find the z-score in column 1, Table A2, locate value in same row, column 4, multiply by 100.

- **No** → **Do you want a z-score or raw score equivalent for a percentile rank?**
  - **Yes** → **Percentile rank greater than 50?**
    - **Yes** → Divide percentile rank by 100, find result in column 3, Table A2. Locate z-score in column 1, same row. z-score will be positive.
    - **No** → Divide percentile rank by 100. Find result in column 4, Table A2. Locate z-score in column 1, same row. z-score will be negative.
  - Then → **Do you want raw score equivalent for z-score?**
    - **Yes** → Multiply z by standard deviation. Add result to mean.
    - **No** → (exit)
  - **No** → (exit)

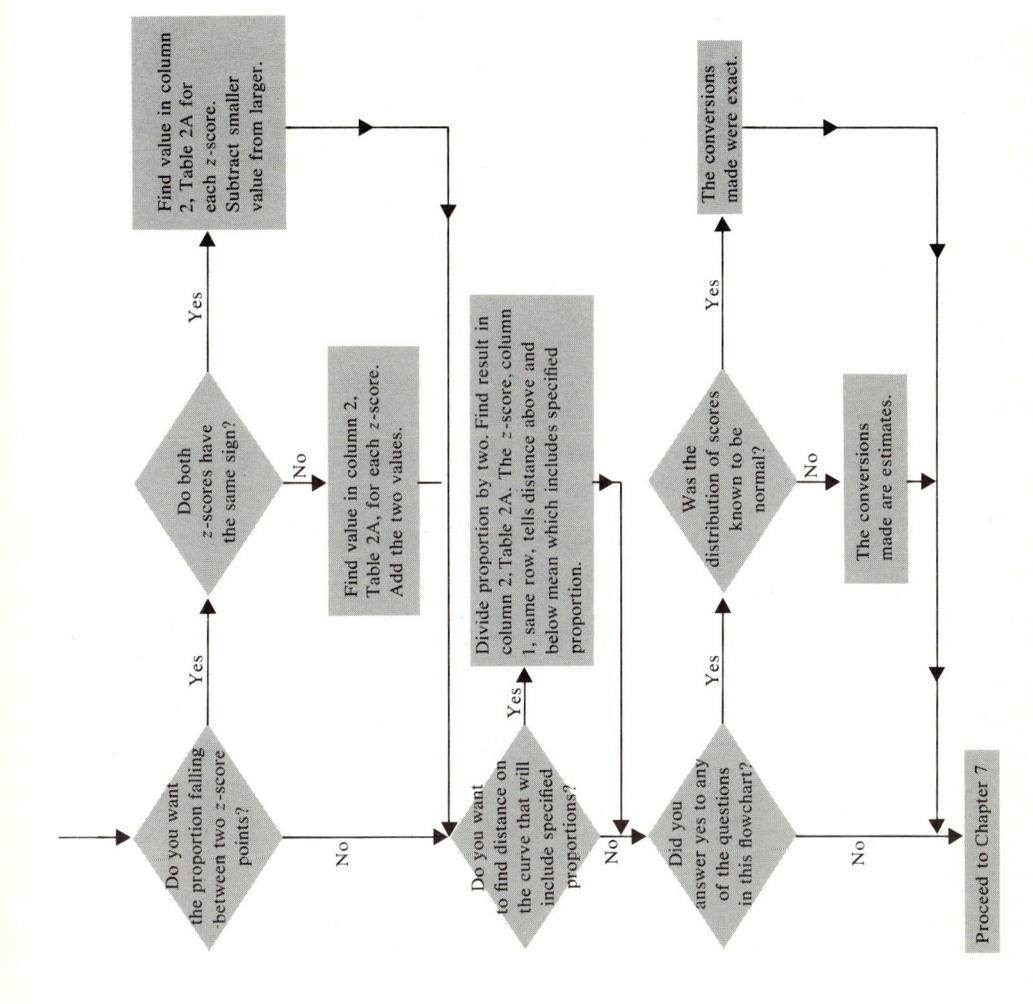

```
                                                    ┌─────────────────┐
                                                    │ Find value in   │
                                                    │ column 2,       │
                                                    │ Table 2A for    │
                                                    │ each z-score.   │
                                                    │ Subtract smaller│
                                                    │ value from      │
                                                    │ larger.         │
                                                    └────────┬────────┘
                                                             │
                                              Yes            │
                                         ┌──────►────────────┘
                                        ╱ ╲
                                       ╱   ╲
                                      ╱ Do  ╲
                                     ╱both   ╲
                                    ╱z-scores ╲                                         ┌──────────────┐
                                   ╱  have the ╲  No    ┌─────────────────┐             │ The          │
                                   ╲ same sign?╱ ─────► │ Find value in   │             │ conversions  │
                                    ╲         ╱         │ column 2,       │             │ made were    │
                                     ╲       ╱          │ Table 2A, for   │             │ exact.       │
                                      ╲     ╱           │ each z-score.   │             └──────▲───────┘
                                       ╲   ╱            │ Add the two     │                    │
                                        ╲ ╱             │ values.         │                    │
                                  Yes    │              └────────┬────────┘                    │
                              ┌───►──────┘                       │                             │
                             ╱ ╲                                 │                             │
                            ╱   ╲                                ▼                             │
                           ╱ Do  ╲                   ┌───────────────────────┐                 │
                          ╱ you   ╲                  │ Divide proportion by  │                 │
                         ╱want the ╲                 │ two. Find result in   │                 │
                        ╱proportion ╲                │ column 2, Table 2A.   │                 │
                       ╱ falling    ╲                │ The z-score, column   │                 │
                       ╲ between two╱                │ 1, same row, tells    │                 │
                        ╲ z-score   ╱                │ distance above and    │                 │
                         ╲ points?  ╱                │ below mean which      │                 │
                          ╲        ╱                 │ includes specified    │                 │
                           ╲      ╱                  │ proportion.           │                 │
                            ╲    ╱                   └───────────▲───────────┘                 │
                             ╲  ╱                                │                             │
                              ╲╱                                 │                            Yes
                               │ No                        Yes   │                         ┌───┴───┐
                               ▼                              ┌──┴──┐                     ╱         ╲
                              ╱ ╲                            ╱       ╲                   ╱  Was the  ╲
                             ╱   ╲                          ╱ Do you  ╲                 ╱distribution╲
                            ╱     ╲                        ╱want to    ╲               ╱ of scores    ╲
                           ╱       ╲                      ╱find distance╲              ╲ known to be  ╱
                          ╱         ╲                     ╲on the curve ╱               ╲ normal?     ╱
                                                           ╲that will   ╱                ╲           ╱
                                                            ╲include   ╱                  ╲         ╱
                                                             ╲specified╱                   ╲   No  ╱
                                                              ╲propor-╱                    │       │
                                                               ╲tions?╱                    ▼       │
                                                                ╲    ╱           ┌──────────────┐  │
                                                                 ╲  ╱            │ The          │  │
                                                                  ╲╱             │ conversions  │  │
                                                                   │ No          │ made are     │  │
                                                                   ▼             │ estimates.   │  │
                                                                  ╱ ╲            └──────┬───────┘  │
                                                                 ╱Did╲                   │         │
                                                                ╱ you  ╲  Yes            │         │
                                           ──►                 ╱ answer ╲────►───────────┤         │
                                                               ╲yes to   ╱                         │
                                                                ╲any of ╱                          │
                                                                 ╲ques- ╱                          │
                                                                  ╲tions╱                          │
                                                                   ╲   ╱                           │
                                                                    ╲ ╱                            │
                                                                     │ No                          │
                                                                     ▼                             │
                                                              Proceed to Chapter 7 ◄───────────────┘
```

159

standard deviations from the mean. Many natural phenomena and many chance occurrences have distributions similar to the normal curve.

The areas associated with various points along the normal curve have been tabled. This table can be used to identify percentile rank equivalents of z-scores or vice versa. The table can also be used to identify areas falling between specified z-scores or the z-scores that mark off specified areas around the mean. These interpretations are exact if a distribution is exactly normal and are good approximations if a distribution is similar to the normal curve.

EXERCISES

1. Find the proportion of the area under the normal curve between the mean and the following z-scores:
   (a) −2.10
   (b) .45
   (c) 1.33
   (d) −1.95
   (e) −.28
   (f) 2.56
   (g) 1.05
2. Given a score located 1.65 standard deviations above the mean in a normal distribution, determine the following:
   (a) area between the mean and the score
   (b) total area to the left of the score
   (c) total area to the right of the score
   (d) percentile rank of the score
3. Given a score located −1.90 standard deviation units below the mean in a normal distribution, determine the following:
   (a) area between the mean and the score
   (b) total area to the left of the score
   (c) total area to the right of the score
   (d) percentile rank of the score
4. Given the following percentile ranks in a normal distribution, find the corresponding z-scores.
   (a) 25
   (b) 92
   (c) 50
   (d) 65
   (e) 10
   (f) 40
5. The following assumes a normal distribution of 500 cases with a mean = 60 and a standard deviation = 10 (you may round off answers):
   (a) What is the percentile rank of a raw score of 45?
   (b) What is the percentile rank of a raw score of 80?
   (c) How many cases would fall below a raw score of 50?
   (d) How many cases would be above a raw score of 75?
   (e) What score falls at the 20th percentile?
   (f) What score falls at the 90th percentile?
   (g) What percentage of the cases are found between 55 and 65?
   (h) Between what raw scores do the middle 50 percent of the cases fall?
   (i) If one wanted to select the top 5 percent on this test, what score would be used as the cut-off?

6. Mary Smith made a score of 656 on the CEEB examination where the mean is 500 and the standard deviation is 100.
   (a) What percent of cases would fall between her score and the mean?
   (b) What is her percentile rank in the population of students taking the exam?
   (c) What percent of cases would fall between scores 425 and 625?
   (d) What is the 85th percentile on this exam?
7. The mean IQ on the WISC test is 100 and the standard deviation is 15. Ten percent of elementary school children have a WISC IQ greater than what value?
8. On a graduate statistics quiz, the mean score was 50 and the standard deviation 5. The professor announced that 15 percent of the class would get a grade of A. What score would be required to get the lowest A?
9. A student received a score of 625 on the CEEB examination (mean = 500, $\sigma = 100$).
   (a) What percent of the comparison group received scores higher than this student?
   (b) How many students out of the comparison group of 1,000 were below this student?
   (c) What is the percentile rank of this student?

ANSWERS

1. (a) .4821 (b) .1736 (c) .4082 (d) .4744 (e) .1103
   (f) .4948 (g) .3531
2. (a) .4505 (b) .9505 (c) .0495 (d) 95
3. (a) .4713 (b) .0287 (c) .9713 (d) 2.87 or 3
4. (a) −.67 (b) 1.41 (c) 0 (d) .39 (e) −1.28
   (f) −.25
5. (a) $z = -1.5$; therefore, PR = 6.68 or 7
   (b) $z = 2$; PR = 98
   (c) $z = -1$; .1587 area below; .1587 × 500 = 79 cases below
   (d) $z = 1.5$; .0668 area above; .0668 × 500 = 33 cases above
   (e) $z = -.84$; therefore, raw score = 52
   (f) $z = 1.28$; raw score = 73
   (g) $z = -.5$ to $+.5$; area = .1915 + .1915 = .3830 or 38 percent of cases
   (h) middle 50 percent between −.67 and +.67 z; or raw scores 53–67
   (i) at $z = 1.645$ the area below is .9500; raw score = 76
6. (a) 44.06 percent
   (b) 94
   (c) $z = -.75$ to $+1.25$; area = .3944 + .2734 = 66.78 percent of cases
   (d) $z = 1.04$; raw score = 604
7. 119.20 (The z-score corresponding to P_{90} is 1.28; therefore, $X = 100 + 15(1.28) = 119.20$. Thus, 10 percent of elementary school children have an IQ of 119.20 or above; 90 percent fall below this score.)
8. 55.20 (A z of 1.04 would set off the upper 15 percent. The raw score corresponding to z of 1.04 is: $X = 50 + 5(1.04) = 55.20$.) Thus, a student would have to get a score of 55.20 or greater to get an A.
9. (a) 10.56 percent (b) 894 (c) 89

CHAPTER SEVEN
CORRELATION

OBJECTIVES

After completing this chapter the reader should be able to:

1. Distinguish between univariate and bivariate distributions.
2. Recognize situations in which a correlation index would be an appropriate statistic.
3. Interpret a scattergram.
4. Describe the difference between a positive and a negative correlation.
5. Describe the relationship between the value of a correlation index and the strength of a relationship between two variables.
6. Describe how the extent of the similarity of paired z-scores determines the value of the Pearson product-moment correlation coefficient.
7. Use paired z-scores to compute a Pearson correlation coefficient.
8. Use paired raw scores to compute a Pearson correlation coefficient.
9. List the assumptions of the Pearson correlation coefficient.
10. Describe factors that influence the correlation coefficient.
11. Interpret a Pearson correlation coefficient.
12. Interpret a coefficient of determination.

Correlation

Thus far we have focused on those statistical procedures used for describing single variables or for analyzing what are called *univariate* distributions. We have computed measures of central tendency and variability, but always only one variable, such as test scores, was involved at one time. However, we also need statistical procedures to investigate relationships that may exist between two variables (in a population).

Is there a relationship between the length of time spent in spelling drill and spelling achievement? Does student achievement increase as schools increase their per pupil expenditures? Is there a relationship between socioeconomic status and participation in school activities? Statistical techniques used for determining relationships between such pairs of variables are referred to as *correlational procedures.* Correlational procedures are used with *bivariate* distributions, that is, distributions where two variables are involved. The bivariate data consist of observations that are paired on some logical basis. Typically, measurements on two variables are available for each member of a population and one determines if there is a relationship between these paired measurements (bivariate data).

We commonly observe that certain variables tend to be related. People who are good golfers are more likely to be good bowlers than those who are poor golfers. A family that spends little on gasoline probably spends little on tires, while a family with high gasoline expenditures probably also has high tire expenditures. When variables are related so that an increase in one is accompanied by a systematic increase or decrease in the other, we say they are correlated. Statistical indexes have been developed that enable us to concisely describe such relationships. These are *correlation indexes.*

A correlation index describes the relationship of two variables in a single population. By convention the two variables are labeled X and Y. We might employ a correlation procedure to identify the relation between intelligence test scores (X) and reading test scores (Y) for a group of third graders. Or we might determine the relationship between height and weight for a group of army recruits. However, correlation may be based on measures of the same variable for different individuals who are paired on some logical basis. For example, we might correlate the IQ scores for husbands (X) with IQ scores of their wives (Y) to see the extent to which bright people tend to marry bright people and low IQ people tend to marry low IQ people. Or we might correlate the heights of fathers (X) with the heights of their eldest sons (Y). Our two options are to measure two characteristics of the same individual or a single characteristic for a matched pair of individuals. In either case we must collect our data so that if we know whose X score we have, we know who its paired Y score belongs to; in other words, our sets of scores must have a common source. If X is Private Brown's height, Y is his weight. If X is Mr. Green's IQ, Y is Mrs. Green's IQ. If two sets of scores do not have a common source we cannot employ correlation. For example, we could not correlate math scores of boys in a class with math scores of girls in the same class since there is no logical basis for pairing the subjects. If we wish to correlate math and reading scores for a class of 25 pupils, but

164 DESCRIPTIVE STATISTICS

find that two pupils have not taken the math test and three other pupils have not taken the reading test, we can only correlate scores for 20 pupils for whom both scores are available.

Exercises

Which of the following sets of scores could be correlated?
1. Reading test scores for this year's fourth graders with reading test scores for last year's fourth graders.
2. Reading test scores for this year's fifth graders with the reading test scores the same pupils had last year.
3. The social adjustment scores of identical twins.
4. The number of dates the girls in a sorority have with their grade-point average.
5. The numbers of dates sorority girls have with the number of dates nonsorority girls have.

Solutions

Numbers 2, 3, and 4 could be correlated.

Scattergrams

A useful way to show correlation visually is to construct a scatter diagram or scattergram. A *scattergram* is a graph in which a single dot is used to locate each individual on two dimensions. The pattern formed by the dots illustrates the correlation. For example, 12 members of a jury have indicated their years of formal education, and their annual salary to the nearest thousand dollars. Are these two variables correlated in this group?

	YEARS OF EDUCATION	ANNUAL INCOME
Juryman A	8	9,000
Juryman B	12	14,000
Juryman C	14	17,000
Juryman D	10	9,000
Juryman E	8	11,000
Juryman F	16	20,000
Juryman G	17	23,000
Juryman H	12	24,000
Juryman I	12	30,000
Juryman J	13	14,000
Juryman K	16	18,000
Juryman L	12	10,000

To determine if a relationship exists between these two variables, the data above may be graphed in a scattergram.

To construct a scattergram we first lay out the scale for one variable on the abscissa and the scale for the other on the ordinate. By convention the variable on the abscissa is labeled X and the variable on the ordinate is labeled Y. In Figure 7.1 we have years of education as X and annual salary as Y.

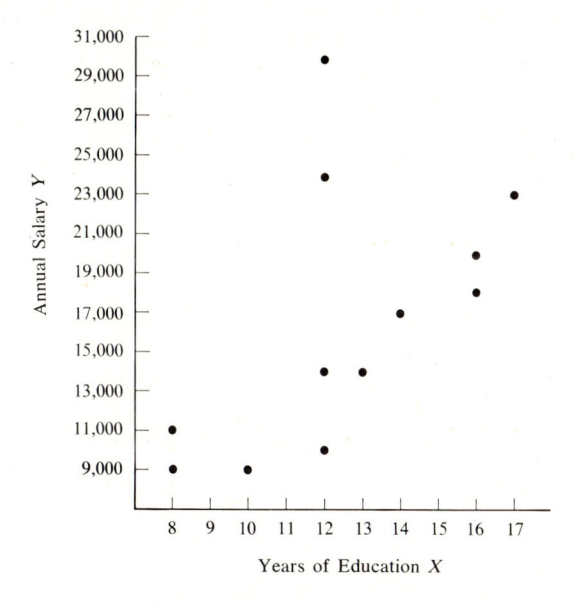

Figure 7.1 Scattergram showing Education and Income of 12 Jurymen

For Juryman A we find his years of education (8) and follow the vertical line from that number until we come to the horizontal line representing his salary. A dot at the point represents Juryman A on both variables. A dot at the junction of the vertical line for 12 years and the horizontal line for $14,000 represents Juryman B, and so on.

In the scattergram we can see that there is definitely a relationship between the two variables. Those with more education tend to have the higher salaries and those with less education tend to have the lower salaries. In other words, the two variables are correlated.

When high scores on one variable are associated with high scores on a second variable and low scores on the one are associated with low scores on the other, the variables are said to be positively correlated, or to show *positive correlation*. Figure 7.1 illustrates a positive correlation between the variables, years of education and annual salary. As a further example, a positive correlation has been observed between intelligence test scores and

166 DESCRIPTIVE STATISTICS

reading test scores. That is, those students scoring high on the IQ test tend to score high on the reading test. Those receiving low scores on the IQ test tend to receive low scores on the reading test.

When high scores on one variable are associated with low scores on a second variable, and vice versa, the variables are said to be *negatively correlated*. One would expect to find a negative correlation between alcohol consumption and performance on a manual dexterity test. That is, as alcohol consumption increases, the scores on the dexterity test would decrease. Another example of negative correlation is the relationship between fatigue and sensory acuity; that is, as fatigue increases sensory acuity tends to decrease. When variables show negative correlation there is an inverse relationship, so that individuals high or above average on one variable tend to be low or below average on the second variable. Figure 7.2 is a scattergram showing a negative correlation.

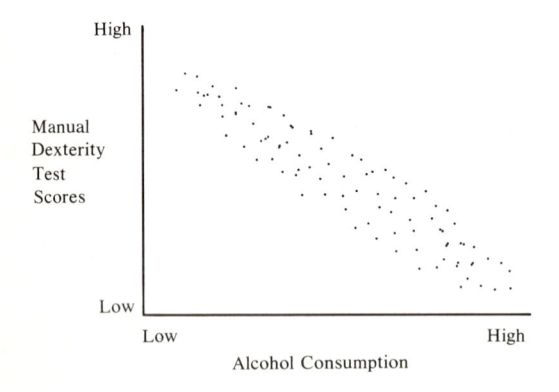

Figure 7.2 Negative Correlation

In a positive correlation the dots in the scattergram spread from lower left to upper right (see Figure 7.1). In a negative correlation the dots spread from upper left to lower right (see Figure 7.2).

Some variables show no tendency to vary or change together. That is, individuals scoring high on one variable tend to score neither systematically high nor low on the other variable. In this case, one reports that there is no correlation between the variables. For example, height and grade point average for a group of male college seniors would not be expected to bear any systematic relationship to each other and thus would show no correlation. When there is no correlation the dots in the scattergram are spread in a random manner.

Exercises

1. Would you expect the correlation between the following variables to be (i) positive; (ii) negative; (iii) near zero?
   (a) the IQ's of oldest and youngest siblings
   (b) chronological age and mental age in elementary school children
   (c) chronological age and IQ
   (d) grades in chemistry and algebra
   (e) extent of fatigue and performance on a speed test
   (f) horsepower of car and miles-per-gallon of gasoline
   (g) scores on a social studies test and performance on the 50-yard dash for elementary school boys

2. Below are test scores in social studies (X) and reading (Y) for ten elementary school students. Construct a scattergram of these scores.

STUDENT	SOCIAL STUDIES	READING	STUDENT	SOCIAL STUDIES	READING
A	83	93	F	84	94
B	80	91	G	82	90
C	89	98	H	83	92
D	88	96	I	85	94
E	86	95	J	87	97

3. What does the scattergram indicate about the correlation between social studies and reading for these students?

Solutions

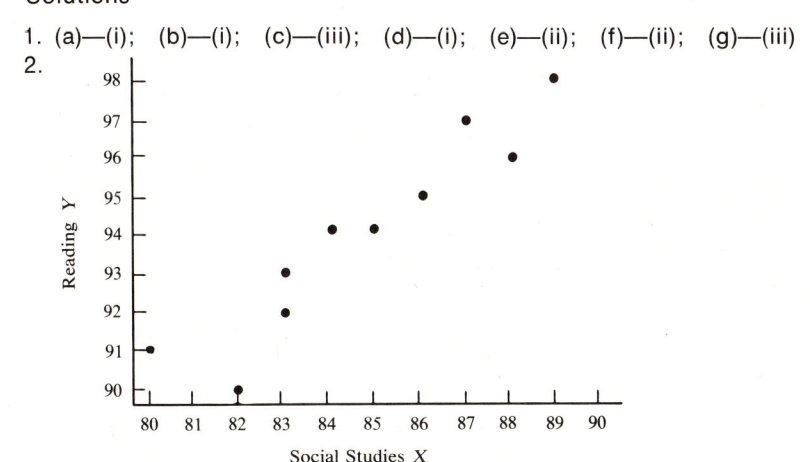

1. (a)—(i); (b)—(i); (c)—(iii); (d)—(i); (e)—(ii); (f)—(ii); (g)—(iii)

2.

3. It shows a high positive correlation between these two variables.

168 DESCRIPTIVE STATISTICS

Strength of Relationship

One is interested not only in the *direction* of the relationship between variables (positive or negative) but also in the *magnitude* or strength of the relationship. A numerical index called the *coefficient of correlation* expresses the degree or magnitude of the relationship. The numerical index 1.0 is the highest possible value that the correlation coefficient assumes and it indicates perfect relationship between the variables. A perfect positive relationship (+1) indicates a direct relationship where each individual is as high or as low on one variable as he is on the other. In other words, each subject has the same z-score on X as his z-score on Y. Figure 7.3 is a scattergram of a perfect positive correlation. In a perfect correlation all the dots in the scattergram can be connected with a single straight line. An example of a perfect positive relationship would be the heights of subjects measured in inches and the heights of these subjects measured in centimeters.

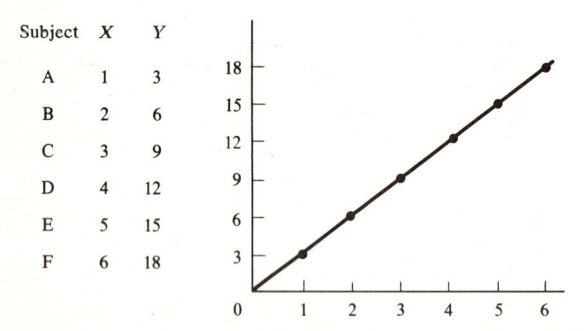

Subject	X	Y
A	1	3
B	2	6
C	3	9
D	4	12
E	5	15
F	6	18

Figure 7.3 Perfect Positive Correlation

A correlation coefficient of −1.0 indicates a perfect negative relationship. This means that the two variables bear an inverse relationship to each other so that the highest z-score on one variable is associated with the lowest z-score on the other, and so forth. An example of a perfect negative relationship is that between the volume of a gas and the pressure applied when temperature is held constant. As pressure increases the volume of the gas decreases. Figure 7.4 shows a perfect negative correlation. The dots can be connected with a single straight line as was the case with a perfect positive correlation. Of course, the straight lines run in different directions.

Perfect correlations are not generally found among educational and psychological variables. There is a perfect relationship between the radius and the circumference of a circle, or in the physical sciences, a perfect relationship exists between the volume of a gas and its temperature; that is, at a constant pressure the volume of a gas varies directly with the temperature applied.

Subject	X	Y
A	1	12
B	2	10
C	3	8
D	4	6
E	5	4
F	6	2

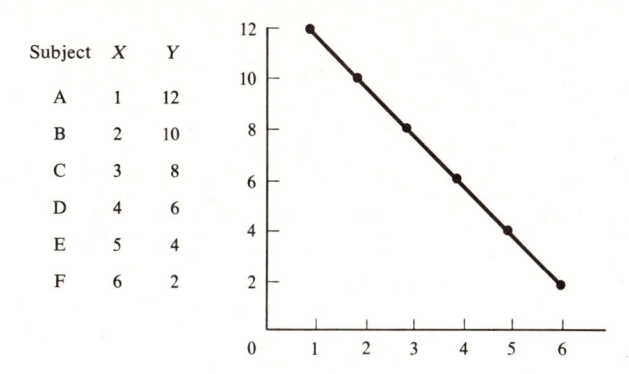

Figure 7.4 Perfect Negative Correlation

As a positive relationship between variables becomes less close, the correlation coefficient assumes a value smaller than 1.0. Students with high intelligence tend to make high grades, but there are exceptions because of several factors other than intelligence that affect school grades. Although still positive the correlation coefficient between these two variables is less than 1.0. Other variables may show a negative relationship that is less than perfect; in such cases the correlation coefficient would have a negative sign but between 0 and -1.0. A negative relationship would be expected between the number of truancies among boys and their scholastic performance, but the relationship would certainly not be perfect. When there is no relationship between the variables, the coefficient is 0.0.

Thus a coefficient of correlation indicates both the direction and the strength of relationship between two variables. The direction of relationship is indicated by the sign of the coefficient (+ or −) and the strength of relationship is represented by the absolute size of the coefficient, that is, how far the coefficient is from 0 and how near it is to $+1.0$ or -1.0.

It must be pointed out that the strength of relationship is independent of the sign of the correlation coefficient. A correlation of -1.0 represents exactly the same amount of relationship as a coefficient of $+1.0$; the difference is in the direction of the relationship only; that is, inverse rather than direct. Similarly, a coefficient of $+.90$ indicates the same degree of relationship as $-.90$. The strength or extent of the relationship is reflected in the absolute size of the correlation coefficient, not the sign.

Examination of a scatter diagram can reveal both the direction and approximate degree of correlation between two variables. The slope of the plotted points indicates the *direction* of the relationship and the closeness of the points around a sloping line indicates the *extent* of relationship. As we saw in Figures 7.3 and 7.4, if the points all fall exactly on a sloped straight line, the relationship is *perfect*. The direction of the slope distinguishes perfect

170 DESCRIPTIVE STATISTICS

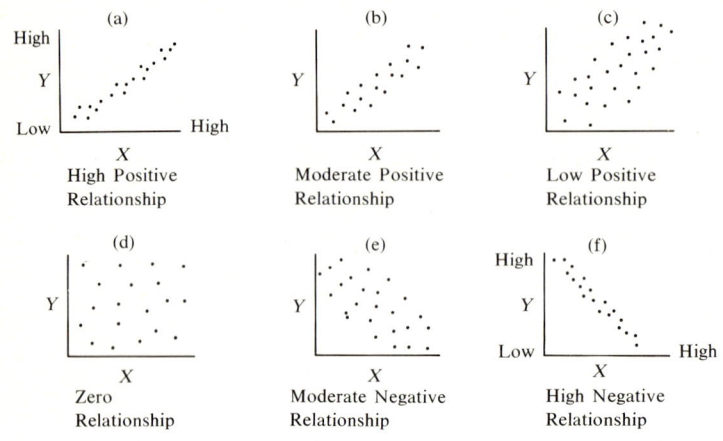

Figure 7.5 Scatter Diagrams Showing Various Degrees and Kinds of Relationship

positive from perfect negative. In Figure 7.5 we have scattergrams showing various correlations.

Exercise

1. Indicate the direction (+ or −) and the approximate extent (high, moderate, or low) of relationship that you would expect between the following variables:
   (a) IQ scores and scores on a reading test for elementary school children
   (b) The floor area of rooms and the square yards of carpet required to cover them
   (c) chronological age and weight of elementary school boys
   (d) scores on a verbal aptitude test and time required to learn a list of nonsense syllables

Solution

1. (a) high positive
   (b) perfect positive
   (c) moderately positive
   (d) moderately negative

Bivariate Frequency Distribution

Another way to show a correlation visually is to construct a bivariate frequency distribution. Values of the X variable are shown on the abscissa

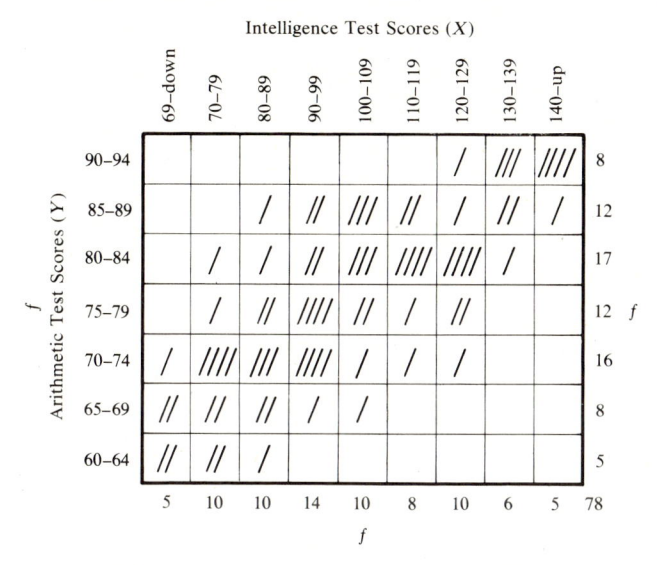

Figure 7.6 Bivariate Frequency Distribution of Intelligence and Arithmetic Test Scores of 78 Students

and values of the Y variable are shown on the ordinate. A tally mark shows each score combination. Bivariate frequency distributions can be constructed with either ungrouped data or with grouped data as shown in Figure 7.6.

One can see the general nature of the relationship by examining the bivariate frequency distribution. In Figure 7.6 the general trend of the tally marks is from the lower left to the upper right, which indicates a positive relationship. That is, high scores on Test X tend to be associated with high scores on Test Y. Since all the tallies do not fall along a single straight line, we know that the relationship is somewhat less than perfect.

Exercises

1. Which correlation coefficient represents the strongest relationship between two variables?
(a) 0.0
(b) −.80
(c) +.60

2. Which scattergram represents each of the following relationships?

_____ positive correlation   _____ zero correlation

_____ perfect negative correlation   _____ negative correlation

172 DESCRIPTIVE STATISTICS

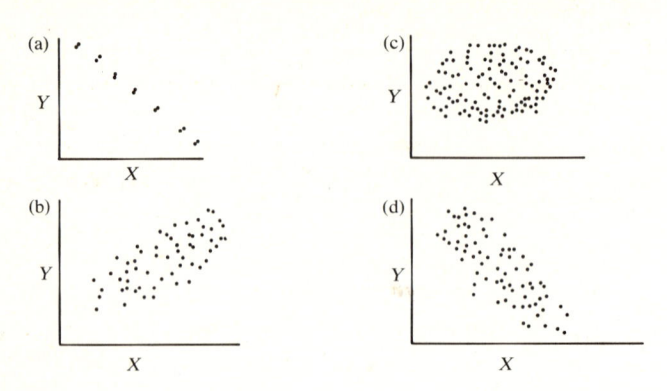

3. A researcher reported the correlation coefficient between two variables to be 1.55. How would you interpret this finding?
4. Sketch a scattergram showing a correlation of 1.00 between variables X and Y.

Solutions

1. (b)
2. (a) perfect negative correlation
   (b) positive correlation
   (c) zero correlation
   (d) negative correlation
3. He has made an error in his computations, since the value of the correlation coefficient cannot exceed 1.00.
4.

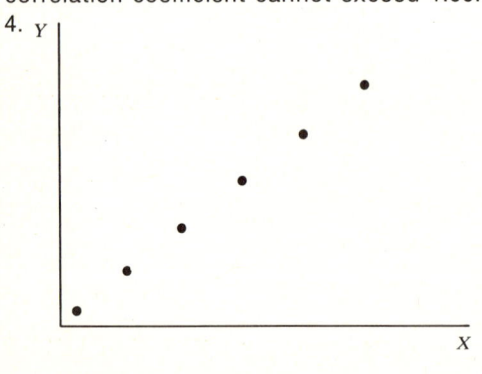

Calculation of Pearson Correlation Coefficient

While inspection of a scattergram furnishes some general information on the relationship between two sets of measures for a given group, a numerical index indicating precisely the degree of relationship is much more helpful. Several correlation indexes have been developed. The most widely used is an index which is employed when both variables are expressed as interval data. The use of this procedure for measuring relationship was first proposed by an English statistician, Karl Pearson. It is called the *Pearson product-moment coefficient of correlation* or simply the *correlation coefficient*. The parameter is symbolized by the Greek letter rho, ρ, and the statistic by a lower case r.[1] Formula (7.1) is the definitional formula for the correlation coefficient:

$$\rho = \frac{\Sigma z_x z_y}{N}$$

where: ρ = Pearson coefficient of correlation for a population
$z_x z_y$ = an individual's z-score on X multiplied by his z-score on Y
N = number of paired scores
Σ = sum of

Formula (7.1)
Pearson ρ
(definition formula)

The Pearson ρ is defined as the mean of the z-score products, since we compute it by converting the X- and Y-score values to z-scores, multiplying each z_x by its z_y partner, summing these products, and dividing by the number of pairs.

You recall that a z-score indicates relative position in a population by showing the deviation of each score from the mean in standard deviation units. The coefficient of correlation then shows the extent to which individuals have the same relative position (z-scores) on *two* variables.

When computing a coefficient of correlation, it is not necessary that the X and Y variables be measured on the same type of scale. Since raw score values are expressed as z-scores, individuals' relative positions in the distributions will be the same regardless of the measuring units used. For example, one can correlate test scores and weight or test scores and height. Furthermore, the height could be measured in inches, feet, or meters and the coefficient of correlation would be the same.

Let us illustrate correlation as the mean of the z-score products.

In Table 7.1 we have spelling and phonics test scores for five sixth grade girls. The raw scores for phonics are shown in column X, the raw scores for spelling are shown in column Y. Looking at the raw scores we see that there

---

[1] Most texts use the symbol r when introducing the Pearson product-moment correlation coefficient and this coefficient is commonly called the "Pearson r." However, in this text we have made it a practice to introduce indexes for population parameters such as μ for the mean and σ for the standard deviation first and to reserve sample indexes such as $\bar{X}$ and s for those cases where a sample statistic is used as an estimate of a population parameter. We will use the Latin r instead of the Greek ρ when a sample product-moment correlation index is being used as an estimate of the population parameter ρ.

TABLE 7.1

	X PHONICS SCORE	Y SPELLING SCORE	x	y	z_x	z_y	PRODUCT $z_x z_y$
Ann	26	14	12	6	+1.5	+1.5	2.25
Bess	18	10	4	2	+.5	+.5	.25
Carol	14	8	0	0	0	0	0
Dora	10	6	−4	−2	−.5	−.5	.25
Emma	2	2	−12	−6	−1.5	−1.5	2.25

$\Sigma X = 70 \quad \Sigma Y = 40 \quad \Sigma x^2 = 320 \quad \Sigma y^2 = 80 \quad \Sigma z_x z_y = 5.00$
$\mu_x = 14 \quad \mu_y = 8$
$\sigma_x = 8 \quad \sigma_y = 4$

$$\rho_{xy} = \frac{\Sigma z_x z_y}{N} = \frac{5}{5} = +1.0$$

is a relationship between the two sets of scores. Let us convert the raw scores to z-scores by finding the deviation of each score from its mean and dividing by the standard deviation. Thus,

$$z_x = \frac{X - \mu_x}{\sigma_x} \quad \text{or} \quad \frac{x}{\sigma_x}$$

$$z_y = \frac{Y - \mu_y}{\sigma_y} \quad \text{or} \quad \frac{y}{\sigma_y}$$

$$\left(\text{or} \quad z_x = \frac{X - \bar{X}}{s_x} \quad z_y = \frac{Y - \bar{Y}}{s_y}\right)$$

For example, Ann's z-score in phonics (X) is found as follows: $z_x = (26 - 14)/8 = 1.5$. When all the scores have been converted to z-scores (columns z_x and z_y), the relationship between X and Y becomes even more apparent. Each girl has the same z-score on both variables, that is, their relative positions are the same on the two measures. Therefore, there is a perfect positive correlation between these two variables for these five girls.

The Pearson product-moment correlation procedure provides a single number, the correlation coefficient, to indicate the degree of relationship. We multiply each girl's z_x-score in phonics by her z_y-score in spelling to get the products shown in the last column in Table 7.1. Summing these products we get +5.0. Dividing the sum by the number of girls (5) we have +1.0 as our correlation coefficient. It indicates that there is a perfect positive relationship between the two sets of scores. A coefficient of 1.00 is the maximum positive correlation and is obtained only when the paired z-scores are exactly the same on the two variables. Note that in a situation where

$\rho = 1.00$, the sum of the cross-products of the paired z-scores ($\Sigma z_x z_y$) is equal to N, the number of pairs.

We also have arithmetic scores for the same girls. Let us compare these scores with the spelling scores (Table 7.2). Looking over the raw scores we can see that there is a positive relationship between the scores. The two girls who did well in spelling are also doing well in arithmetic and the two who did poorly in spelling are doing poorly in arithmetic. However, the relationship is not as strong as the relationship between spelling and phonics. When we look at the girls' z-score positions on the two variables in Table 7.2 we see that there is some discrepancy in their relative position on the two variables. However, the girls who are above average on one variable are also above average on the other variable and those who are below average on the one variable are below average on the other. Therefore, all the z-score products will be positive and hence the ρ will be positive but will be less than 1.00.

TABLE 7.2

	X ARITHMETIC SCORE	Y SPELLING SCORE	x	y	z_x	z_y	PRODUCT $z_x z_y$
Ann	20	14	3	6	.5	1.5	.75
Bess	26	10	9	2	1.5	.5	.75
Carol	17	8	0	0	0	0	0
Dora	8	6	−9	−2	−1.5	−.5	.75
Emma	14	2	−3	−6	−.5	−1.5	.75

$\Sigma X = 85 \quad \Sigma Y = 40 \quad \Sigma_x^2 = 180 \quad \Sigma_y^2 = 80 \quad \Sigma z_x z_y = 3.00$

$\mu_x = 17 \quad \mu_y = 8$

$\sigma_x = 6 \quad \sigma_y = 4$

$\rho_{xy} = \dfrac{\Sigma z_x z_y}{N} = \dfrac{3.0}{5} = +.60$

Computing the Pearson ρ we have $+.60$ which tells us there is a positive relationship between the variables arithmetic and spelling but it is not a perfect relationship. A correlational relationship of this magnitude would be described as moderate and positive.

The computation of the correlation between days absent and spelling scores is shown in Table 7.3. When we look at the girls' z-scores on the two variables we see that they vary not only in size but also in sign. The one exception is Carol, who scored at the mean on both variables. The girls who are above the mean on one variable are below the mean on the other variable,

176 DESCRIPTIVE STATISTICS

TABLE 7.3

	X DAYS ABSENT	Y SPELLING SCORES	x	y	z_x	z_y	PRODUCT $z_x z_y$
Ann	2	14	−1	6	−.5	1.5	−.75
Bess	0	10	−3	2	−1.5	.5	−.75
Carol	3	8	0	0	0	0	0
Dora	6	6	3	−2	1.5	−.5	−.75
Emma	4	2	1	−6	.5	−1.5	−.75

$\Sigma X = 15 \quad \Sigma Y = 40 \quad \Sigma_x^2 = 20 \quad \Sigma_y^2 = 80 \quad \Sigma z_x z_y = 3.00$
$\mu_x = 3 \qquad \mu_y = 8$
$\sigma_x = 2 \qquad \sigma_y = 4$
$\rho_x = \dfrac{\Sigma z_x z_y}{N} = \dfrac{-3.00}{5} = -.60$

but not to the same extent. Therefore, the z-score products are all negative or zero and the ρ is negative.

The correlation between spelling and days absent is −.60. This tells us that girls with high scores on days absent tend to have low scores in spelling. This is a negative relationship but not a perfect negative relationship.

In Table 7.4 is illustrated a perfect negative relationship by using scores on an antisocial attitudes scale and the spelling scores.

TABLE 7.4

	X ANTISOCIAL ATTITUDES	Y SPELLING SCORES	x	y	z_x	z_y	PRODUCT $z_x z_y$
Ann	10	14	−60	6	−1.5	1.5	−2.25
Bess	50	10	−20	2	−.5	.5	−.25
Carol	70	8	0	0	0	0	0
Dora	90	6	20	−2	.5	−.5	−.25
Emma	130	2	60	−6	1.5	−1.5	−2.25

$\Sigma X = 350 \quad \Sigma Y = 40 \quad \Sigma_x^2 = 8,000 \quad \Sigma_y^2 = 80 \quad \Sigma z_x z_y = -5.00$
$\mu_x = 70 \qquad \mu_y = 8$
$\sigma_x = 40 \qquad \sigma_y = 4$
$\rho_{xy} = \dfrac{\Sigma z_x z_y}{N} = \dfrac{-5}{5} = -1.00$

Note that the pair of z-scores for each girl are the same value but opposite in sign. That is, each z_x has the same numerical value as the z_y with which it is paired, but they differ in sign ($z_x = -z_y$). This means that each girl is as high in antisocial attitudes as she is low in spelling. Therefore, the sum of the z-score products will assume a maximum absolute value, but will have a negative sign. This results in a correlation coefficient of -1.0, which represents the maximum negative correlation.

We can illustrate a near-zero correlation using the number of successful pull-ups and spelling scores for the five girls in our example, as shown in Table 7.5. Note that there is little tendency toward agreement between the pull-up and spelling z-scores. The sum of the z-score products is near zero and the Pearson ρ is near zero.

TABLE 7.5

	X PULL-UPS	Y SPELLING SCORES	x	y	z_x	z_y	PRODUCT $z_x z_y$
Ann	5	14	−2	6	−.5	1.5	−.75
Bess	9	10	2	2	.5	.5	.25
Carol	13	8	6	0	1.5	0	0
Dora	1	6	−6	−2	−1.5	−.5	.75
Emma	7	2	0	−6	0	−1.5	0

$\Sigma X = 35 \quad \Sigma Y = 40 \quad \Sigma_x^2 = 80 \quad \Sigma_y^2 = 80 \quad \Sigma z_x z_y = .25$

$\mu_x = 7 \quad \mu_y = 8$

$\sigma_x = 4 \quad \sigma_y = 4$

$\rho_{xy} = \dfrac{\Sigma z_x z_y}{N} = \dfrac{.25}{5} = .05$

These relationships in z-score position are illustrated in Figure 7.7 by means of a graph in which the axes represent z-scores on the two variables. You recall that z-scores have means of 0 and standard deviations of 1; the lines have been drawn through the means of the two axes.

Let us see how the *sign* of the correlation coefficient is determined. If the pair of z-scores falls in the upper right-hand quadrant of Figure 7.7, then both z_x and z_y are positive and their product will be positive. A positive contribution is made to the total sum of cross-products ($\Sigma z_x z_y$). If the pair falls in the lower left-hand quadrant, their product will also be positive since a negative z_x times a negative z_y will result in a positive $z_x z_y$ and will also make a positive contribution to the sum of the cross products. If the pair of z-scores

178 DESCRIPTIVE STATISTICS

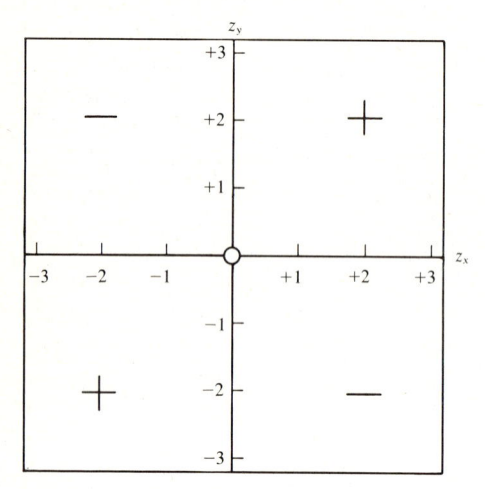

Figure 7.7 Standard (z) Scores Arranged into Quadrants

fall in either the upper left or the lower right, their product will be negative since one z-score is positive and the other is negative ($- \times + = -$). When the cross-products are summed, a negative quantity would thus be contributed to the total.

One can also see how the *size* of the coefficient is determined. If both z-scores in each pair are equal ($z_x = z_y$) in size or nearly equal, the sum of cross-products will be large and the coefficient will be large. The maximum positive value occurs when the signs of the z's are the same (both positive or both negative, as in the upper right and lower left quadrants) and when each z_x exactly equals in value its paired z_y. Exactly equal z-scores will result in a perfect positive correlation which would be indicated on a scattergram by a 45-degree straight line. (See Figure 7.8.) When the relative position on $X(z_x)$ is not equal to the relative position on $Y(z_y)$, then the $\Sigma z_x z_y$ will be lowered in size and the coefficient of correlation will be less than 1.00.

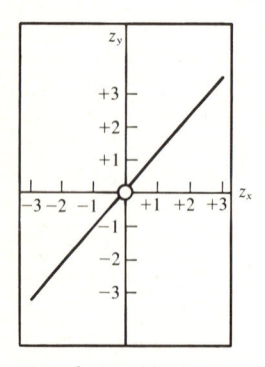

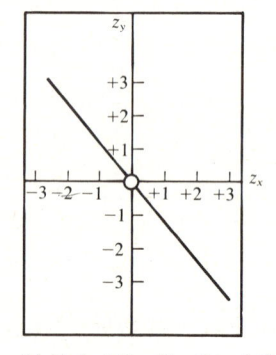

(a) Perfect Positive Correlation (b) Perfect Negative Correlation

Figure 7.8 z-Scores Arranged To Show Perfect Correlation

The maximum negative correlation coefficient of -1.00 occurs when $z_x = -z_y$. That is, the pair of z-scores is equal in absolute size but is different in sign. On a scattergram all points would fall exactly on a straight line but the line would run in a different direction to what it did in a perfect positive correlation. (See Figure 7.8.)

As discrepancy increases in the relative positions of the pair of z-scores (size and sign), then the algebraic sum of the cross-products is lowered and hence the correlation coefficient is lower. It will fall between the two extremes (-1.0 and $+1.0$).

Of course, if the z's are scattered over all four quadrants, that is, if some are positive and some are negative, then the negative cross-products will equal or nearly equal the positive cross-products. As a result the sum of the cross-products will be zero or near zero and the coefficient correlation will be zero or near zero.

Thus, the correlation coefficient may range from 1.0 (perfect positive correlation) through 0.0 (no correlation) to -1.0 (perfect negative correlation). Most correlations reported in education and psychology are less than ± 1.00, indicating that the relationship between behavioral variables is usually less than perfect.

Exercise

1. Below are listed the paired z-scores for a group of students. Find the correlation coefficient for the X and Y variables.

STUDENT	z_x	z_y
A	0	1.0
B	1.5	1.0
C	0	$-.5$
D	.5	$-.5$
E	.5	$-.5$
F	-2.0	1.0
G	-1.0	1.0
H	0	-1
I	1.0	-2
J	$-.5$	.5

Solution

1. $\rho = \dfrac{-4.25}{10} = -.425$

Computation of the Pearson Correlation Coefficient with Raw Scores We have seen that the correlation coefficient is defined as the mean

180 DESCRIPTIVE STATISTICS

of the z-score products. While this z-score interpretation facilitates our understanding of the concept of correlation, the definitional formula using z-scores is not recommended for computational purposes, especially if the number of cases is large. It is too laborious and time-consuming, because each score in the pair of measures must be converted to a z-score which, of course, requires prior computation of the mean and standard deviation of each distribution.

Other formulas have been derived from the basic definitional formula. One of the most convenient allows us to work directly with the original raw scores without the necessity of finding the means and standard deviations.

$$\rho_{xy} = \frac{N \Sigma XY - (\Sigma X)(\Sigma Y)}{\sqrt{[N \Sigma X^2 - (\Sigma X)^2][N \Sigma Y^2 - (\Sigma Y)^2]}}$$

where: ΣX = sum of the raw X scores
ΣY = sum of the raw Y scores
ΣXY = sum of the products of each X times each Y
ΣX^2 = sum of the squares of each X-score
ΣY^2 = sum of the squares of each Y-score
$(\Sigma X)^2$ = the square of the total sum of X-scores
$(\Sigma Y)^2$ = the square of the total sum of Y-scores
N = the number of paired scores

Formula (7.2)
Pearson ρ
(Raw Score formula)

Table 7.6 illustrates the computation of the correlation coefficient using this raw score formula. The data are the arithmetic and spelling scores from our previous example.

TABLE 7.6 Computation of the Correlation Coefficient Using Raw Scores

	X ARITHMETIC	Y SPELLING	XY	X^2	Y^2
Ann	20	14	280	400	196
Bess	26	10	260	676	100
Carol	17	8	136	289	64
Dora	8	6	48	64	36
Emma	14	2	28	196	4
Σ	85	40	752	1,625	400

$$\rho = \frac{5(752) - (85)(40)}{\sqrt{[5(1{,}625) - 85^2][5(400) - 40^2]}}$$

$$= \frac{3{,}760 - 3{,}400}{\sqrt{(8{,}125 - 7{,}225)(2{,}000 - 1{,}600)}}$$

$$= \frac{360}{\sqrt{(900)(400)}}$$

$$= \frac{360}{600}$$

$$= .6$$

Notice that we obtained the same correlation coefficient that we did when using the z-score formula in Table 7.2. [Formulas (7.1) and (7.2) are mathematically equivalent.]

Normally the raw score formula is much easier to calculate than the z-score formula. This is not apparent in our example because the means and standard deviations in the example are whole numbers. In most cases z-score values are mixed decimal numbers and the calculation is more difficult.

Today the Pearson ρ is seldom calculated with pencil and paper. Most universities have computer programs for this purpose and the instruction manuals for most pocket and desk calculators include instructions for calculating ρ.

Exercise

Compute the Pearson ρ for the following scores: (a) with the definition formula (7.1); (b) with the raw score formula (7.2).

SUBJECT	X	Y	SUBJECT	X	Y
A	2	4	D	5	8
B	4	10	E	7	14
C	4	8	F	8	16

Solution

(a)

X	Y	z_x	z_y	$z_x z_y$
2	4	−1.5	−1.5	2.25
4	10	−.5	0.0	0.00
4	8	−.5	−.5	.25
5	8	0.0	−.5	0.00
7	14	1.0	1.0	1.00
8	16	1.5	1.5	2.25
				5.75

182 DESCRIPTIVE STATISTICS

$$\rho = \frac{\Sigma z_x z_y}{N}$$

$$\rho = \frac{5.75}{6}$$

$$\rho = .96$$

(b)

X	Y	XY	X^2	Y^2
2	4	8	4	16
4	10	40	16	100
4	8	32	16	64
5	8	40	25	64
7	14	98	49	196
8	16	128	64	256
$\Sigma = 30$	60	346	174	696

$$\rho = \frac{N \Sigma XY - (\Sigma X)(\Sigma Y)}{\sqrt{[N \Sigma X^2 - (\Sigma X)^2][N \Sigma Y^2 - (\Sigma Y)^2]}}$$

$$\rho = \frac{6(346) - (30)(60)}{\sqrt{[6(174) - (30)^2][6(696) - (60)^2]}}$$

$$\rho = \frac{6(346) - (30)(60)}{\sqrt{(1{,}044 - 900)(4{,}176 - 3{,}600)}}$$

$$\rho = \frac{2{,}076 - 1{,}800}{\sqrt{(144)(576)}}$$

$$\rho = \frac{276}{\sqrt{82{,}944}}$$

$$\rho = \frac{276}{288}$$

$$\rho = .96$$

Assumptions of the Pearson ρ

The Pearson ρ is a meaningful index of the relationship between two variables if the data meet certain underlying assumptions. The basic assumptions of the Pearson ρ are:

1. The relationship is *linear*.
2. The two distributions are similar in shape.
3. The scattergram is *homoscedastic*.

1. A linear, or rectilinear, relationship is one where the plotted X and Y values tend to scatter along a straight line rather than along a curved line.

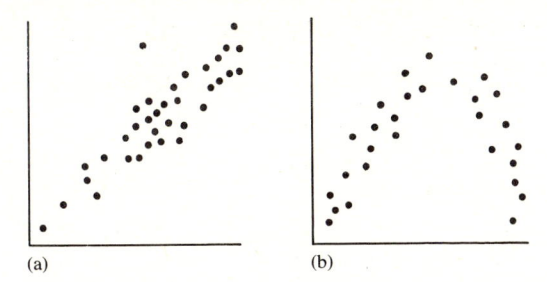

Figure 7.9 Linear (a) and Curvilinear (b) Relationships

Scattergrams where a curved line fits the data better than a straight line are described as *curvilinear*. Figure 7.9 shows linear and curvilinear relationships.

The most common curvilinear relationships follow a curve with a single arc. For example, if we were plotting a measure of efficiency of performance on certain tasks (Y) against anxiety (X), we might find that performance increases in efficiency as anxiety increases *up to a certain level*. Thereafter, increases in anxiety result in a decline in the efficiency of performance. The graph would look like scattergram (b) in Figure 7.9. Similarly, the relationship between physical strength and age is curvilinear. As age increases, physical strength increases up to adulthood, remains stable for a number of years, but then decreases as old age is reached.

The Pearson correlation coefficient is inappropriate to describe such curvilinear relationships. The assumption of linearity is built into the rationale of the Pearson ρ and its use is restricted to linear variables. Other correlational techniques are available for determining the degree of curvilinear correlation.[2]

Inspection of a scattergram is the easiest guide to use in determining linearity. As long as the plotted points tend to fall in a relatively straight pattern, one can assume linearity and use the Pearson correlation coefficient as an index of relationship. If the data, however, show a definite curvilinear trend, the Pearson coefficient should *not* be used because the degree of relationship will be greatly underestimated. Fortunately many of the variables in education and psychology show a linear relationship and one can often safely use the Pearson ρ.

2. The Pearson ρ is based on the assumption that the two distributions X and Y are similar in shape. If the shapes of the two distributions are very different, for example, if one is markedly skewed to the right and the other is

---

[2] The reader is referred to N. M. Downie and R. W. Heath, *Basic Statistical Methods*, 3rd ed. (New York: Harper & Row, 1970), Chapter 8.

markedly skewed to the left, the Pearson ρ will underestimate the relationship between variables.

3. Another assumption underlying the Pearson ρ is that the width of the pattern of dots is the same in all parts of the scattergram. This is called the *assumption of homoscedasticity*. Scatterdiagram (a) in Fig. 7.10 illustrates a

(a) Homoscedastic Relationship

(b) Nonhomoscedastic Relationship

(c) Relationship Approximating Homoscedasticity

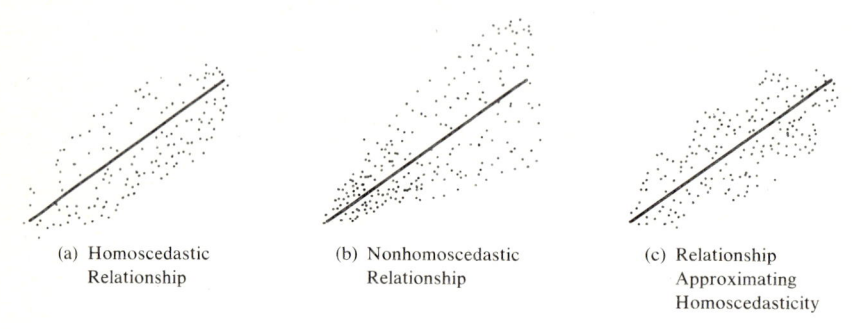

Figure 7.10 Homoscedastic and Nonhomoscedastic Scattergrams

homoscedastic relationship. Scatterdiagram (b) is a nonhomoscedastic relationship as the points are close together at the lower left part of the diagram but far apart at the upper right part. A Pearson ρ would underestimate the relationship in the denser part of the scattergram and overestimate it in the more scattered area. Figure 7.10(c) is not homoscedastic inasmuch as it has wide and narrow portions. However, the wide and narrow portions are scattered along the line and the figure is near enough homoscedastic that one would probably decide to use the Pearson ρ, but to interpret it with caution. There are more sophisticated ways to determine homoscedasticity and linearity than inspection of a scattergram but these are outside the scope of this text.

Factors Influencing the Correlation Coefficient

When interpreting a correlation coefficient one should always consider the nature of the population in which the two variables were observed. The correlation coefficient observed between two variables will vary from one population to another because: (a) the basic relationship is different in different populations, or (b) the variability in the populations differs, or (c) the correlation of the two variables is influenced by their relationship with a third variable.

(a) The inherent relationship of variables may differ from population to population. Among humans between the ages of 10 and 16 physical prowess and chronological age are highly correlated. Among humans between the ages of 20 and 26 these two variables are not correlated. Among children the variables mental age and chronological age are positively correlated, among the middle-aged there is no correlation between these two variables; among the elderly they are somewhat negatively correlated.

(b) When a population is heterogeneous in the variables of concern we expect to observe a higher correlation than when a population is homogeneous in these variables. For example, in a general population of college men we expect a positive correlation between height and success in basketball because the taller boys will tend to do better at the game. In a professional basketball team we would not expect such a relationship. The members of a professional team are all tall and all very good at the game. In a group that is very homogeneous we would not observe the correlation that exists in the population at large. A test of manual dexterity may be highly correlated with machinist skills in the general population but among veteran machinists we would find little correlation between the two variables as the veteran machinists are all highly skilled at their trade.

(c) We may find a correlation between two variables not because there is an intrinsic relationship between these variables but because they are both related to a third variable. For example, if we correlate the average teacher's salary for each of the last 20 years and the dollar value of hard liquor sold during each of these years we get a high correlation. This does not mean that as soon as teacher's salaries are raised they spend the extra money on booze. We observe a correlation between the two variables because each of them is highly correlated with a third variable, general inflation.

Figure 7.11 illustrates another example of how a correlation coefficient may vary from one population to another. If we graph shoe size and reading vocabulary scores for all the children in an elementary school we get a long

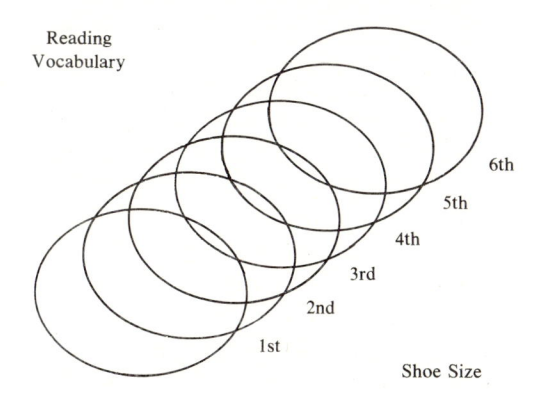

Figure 7.11 Relationship between Shoe Size and Reading Vocabulary for Single Grade Levels and Combined Grade Levels

thin scattergram showing a high correlation. If we look at any one grade we get a circular scattergram which shows 0 correlation. Both shoe size and reading vocabulary are correlated with chronological age. When we measure children of the same age on these variables we get no correlation between these two variables. When we correlate the same variables in a population which includes a wide range of ages we get a correlation between shoe size and reading vocabulary, because these variables both correlate with chronological age.

Thus we see that it is meaningless to refer to *the* correlation between any two variables, apart from a description of the population involved. The relationship between any two characteristics may vary with the population studied.

Interpreting a Correlation Coefficient

The correlation coefficient indicates the direction and strength of the relationship between two variables. Although expressed as a decimal fraction, the coefficient is not to be interpreted as a percent. A coefficient of .70 does *not* indicate a 70 percent relationship between two variables X and Y. Furthermore, the coefficient cannot be considered as directly proportional to the strength of relationship. One could *not* say that a coefficient of .90 represents exactly twice as much relationship as a coefficient of .45.

Valid interpretations concerning proportion or percent can be made with an index called the *coefficient of determination*. This is the square of the Pearson coefficient (ρ^2). It represents the proportion of the variance of Y that may be accounted for by its linear relationship with X or vice versa. Thus, a ρ of .70 tells us that .49 or 49 percent of the Y variance can be accounted for by its correlation with X; the other 51 percent of the variability cannot be accounted for through its relationship with X. If there is a ρ of .80 between IQ and achievement for a group of elementary school children, 64 percent of the variance in achievement is accounted for by intelligence; the other 36 percent can be attributed to other factors.

Another caution is that a correlation does not necessarily indicate a cause-and-effect relationship between the two variables. Correlation must not be interpreted to mean that one variable is causing the scores in the other variable to be what they are. Frequently there are other factors that influence both of the variables under consideration. For example, we may find that there is a negative correlation between measures of anxiety and measures of intelligence. It should not be interpreted that there is a causative relationship between anxiety and intelligence; that is, pupils are anxious because they are unintelligent, or that pupils appear unintelligent because they are anxious. It might be that there are other underlying characteristics of individuals that tend to make some appear unintelligent and anxious, and others intelligent and not anxious. Interpretation of such a correlation is difficult without experimental confirmation. It is recommended in many instances that a relationship study be supplemented by an experimental study in which the alternative interpretations can become hypotheses for experimentation. For example, the relationship between anxiety measures and intelligence measures could be investigated experimentally by deliberately inducing anxiety in a testing situation and determining the effect on intelligence test scores.

Perhaps the fallacy of interpreting a ρ in terms of a cause-and-effect relationship can most readily be seen by means of a ludicrous example. If one took a sample of cities, one would probably find a correlation between church membership and the amount of beer sold in the city. One would certainly not interpret this as indicating that church membership causes more beer to be sold or that increased beer sales lead to increased church membership. One could only say that some other factor or factors account for both of these phenomena. In this case size of city is a third factor that could account for both of the other variables. Larger cities would have more church members and more beer sold than smaller cities.

Thus, no conclusions about causality are justified on the basis of correlational data alone. In the absence of experimental control of other relevant factors, one cannot assume that variable X causes variable Y or that Y causes X. Correlation indicates only that the two variables tend to change together in a systematic way.

SUMMARY

In this chapter we looked at bivariate distributions and the question of correlation. Correlation refers to the extent to which two variables are related in a population. Correlation may be illustrated graphically by means of a scattergram where the values of the two variables X and Y for each member are plotted as points. Inspection of a scattergram indicates the direction and extent of relationship between the variables. If high scores on one variable are associated with high scores on the other and low scores are associated with low scores, a positive correlation is indicated. The plotted points run from lower left to upper right. If high scores on one variable are associated with low scores on the other and vice versa, a negative correlation is indicated. The plotted points run from upper left to lower right. If the points are scattered in a random fashion all over the graph, this is indicative of an absence of correlation between the variables.

The closeness of the plotted points to a straight line indicates the degree of correlation. When the relationship is perfect (an increase on X is accompanied by the same degree of increase or decrease on Y), all of the plotted

188 DESCRIPTIVE STATISTICS

points fall on a straight line. As the degree of relationship becomes less, the more removed the points are from a straight line.

Correlation is expressed quantitatively by means of an index called the coefficient of correlation. The best known correlation coefficient is the Pearson, which is defined as the mean of the z-score products of the X and Y variables. The coefficient assumes a value between 0 and ± 1.00. A coefficient of 1.00 indicates perfect positive correlation and occurs only when the relative positions on X and Y are identical (every $z_x = z_y$). If the paired z-scores are equal in magnitude but opposite in sign ($z_x = -z_y$), the coefficient is -1.00, which indicates perfect negative correlation. When the relative positions on the two variables fall between these two extremes, the coefficient is between ± 1. If there is no correlation at all between the variables, the correlation coefficient is 0.00.

The coefficient was calculated using both the z-score formula and a formula based on the raw scores from X and Y. A correlation coefficient shows the extent to which two variables are related in the population under consideration. The correlation observed between two variables may vary from one population to another because (a) the basic relationship may be

FLOWCHART 7

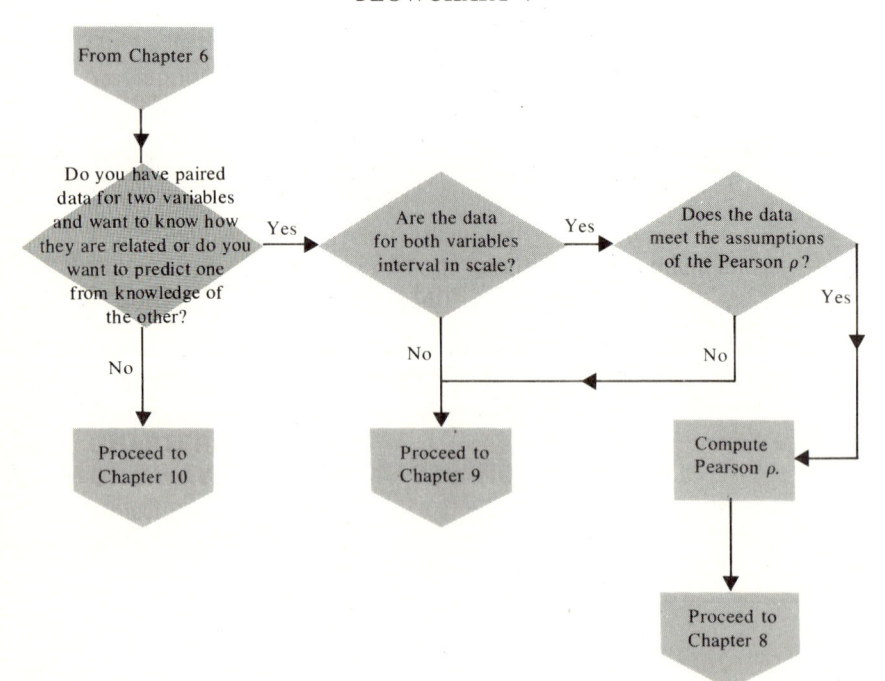

From Chapter 6

Do you have paired data for two variables and want to know how they are related or do you want to predict one from knowledge of the other? — Yes → Are the data for both variables interval in scale? — Yes → Does the data meet the assumptions of the Pearson ρ? — Yes → Compute Pearson ρ. → Proceed to Chapter 8

No → Proceed to Chapter 10

No → Proceed to Chapter 9

No →

different, (b) the variability in the populations may differ, or (c) the correlation of the two variables is influenced by a third variable that has systematically different values in the two populations. Therefore, the population studied should always be specified when reporting a correlation coefficient.

Several points were made about the interpretation of a coefficient of correlation. A coefficient does not indicate the percent of relationship between two variables. However, the square of the coefficient (an index called the coefficient of determination) does indicate the proportion of the variance on one variable that may be accounted for by its relationship with the other. Finally, we stressed that a correlation between two variables is not evidence of a causal relationship between these variables.

EXERCISES

1. Can correlation procedures be used to analyze univariate distributions?
2. What is a correlation index?
3. What are the available options for selecting variables to be described by a correlation index? Give an example of each.
4. (a) When do variables show positive correlation?
   (b) When do variables show a negative correlation?
5. Describe the relationship shown by these scattergrams. Then estimate the correlation coefficients.

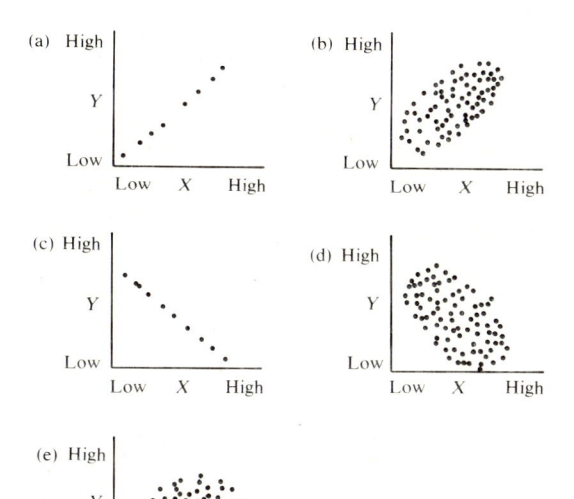

190 DESCRIPTIVE STATISTICS

6. Write true or false for the following statements.
   (a) A coefficient of correlation indicates both the direction and the extent of relationship between two variables.
   (b) The sign of the coefficient indicates the extent of relationship.
   (c) The direction of the relationship is indicated by the absolute size of the coefficient.
   (d) The extent of relationship is dependent on the sign of the correlation coefficient.
7. Given the following graphs, describe the correlation between X and Y. Coordinates are in terms of z-scores.

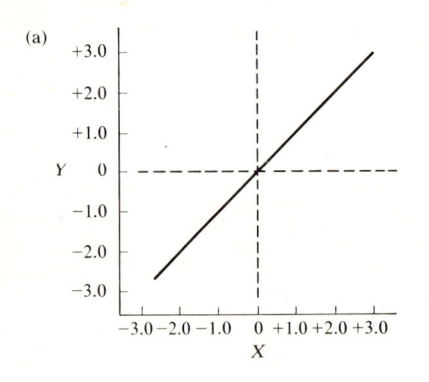

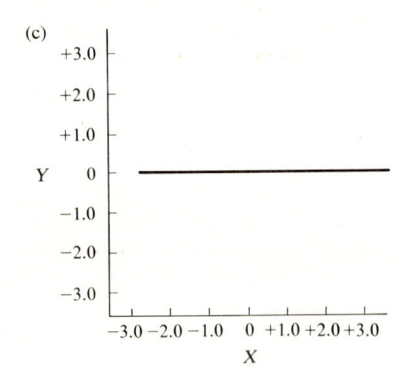

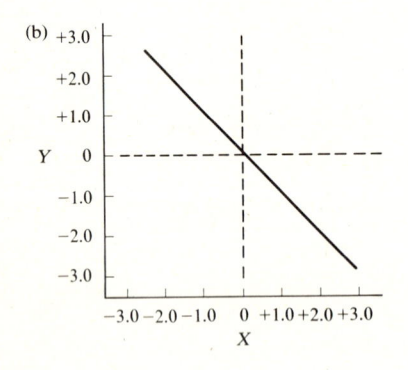

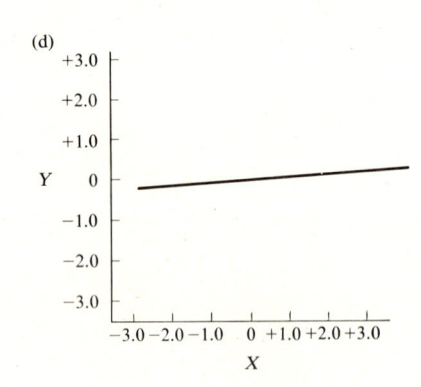

8. The following data are scores obtained by ten students on an Abstract Reasoning Test and their cumulative grade-point average in philosophy. Calculate the Pearson ρ for these data.

STUDENT	ABSTRACT REASONING	GRADE-POINT AVERAGE
A	15	1.5
B	20	2.5
C	30	3.0
D	35	2.0
E	25	3.0
F	40	3.5
G	35	4.0
H	5	1.0
I	15	2.0
J	10	2.5

9. For the following students in Music 101 last semester we have their scores on an Auditory Discrimination Test and the number of Ear-Training Exercises they passed. Construct a scattergram for the data. Would a Pearson ρ be meaningful for this data? Explain your answer.

STUDENT	AUDITORY TEST	EXERCISES PASSED
Tom	35	9
Fred	32	9
Harry	6	3
Ruth	11	4
Joan	25	6
Ron	18	10
Sally	7	4
Mary	5	2
Helen	23	8
Gale	32	6
Sam	17	5
Dave	3	1
Evelyn	31	5
Joe	16	8
Susan	39	5
Emma	22	10
Don	26	10
Bob	28	5
Janet	19	7
Ann	25	4
Dora	12	3
Mike	14	9
Ted	13	7
Steve	6	3
John	34	7
Sara	5	2
Lisa	2	1
Cheryl	10	4

192   DESCRIPTIVE STATISTICS

10. Interpret the following correlation coefficients.
    (a) +.30   (b) −.60   (c) +.71   (d) +.86
11. A researcher demonstrated a correlation of +.60 between teacher attire and student academic performance, across 150 grade schools in his state. He concluded that encouraging teachers to be properly attired will increase academic performance. Comment on his conclusion.
12. A researcher demonstrated a correlation −.45 between hours of television watched and scholastic achievement, across 200 high schools in his state. He concluded that television viewing is detrimental to achievement in school. Comment on his conclusion.

ANSWERS

1. No. Correlation procedures are used to analyze bivariate distributions.
2. An index that describes the relationship of two variables in a single population.
3. We can measure two characteristics in the same individual, for example, height (X) and weight (Y) among army recruits. We can measure a single characteristic for a matched pair of individuals: IQ scores for husbands (X) and IQ scores of their wives.
4. (a) When high scores on one variable are associated with high scores on a second variable and low scores on the one are associated with low scores on the other.
   (b) When high scores on one variable are associated with low scores on a second variable, and vice versa.
5. (a) perfect positive +1.0   (b) positive +.75   (c) perfect negative −1.0
   (d) negative −.75   (e) no correlation .0
6. (a) True   (b) False   (c) False   (d) False
7. (a) positive correlation; $\rho = +1.0$ since angle is 45°
   (b) negative correlation; $\rho = -1.0$ since angle is 45°
   (c) no correlation; $\rho = .0$ since line is flat
   (d) positive correlation but ρ is small since angle is small
   There is a slight positive correlation between the Mathematic Aptitude score and the grade in Geometry.
8.

X	Y	XY	X^2	Y^2
15	1.5	22.5	225	2.25
20	2.5	50	400	6.25
30	3.0	90	900	9
35	2.0	70	1,225	4
25	3.0	75	625	9
40	3.5	140	1,600	12.25
35	4.0	140	1,225	16
5	1.0	5	25	1
12	2.0	24	144	4
10	2.5	25	100	6.25
$\Sigma = 227$	25.0	641.5	6,469	70

$$\rho = \frac{N \Sigma XY - (\Sigma X)(\Sigma Y)}{\sqrt{[N \Sigma X^2 - (\Sigma X)^2][N \Sigma Y^2 - (\Sigma Y)^2]}}$$

$$= \frac{10(641.5) - (227)(25)}{\sqrt{[10(6,469) - (227)^2][10(70) - (25)^2]}}$$

$$= \frac{6,415 - 5,675}{\sqrt{(64,690 - 51,529)(700 - 625)}}$$

$$= \frac{740}{\sqrt{(13,161)(75)}}$$

$$= \frac{740}{\sqrt{987,075}}$$

$$= \frac{740}{993.52}$$

$$= .74$$

9.

EXERCISES PASSED

	1	2	3	4	5	6	7	8	9	10
35–40					1				1	
30–34					1	1	1		1	
25–29				1	1	1				1
20–24								1		1
15–19					1		1	1		1
10–14			1	11			1		1	
5–9		11	11	1						
0–4	11									

No, a Pearson would not be meaningful since the distribution is nonhomoscedastic. A Pearson ρ would underestimate the relationship in the denser part of the scattergram and overestimate it in the more scattered area.

10. (a) A ρ of +.30 tells us that 9 percent [100(.30²)] of the variability in Y can be accounted for by its correlation with X; the other 91 percent of the variability cannot be explained by Y's relationship with X.
(b) 36 percent of the variability in Y can be accounted for by its correlation with X.
(c) About 50 percent of the variance in Y is accounted for by X.
(d) About 74 percent of the variance in Y is accounted for by X.

11. The researcher has no justification for inferring a causal relationship merely on the basis of correlational evidence. Teacher attire and student academic performance could very well be a function of some other variable.
12. The researcher is inferring a causal relationship on the basis of correlational evidence alone. Experimental evidence is the basis upon which cause-and-effect relationships are determined.

CHAPTER EIGHT
PREDICTION

OBJECTIVES

In the last chapter we discussed correlation, a procedure designed to show the relationship between two measures, X-scores and Y-scores. In this chapter, we shall consider the closely related topic of prediction. After completing this chapter, the reader should be able to:

1. Recognize situations in which linear prediction is used.
2. Explain the rationale for prediction.
3. Explain the meaning of the constants β and α in the equation for a straight line.
4. Calculate the constants β and α.
5. Develop a prediction or regression equation of Y on X from the given data.
6. Develop a prediction or regression equation of X on Y from the given data.
7. Compute predicted scores using the regression equation.
8. Define error of prediction.
9. Define "best-fitting" line.
10. Identify the least squares criterion for the best-fitting line.
11. Plot the regression line of Y on X on a graph.
12. Plot the regression line of X on Y on a graph.
13. Use the regression lines to predict Y from X or X from Y.
14. State the relationship between the magnitude of the correlation and the slope of the regression lines.

15. Define standard error of estimate.
16. Compute a standard error of estimate from the given data.
17. Interpret the meaning of standard error of estimate in a prediction situation.
18. Write the regression equation in standard score form.
19. Compute the predicted standard score on Y based on knowledge of the standard score on X.
20. Interpret the phenomenon of regression toward the mean.

Prediction

When two variables are correlated we can make estimates or predictions about what an individual's score will be on one variable if we know his score on the other variable. For example, knowing that IQ and academic performance are positively correlated we would predict that a student who has an above-average IQ is likely to be above average in his grades in school. This does not mean that every student with an above-average IQ will have above-average grades. However, since the two variables are correlated, if we predict that those with above-average IQ's will be above average in grades and those with below-average IQ's will be below average in grades we will be right more often than we are wrong.

High school grade-point average can be used as a predictor of academic performance in college. In this case, one knows high school average, the *predictor* variable, and wishes to use this to gain information about college grade-point average, the *predicted variable*. Admissions officers at colleges and others involved in selection processes depend on such prediction techniques. They must predict future performance when only scores on a selected predictor are known. Prediction is, of course, not perfect because the correlation between the variables involved is less than perfect. Some error is always present in prediction, but it is less than would be involved if one made just blind predictions without benefit of correlation studies.

Although the goal is to predict a second variable when one knows only the first, such prediction is in practice only possible in situations where previous studies have been conducted with groups of subjects on whom both measures were available. The correlation between variables has been established with these subjects. The findings from these studies may be applied later to similar groups on whom only one (predictor) measure is available. For example, the Scholastic Aptitude Test (SAT) is used to predict college performance in the freshman year. This is possible because studies were made in which the SAT was administered to a sample and then a follow-up study obtained the grade-point average of these individuals at the end of the freshman year. The correlation between SAT and grade-point average was then established. Knowing this correlation it has been possible to make predictions of scholastic achievement in the freshman year from the SAT performance of other similar groups.

In order to make such predictions one must assume that the new group and the old group are both samples from the same population and, therefore, the correlation between the two variables in the new sample is the same as the correlation in the original sample except for sampling error. Whenever that assumption is well justified the predictions are valid.

A memorable example of the practical usefulness of prediction occurred during World War II when the U.S. Army Air Corps was concerned with the high failure rate among pilot trainees. Only 25 percent of the trainees earned their wings. The Air Corps employed psychologists to seek measures that correlated with pilot success. The psychologists found that certain measures such as tests of spatial orientation and instrument reading were correlated with pilot performance and could be used to predict which candidates were most likely to succeed. The Air Corps began administering these tests to applicants for pilot training and selected only those who did well on these tests. The result was an increase of the success rate in pilot school from 25 percent to 75 percent![1]

Using the best prediction procedure they could devise the Army Air Corps were not able to increase the pilot candidate success rate to 100 percent. They were looking for measures that would enable them to make reasonable, although not infallible, predictions. The use of the predictor scores could not insure that every candidate would win his wings but it did triple the number of successes. Except in cases where the correlation between variables is known to be perfect the prediction process cannot guarantee that every individual will perform as predicted. It can, however, provide a reasonable estimate which can then be used as the basis for action.

The accuracy of our prediction depends in part on the strength of the correlation. If the correlation between variables is perfect ($+1.0$ or -1.0) we can make perfectly accurate predictions. Heights in inches are perfectly positively correlated with heights in centimeters. If we know a person's correct height in centimeters we can identify his correct height in inches. On the Amtrack line between Chicago and New Orleans each city's distance from Chicago is perfectly negatively correlated with its distance from New Orleans ($\rho = -1.0$). Therefore, if we know the rail distance of any city from Chicago we also know its exact rail distance from New Orleans.

When the correlation between variables is less than perfect our predictions are good estimates rather than exact predictions. The higher the correlation the more accurate our predictions will be.

---

[1] John C. Flanagan, "Scientific Development of the Use of Human Resources: Progress in the Army Air Forces," Science, **105**, 57–60 (1947).

198 DESCRIPTIVE STATISTICS

Predicting z-Scores

The basic procedure involved in using a Pearson ρ correlation coefficient for prediction can be more easily explained when we are predicting z-scores. By convention the variable we are predicting is labeled Y and the variable we are using to predict is labeled X. It has been demonstrated mathematically that the predicted z-score for Y can be calculated by multiplying the z-score for X by the correlation coefficient (ρ_{XY}).

$$z_{\tilde{Y}} = z_X \rho_{XY}$$

where: $z_{\tilde{Y}}$ = the predicted z-score for the Y variable

z_X = the z-score we are using to predict

ρ_{XY} = the correlation between X and Y

[The symbol tilde (~) is used to distinguish predicted scores from other scores.] Formula (8.1)
Prediction of $z_{\tilde{Y}}$ from z_X

For example, if we know that in a particular population the correlation between reading test scores and spelling test scores is +.7, we would predict that an individual in that population with a z-score of +1.0 in reading can be expected to have a z-score of (1.0)(.7) = .7 in spelling. We would predict that a student with a z-score of −2.0 in reading can be expected to have a z-score of (−2.0)(.7) = −1.4 in spelling. When a correlation is less than perfect our predictions are good estimates rather than absolute facts. Formula (8.1) can be used in either direction, that is, we can use it to predict reading z-scores from spelling z-scores as well as to predict spelling z's from reading z's.

The relation between the predictor (X) z-score and the predicted (Y) z-score is a function of both the strength and the sign of the correlation coefficient. When the correlation is positive we predict Y z-scores with the same sign as their paired X z-scores. When the correlation is negative we predict z-scores of the opposite sign.

If the two variables are perfectly positively correlated ($\rho = +1.0$), we predict that an individual will have the same z-score on the predicted variable Y that he has on the predictor variable X. If the correlation is a perfect negative ($\rho = -1.0$), we predict the same z-score with the opposite sign. If the correlation is positive but less than +1.0, we predict Y z-scores of the same sign but nearer zero than the X z-scores. With a negative correlation less than perfect (between 0 and −1.0), the predicted Y z-scores are opposite in direction and nearer zero than the X z-scores. When the correlation is high (near +1.0 or −1.0) we predict Y z-scores that are only slightly nearer the mean than the X z-scores. When the correlation is low all our predicted z-scores are near the mean. The strength of the correlation determines how far from the mean our predicted scores will be. If the correlation is 0 all our predicted Y-scores will be at the Y mean.

Exercises

1. If we know that in a particular class the correlation between arithmetic scores and reading scores is $\rho = .8$, what arithmetic z-score do we predict for a student with (a) a z-score of 1.5 in reading, or (b) a z-score of $-.6$ in reading? (c) What reading z-score do we predict for a student who is 2 standard deviations above the mean in arithmetic?

2. A principal found a correlation of $\rho = -.3$ between times tardy and reading achievement. What reading z-scores do we predict (a) for a student with a times tardy z-score of 3.0, (b) for a student with a times tardy z-score of $-.6$?

Solutions

1. (a) 1.2   (b) $-.48$ (c) 1.6
2. (a) $-.9$ (b) .18

Regression toward the Mean

Regression is an interesting phenomenon that is observed when there is less than perfect relationship between the variables involved in prediction. It refers to the fact that the predicted score will be closer to the mean of the population than is the predictor score. If one selects a number of individuals will tend to be closer to the mean on the Y variable than they are to the overall mean on the X variable. For example, if one takes a number of subjects who are superior on an IQ test (above 140), one will find that while most of these subjects are also above average on an achievement test, only a very few will be as far above average in achievement as they are in intelligence. Similarly, if one selects a group with low scores on the intelligence test, one will find that, as a group, their scores on the achievement test will lie closer to the mean than did their intelligence test scores. Conversely, if a group of subjects each with a z-score of +2.0 in achievement is given an intelligence test, one will not expect them to average +2.0 in intelligence, but will expect them to fall somewhere between +2.0 and 0 (the mean z-score). That is, they will be closer to the mean performance in intelligence than they are in achievement.

Thus, unless two variables are perfectly correlated, there is a tendency for a group scoring at a given level above or below the mean on the first variable to be closer to the mean on the second variable. This effect on scores is called the *regression effect* and since the regression is always toward the mean of the second variable, it is called *regression toward the mean*. Regression is an inherent part of prediction: the predicted Y-scores are closer to the mean than the X-scores.

The extent of correlation between two variables determines how much regression will occur. If the correlation is perfect ($\rho = +1.00$ or -1.00), every measure in the first distribution is paired with another with the same relative position in the second distribution and there would be *no* regression. Regression occurs only when two variables are less than perfectly correlated.

If the correlation is high (but not perfect), there is a slight tendency for the mean score of a group selected on the first variable to move toward the mean of the second variable. If the correlation is low, the tendency would be more pronounced for moves toward the mean of the second distribution.

If there is 0 correlation, there is complete regression to the mean. That is, when $\rho = 0$ and a group is selected because of similar performance on the first variable, their scores on the second variable will have a mean which would coincide with the mean of the entire unselected group. For example, if we have distributions of height and typing speed and select those individuals who are extremely tall, we would find that their mean typing speed would correspond to the overall mean speed for the group.

This phenomenon was first described by Sir Francis Galton who used the term "regression" in his hereditary studies of height to refer to the phenomenon that heights of progeny tend to move toward the mean height of the population. That is, the offspring of exceptionally tall parents are tall but not as tall as their parents and those of exceptionally short parents are short but not as short as their parents. The heights of offspring move "back" toward the general population mean height, a phenomenon which Galton termed "regression toward the mean."

Regression will inevitably occur whenever the correlation is less than perfect. Not too many years ago a respected military leader noted that exceptionally bright children as a group were well above average in school achievement but their mean in school achievement was not as extreme as their IQ. His charge that American schools were failing the very bright children was widely publicized. What he failed to realize was that as long as the correlation between IQ and achievement is less than perfect those who are exceptional on the one variable will, as a group, be less exceptional on the other variable.

Let us look at the effect of regression on the accuracy of prediction. It is easier to see this when the scores are expressed as standard scores. We have looked at the prediction formula using z-scores: $z_{\hat{Y}} = \rho z_X$. Therefore, if $\rho = 0$, then $z_{\hat{Y}} = 0$ ($z_{\hat{Y}} = 0 \cdot z_X = 0$). Thus, when $\rho = 0$, regardless of the score on X, the best prediction for the Y-scores is the mean of the Y's. There is complete regression to the mean. This is why we said earlier that a $\rho = 0$ has no predictive power. If one wanted to know Y, it would not help to know the value of X. One might as well predict the mean of the Y's for everyone; that would be the best guess.

On the other hand, if $\rho = \pm 1.00$, $z_{\hat{Y}} = \rho z_X = 1.0(z_X) = z_X$. This means that the predicted z-score on Y is the same as the z-score on X. The relative position is exactly the same on the two variables. Thus, the prediction would be perfect and there would be no error. There is no regression toward the mean. At values of ρ between these two extremes, there is regression toward the mean; the amount of regression being inversely related to the extent of the correlation.

The Regression Line

When we show the predicted (Y) z-scores on a bivariate graph they will fall on a single straight line. This is true because for each value of X, we multiply by a constant, the value of ρ, to get the predicted Y values. This line connecting the predicted scores is defined as the *regression line*.

Let us use the z-scores from Table 7.2 to illustrate a regression line. The correlation was .6 so the predicted z-scores on spelling will be the z-scores on arithmetic multiplied by .6.

	ARITHMETIC z-SCORE	PREDICTED SPELLING z-SCORE
Ann	.5	.3
Bess	1.5	.9
Carol	.0	.0
Dora	−1.5	−.9
Emma	−.5	−.3

Figure 8.1 shows the predicted spelling z-scores (z_y) as dots along with the regression line connecting them. Note that with $\rho = .6$ the regression line goes up .6 on the Y-axis for every one unit it goes across on the X-axis. We also show the actual spelling z-scores from Table 7.2 with asterisks.

Because of the way the regression line is defined if we take the distance on the Y-axis from it to each Y-score, square those distances and sum them, we have a value that is less than the sum of the squared distances from the Y-scores to any other straight line that might be drawn through the scattergram. This concept is sometimes used to define the regression line. For this reason the regression line is sometimes referred to as the line of least squares. It is the single straight line that best fits the general shape of the scattergram and is the line that will result in the least error in prediction.

202 DESCRIPTIVE STATISTICS

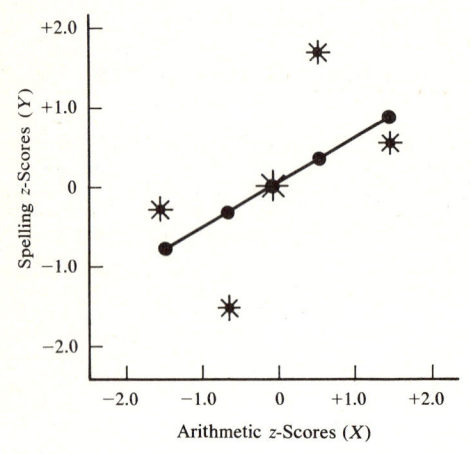

Figure 8.1 Regression Line for Data of Table 7.2

Criterion for Best Prediction

Let us now consider the concept of errors of prediction. When we predict Y values from X, the predicted values of Y will fall along a straight line, but the actual observed values of Y may not fall on this line. That is, the predicted values of Y will not necessarily be equal to the observed values of Y. Figure 8.2 illustrates a linear relationship between X and Y with the "line of best fit" drawn. The observed values of Y corresponding to the X values are plotted as points. The line of best fit is used to predict the values of Y from the known values of X. For example, using the straight line in Figure 8.2 one can see that the best estimate of the Y-score for an individual with an X-score of 5 is 6; however, the individual with an X of 5 actually obtained a score of 9 (see plotted point). The straight line indicates that the best

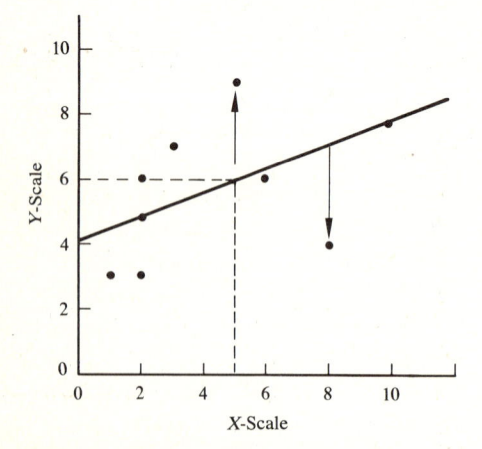

Figure 8.2 Errors of Prediction

estimate of a Y-score for one with an X of 8 is approximately 7; however, the actual Y-score observed was 4. In fact, this discrepancy between the observed Y and the predicted value of $\tilde{Y}$ represents error, the error of prediction. Thus, the error of prediction is defined as the difference between Y and $\tilde{Y}$; it is symbolized as $e = Y - \tilde{Y}$. Of course, these errors might be positive or negative depending on whether the recorded Y-scores are larger or smaller than the predicted score. In Figure 8.2, where $X = 5$, the predicted $\tilde{Y}$ is 6; however, the observed Y is 9. Thus, the prediction error e is $9 - 6 = 3$.

Least Squares Criterion

The aim is to find a prediction line and equation that will make the errors of prediction as small as possible. A first thought might be to find the line so that the algebraic sum of the errors is zero; that is, the line where the sum of all positive errors is exactly equal to the sum of the negative errors. It happens, however, that there are an infinite number of different straight lines that would satisfy this criterion so this procedure would not be adequate. The solution to the problem is the same as was used in dealing with deviations from the mean when computing variance and standard deviation. You recall that the deviations were squared in order that they all become positive. In this case, all of the errors (+ and −) will be squared and a prediction line drawn for which the *sum of the squared errors* (Σe^2) is *smaller* than for any other line that could be used. This is called the *least squares* criterion and it serves as the basis for determining the "line of best fit" for the data. Figure 8.3 illustrates a line of least squares with the vertical distances from the line indicated. This is the straight line about which the sum of the squared distances (errors) is least or at a minimum. This straight line which results in the least error in prediction is called a *regression line*.

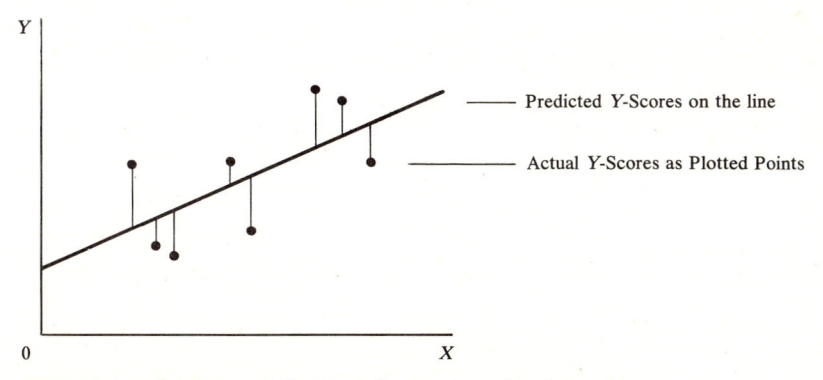

Figure 8.3 Distances of Obtained Scores from Predicted Scores

Exercises

1. (a) What is a regression line?
   (b) Why do all the predicted scores fall on a single line?

2. A high school counselor has Mechanical Reasoning Test z-scores for the following students. On the basis of these scores he wishes to predict their performance on a Spatial Relations Test. From previous research he knows that the correlation between these tests is +.50. Predict the Spatial Relations Test z-scores and graph the regression line.

STUDENT	MECHANICAL REASONING TEST z-SCORES
Bob	1.5
John	.5
Fred	.0
Jim	−1.0
Mike	−1.5

3. We know that family expenditures on gasoline are highly, but not perfectly, correlated with expenditures on tires. We identify the ten families in Barstow, California who have spent the most on gasoline. (a) Would we expect their tire expenditures to be above average? (b) Would we expect their tire expenditures to be as different from the average as their gasoline expenditures?

4. At the beginning of the school year a principal identified 20 fourth graders who had scored very low on a standardized reading test and assigned them to a special program. Their mean on this test (X) was 1.8 standard deviations below the fourth grade population mean. At the beginning of the next year they were given another standardized reading test (Y). This time they were only 1.2 standard deviations below the fifth grade population mean. The principal is delighted with this evidence of success for the special program. Comment.

Solutions

1. (a) A line connecting the predicted (Y) scores.
   (b) Because in order to obtain the predicted (Y) scores each value of X is multiplied by a constant.

2.

STUDENT	MECHANICAL REASONING TEST z-SCORE	PREDICTED SPATIAL RELATIONS z-SCORE
Bob	1.5	.75
John	.5	.25
Fred	.0	.0
Jim	−1.0	−.5
Mike	−1.5	−.75

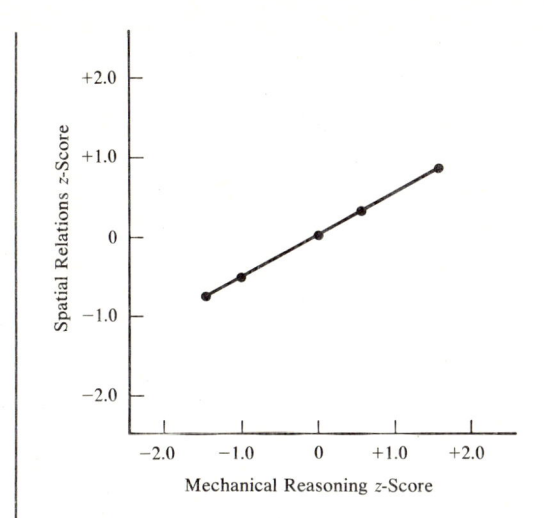

[Graph: Spatial Relations z-Score vs Mechanical Reasoning z-Score, showing a positive linear relationship]

3. Their tire expenditures will be very high but not as exceptionally high as their expenditures on gasoline.
4. Since scores on the two tests are not perfectly correlated, the apparent improvement could easily be due to regression.

Prediction with Raw Scores

Practical prediction situations usually require prediction involving raw scores instead of z-scores. These raw scores will not have identical means and standard deviations, as z-scores do. To make allowances for this two values Beta (β) and Alpha (α) are used.

Beta To adjust for the difference in the standard deviations, the correlation coefficient is multiplied by the standard deviation of the Y variable divided by the standard deviation of the X variable to indicate the slope of the regression line for the raw scores. The symbol β (the Greek letter beta) is used for this value. The subscripts show first the variable being predicted and then the predictor variable.

$$\beta_{yx} = \rho_{xy} \frac{\sigma_y}{\sigma_x}$$

where: β_{yx} = the slope of the raw score regression line for predicting Y given X

ρ_{xy} = the correlation of X and Y

σ_x = the standard deviation of X

σ_y = the standard deviation of Y Formula (8.2)
Slope of Raw Score Regression Line (definition formula)

206 DESCRIPTIVE STATISTICS

For example, we want to predict grammar test (Y) scores from vocabulary test scores (X) with a correlation between the two tests of .5. The grammar test has a mean of 24 and a standard deviation of 3. The vocabulary test has a mean of 10 and a standard deviation of 2. The slope of the raw score regression line is:

$$\beta_{yx} = \rho_{xy}\frac{\sigma_y}{\sigma_x} = .5\left(\frac{3}{2}\right) = .75$$

This tells us that for every unit across the abscissa the regression line moves .75 of a unit along the ordinate. This is illustrated in Figure 8.4.

Beginning with the intercepts of the two means, 10 and 24, for each unit the regression line moves to the right it moves up .75 of a unit. For each unit it moves to the left it goes down .75 of a unit.

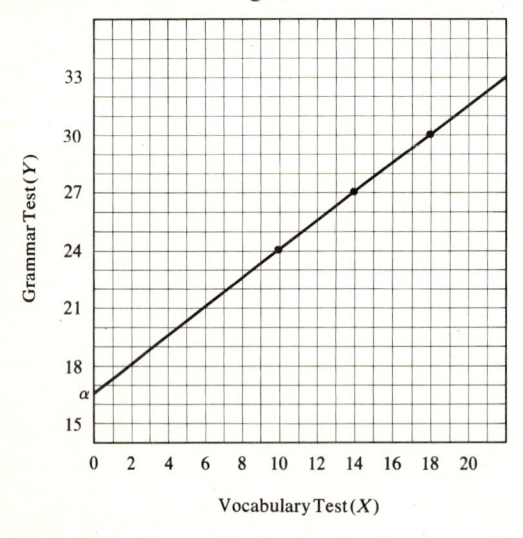

Figure 8.4 Slope of the regression line for predicting grammar scored from vocabulary scores.

β_{yx} can be computed directly from the raw scores without calculating the correlation coefficient and the two standard deviations through the use of Formula (8.3).

$$\beta_{yx} = \frac{N \Sigma XY - (\Sigma X)(\Sigma Y)}{N \Sigma X^2 - (\Sigma X)^2}$$

where: β_{yx} = the slope of the regression line for predicting Y given X
N = number of subjects in the sample
ΣXY = sum of the product of the X and Y scores for each individual
ΣX = sum of the X-scores for all individuals
ΣY = sum of the Y-scores for all individuals

ΣX^2 = sum of the squares of the X-scores for all individuals

Formula (8.3)
Beta (computation formula)

Alpha In order to predict Y raw scores from X raw scores, we need to know the position of the regression line in relation to the Y-axis. This is accomplished by determining the point on the Y-axis where the regression line crosses zero on the X-axis. This point is called *alpha* (α). In Figure 8.4 alpha is 16.5 as that is the point on the Y-axis where the regression line intercepts zero on the X-axis. Alpha can be computed with Formula (8.4).

$$\alpha = \mu_y - \beta_{yx}\mu_x$$

where: α = the Y-intercept
μ_y = the mean of the Y-scores
μ_x = the mean of the X-scores
β_{yx} = the slope of the raw score regression line

Formula (8.4)
Alpha

In our example:

$$\alpha = 24 - (10)(.75) = 16.5$$

Once the values of β_{yx} and α have been established it is possible to predict Y raw scores ($\tilde{Y}$) directly from X raw scores using Formula (8.5).

$$\tilde{Y} = \beta_{yx}X + \alpha$$

where: $\tilde{Y}$ = the predicted raw score
X = the predictor raw score
β_{yx} = the slope of the raw score regression line
α = the Y intercept

Formula (8.5)
Prediction of Y from X

In our example the equation for predicted scores becomes:

$$\tilde{Y} = (.75)(X) + 16.5$$

For an X score of 12 the $\tilde{Y}$ is

$$(.75)(12) + 16.5 = 9 + 16.5 = 25.5$$

For an X score of 6 the $\tilde{Y}$ is

$$(.75)(6) + 16.5 = 4.5 + 16.5 = 21$$

For each value of X on the abscissa in Figure 8.4 the equation $(.75)(X) + 16.5$ tells us where the regression line is located on the ordinate.

Let us compute β_{yx} and α from the data in Table 8.1 and then use these values to calculate a predicted Y for each value of X.

TABLE 8.1

SUBJECT	VOCABULARY SCORES X	ARITHMETIC SCORES Y	X^2	Y^2	XY
A	10	8	100	64	80
B	9	8	81	64	72
C	8	4	64	16	32
D	5	9	25	81	45
E	6	6	36	36	36
F	4	6	16	36	24
G	3	7	9	49	21
H	2	6	4	36	12
I	2	3	4	9	6
J	1	3	1	9	3
$n = 10$	$\Sigma X = 50$	$\Sigma Y = 60$	$\Sigma X^2 = 340$	$\Sigma Y^2 = 400$	$\Sigma XY = 331$

$$\beta_{yx} = \frac{10(331) - (50)(60)}{10(340) - (50)^2}$$

$$\beta_{yx} = \frac{310}{900} = .344$$

$$\mu_x = \frac{50}{10} = 5$$

$$\mu_x = \frac{60}{10} = 6$$

$$\alpha = \mu_y - \beta\mu_x$$
$$\alpha = 6 - (.344)5$$
$$\alpha = 4.28$$

Having established the value of β_{yx} (.344) and of α (4.28) we can now calculate the predicted arithmetic scores (Y) from the X-scores, vocabulary. Let us begin with the X score of 10:

$$\tilde{Y} = \beta_{yx}X + \alpha$$
$$= (.344)10 + 4.28$$
$$= 3.44 + 4.28$$
$$= 7.72$$

Continuing with each value of X will give us the predicted ($\tilde{Y}$) scores as shown in Table 8.2.

TABLE 8.2 Calculation of Predicted Y Values

PAIRED VALUES X	Y	PREDICTED Y $\tilde{Y}$
10	8	7.72
9	8	7.376
8	4	7.032
5	9	6.00
6	6	6.344
4	6	5.656
3	7	5.312
2	6	4.968
2	3	4.968
1	3	4.624

Exercises

1. As a member of the English department's undergraduate admissions committee, you would like to develop a method for predicting grade-point average (GPA) in upper level English courses on the basis of scores in the department's qualifying examination. Develop your prediction equations from the following data.

STUDENT	QUALIFYING EXAM SCORE (X)	ENGLISH GPA (Y)
A	10	2.0
B	15	3.0
C	8	2.0
D	5	1.5
E	18	3.5
F	12	2.5
G	6	2.0
H	16	3.0
I	17	4.0
J	12	3.0

2. Using the results obtained in Problem 1, would you expect a student who scored 15 on the department's qualifying examination to have an English GPA of 3.0 or above? Explain your answer.
3. Graph a regression line for Problem 1.

210 DESCRIPTIVE STATISTICS

Solutions

1.

STUDENT	X	Y	X^2	Y^2	XY
A	10	2.0	100	4.0	20.0
B	15	3.0	225	9.0	45.0
C	8	2.0	64	4.0	16.0
D	5	1.5	25	2.25	7.5
E	18	3.5	324	12.25	63.0
F	12	2.5	144	6.25	30.0
G	6	2.0	36	4.0	12.0
H	16	3.0	256	9.0	48.0
I	17	4.0	289	16.0	68.0
J	12	3.0	144	9.0	36.0

$n = 10 \quad \Sigma X = 119 \quad \Sigma Y = 26.5 \quad \Sigma X^2 = 1{,}607 \quad \Sigma Y^2 = 75.75 \quad \Sigma XY = 345.5$

$\mu_x = 11.9$

$\mu_y = 2.65$

$\beta_{yx} = \dfrac{n \Sigma XY - \Sigma X \Sigma Y}{n \Sigma X^2 - (\Sigma X)^2}$

$= \dfrac{10(345.5) - (119)(26.5)}{10(1{,}607) - (119)^2} = \dfrac{3{,}455 - 3{,}153.5}{16{,}070 - 14{,}161} = \dfrac{301.5}{1{,}909} = .158$

$\alpha = \mu_y - \beta_{yx}\mu_x = 2.65 - (.158)(11.9) = 2.65 - 1.88 = .77$

$\tilde{Y} = \beta_{yx}X + \alpha = .158X + .77$

2. A student who scored 15 on the qualifying exam would probably achieve a GPA of 3.0 or higher since $\tilde{Y} = .158(15) + .77 = 3.14$.

3.

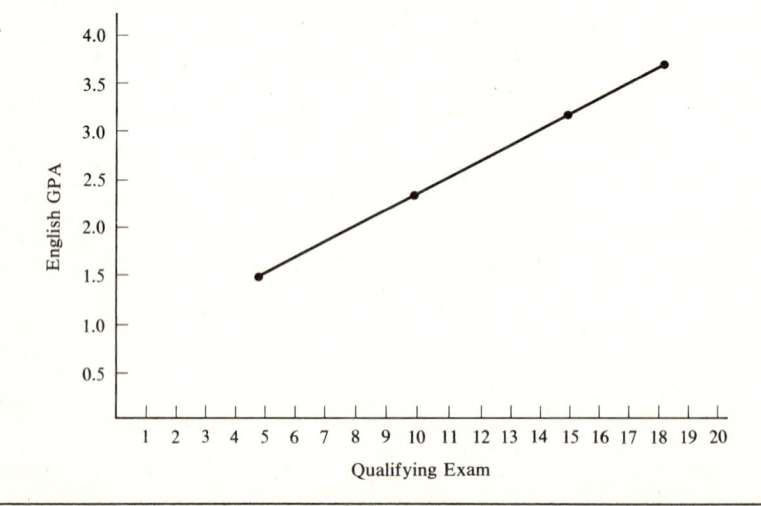

Prediction of X from Y

Thus far we have considered only the prediction of the Y variable from the X variable. However, there may be occasions when the variable previously labeled X is unknown and the variable previously labeled Y is the known quantity. In this case one would predict from a knowledge of Y to the X variable. For example, while typically one predicts achievement (Y) from intelligence test scores (X), one may sometimes wish to predict intelligence (X) from achievement test scores (Y). Thus, it is possible to have two regression lines for any set of paired observations: a regression line of Y on X and a regression line of X on Y. The regression line of Y on X is used when predicting Y from X; the regression line of X on Y is used when predicting X from Y.

The formulas used in the prediction of X from Y are the same as those used in the prediction of Y from X, except that the positions of X and Y are reversed. An X is substituted in the formulas where a Y was formerly used and a Y where an X had been.

Thus, the formula for the regression lines becomes $\tilde{X} = \beta_{xy} Y + \alpha$ where $\tilde{X}$ is the value of X to be predicted from a knowledge of Y, β_{xy} is the slope of the regression line and α is the X-intercept (the value of X when $Y = 0$). Notice that the order of the subscripts has changed too; the variable to be predicted (x) is written first followed by the predictor (y).

The formula for β_{xy}, the slope of the regression line of X on Y, is:

$$\beta_{xy} = \frac{N \Sigma XY - (\Sigma X)(\Sigma Y)}{N \Sigma Y^2 - (\Sigma Y)^2}$$

where β_{xy} = the slope of the regression line when predicting X from Y

N = the number of individuals with X- and Y-scores

ΣXY = sum of the products of the X- and Y-scores for each subject

ΣX = sum of the X-scores

ΣY = sum of the Y-scores

ΣY^2 = sum of the squares of the Y-scores

Formula (8.6)
Beta for $\tilde{X}$ from Y

The formula for a, the intercept constant, is:

$\alpha = \mu_x - \beta_{xy}\mu_y$ where: α = the X-intercept constant (the point where the regression line will cross the X-axis)

μ_x = the mean of the X-scores

μ_y = the mean of the Y-scores

β_{xy} = the slope of the regression line for predicting X from Y

Formula (8.7)
Alpha for $\tilde{X}$ from Y

212 DESCRIPTIVE STATISTICS

Once the values of β_{xy} and α are determined, one has the prediction equation for predicting X values from a knowledge of Y.

$$\tilde{X} = \beta_{xy}Y + \alpha \qquad \text{Formula (8.8)}$$
$$\text{Prediction of } \tilde{X} \text{ from } Y$$

Let us apply these formulas to the data in Table 8.1:

$$\beta_{xy} = \frac{10(331) - (50)(60)}{10(400) - 60^2} \qquad \alpha = \mu_x - \beta_{xy}\mu_y$$
$$= \frac{310}{400} = .775 \qquad = 5 - (.775)6$$
$$= .350$$

Now one is ready to predict from Y to X. If an individual has a Y-score of 8, his predicted X-score is:

$$\tilde{X} = \beta_{xy}Y + \alpha$$
$$= .775(8) + .350$$
$$= 6.55$$

For an individual whose Y-score is 3, $\tilde{X} = .775(3) + .350 = 2.675$. If the individual is at the mean on Y (6), his predicted X-score is the mean on X (5). $\tilde{X} = .775(6) + .350 = 5$. Figure 8.5 shows the two regression lines for the data in Table 8.1. Notice that the regression lines cross at the point representing the mean of the X distribution (5) and the mean of the Y distribution (6). The two regression lines will lie at an angle whenever there is less than perfect correlation between the variables. As the magnitude of the correlation increases, the angle between the lines decreases. That is, the lines become closer and closer until they finally coincide when $\rho = +1.0$ or

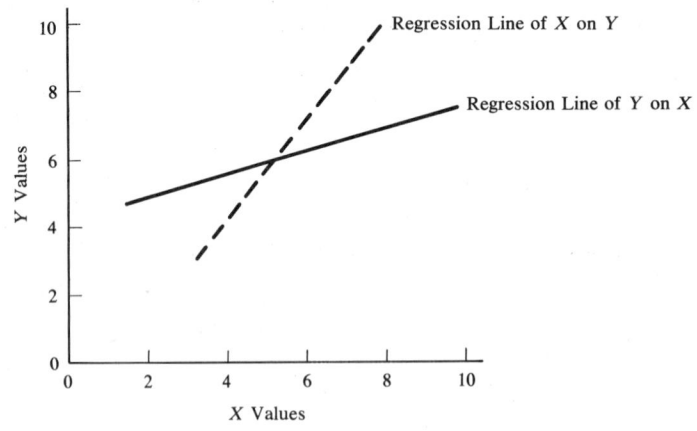

Figure 8.5 Regression Line of Y on X and Regression Line of X on Y for the Data in Table 8.1

−1.0. When $\rho = 0.0$ the two regression lines are perpendicular; notice that each line indicates that, regardless of the value of the predictor, the predicted value of the other variable will always be the mean. This is a result of the fact that a ρ of 0.0 has no predictive value.

Exercises

1. Describe the correlation for these graphed regression lines.

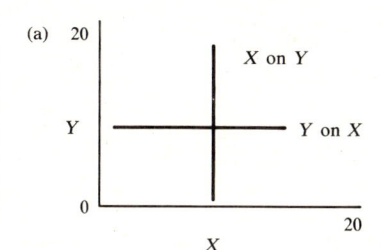

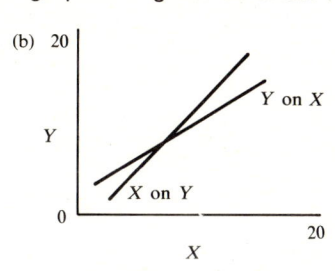

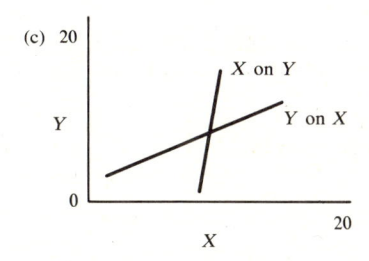

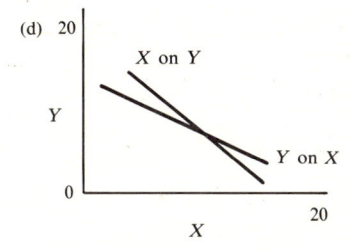

(a), (b), (c), (d) — graphs of regression lines labeled "X on Y" and "Y on X" on axes with X and Y ranging from 0 to 20.

2. As a high school principal, you are interested in the relationship between attitude and achievement, and wish to devise some method by which to predict achievement from a student's score on an attitude scale. You would also like to predict attitude from achievement measures. Using the following data calculate the values needed to predict attitude (Y) from achievement (X) and to predict achievement (X) from attitude (Y). Graph the two regression lines.

STUDENT	ACHIEVEMENT SCORE (X)	ATTITUDE SCORE (Y)
Alice	12	9
Joan	8	6
Tom	14	10
Bill	6	4
Jim	4	2
Susan	10	7
Mike	12	6
Ted	8	4
John	4	3
Carol	2	4

214 DESCRIPTIVE STATISTICS

Solutions

1. (a) zero correlation (b) high positive
   (c) moderate positive (d) high negative

2.

STUDENT	X	Y	X^2	Y^2	XY
Alice	12	9	144	81	108
Joan	8	6	64	36	48
Tom	14	10	196	100	140
Bill	6	4	36	16	24
Jim	4	2	16	4	8
Susan	10	7	100	49	70
Mike	12	6	144	36	72
Ted	8	4	64	16	32
John	4	3	16	9	12
Carol	2	4	4	16	8

$N = 10 \quad \Sigma X = 80 \quad \Sigma Y = 55 \quad \Sigma X^2 = 784 \quad \Sigma Y^2 = 363 \quad \Sigma XY = 522$

$\mu_x = 8.0$
$\mu_y = 5.5$

$$\beta_{yx} = \frac{n \Sigma XY - \Sigma X \Sigma Y}{n \Sigma X^2 - (\Sigma X)^2}$$

$$= \frac{10(522) - (80)(55)}{10(784) - 6{,}400}$$

$$= \frac{5{,}220 - 4{,}400}{7{,}840 - 6{,}400} = \frac{820}{1{,}440} = .569$$

$\alpha = \mu_y - \beta_{yx}\mu_x = 5.5 - (.569)(8.0) = 5.5 - 4.552 = .948$

$\tilde{Y} = \beta_{yx}X + \alpha; \quad \tilde{Y} = .569X + .948$

$$\beta_{xy} = \frac{n \Sigma XY - \Sigma X \Sigma Y}{n \Sigma Y^2 - (\Sigma Y)^2}$$

$$= \frac{10(522) - (80)(55)}{10(363) - 3{,}025}$$

$$= \frac{5{,}220 - 4{,}400}{3{,}630 - 3{,}025}$$

$$= \frac{820}{605}$$

$$= 1.355$$

$\alpha = \mu_x - \beta_{xy}\mu_y = 8.0 - (1.355)(5.5) = 8.0 - 7.4525 = .5475$

$\tilde{X} = \beta_{xy}Y + \alpha; \quad \tilde{X} = 1.355Y + .5475$

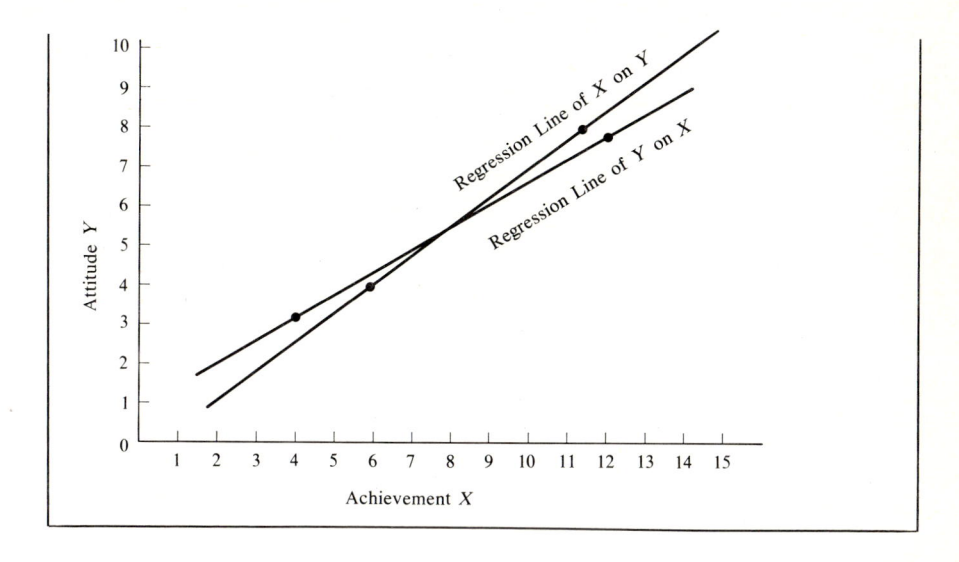

Standard Error of Estimate

Predictions based on prediction equations are never without some error unless the correlation between the variables is exactly +1.00 or −1.00. When the correlation is less than perfect, the prediction equations do enable one to make a more accurate estimate of the score on the predicted variable than would be possible without the equations. However, as we have seen, the actual score received by an individual rarely coincides exactly with the predicted score. The difference between an actual and a predicted score on the predicted variable is the *error of prediction* ($e = Y - \tilde{Y}$). We need some estimate of the amount of error that is involved in prediction.

In the examples of prediction that we have considered so far, a very small number of scores (N) was used. This was done in order to simplify the explanations and the calculations. You must remember, however, that an actual study of this type might involve thousands of subjects. In such a study, a prediction is made of the *mean* score on a second variable (Y) for a whole multitude of individuals who have a given score on the first variable (X). Figure 8.6 illustrates this concept. Notice in Figure 8.6 that there is a subpopulation of Y (GPA) for each given score on X (aptitude test). That is, for any single value on X there is assumed to be a normal distribution of associated Y-scores. It is assumed that each of these subpopulations on Y has a mean falling exactly on the regression line shown. This mean represents the predicted Y-score ($\tilde{Y}$) for that X value. For example, assume that a grade-point average (GPA) of 1.5 is predicted for an aptitude test score of 90. This does not mean that every individual who scores 90 on the test will

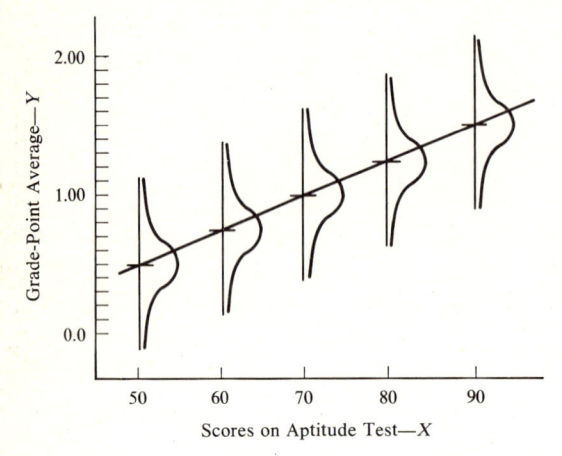

Figure 8.6 Distribution of Y-Scores for Subgroups having Given Scores on X

receive a GPA of 1.5. But rather it means that 1.5 is the predicted mean GPA for a subpopulation of subjects who score 90 on the aptitude test. However, you can see that there is variability about the predicted mean GPA. Some students with a score of 90 will get higher GPA's and some will get lower. Thus, in prediction, it is very important to get an estimate of the amount of *variability* observed in the Y-scores for persons with a given X-score. That is, one would want to determine the variability to be expected in the GPA's of all the subjects who had an aptitude score of 90.

How accurate the prediction is depends on how much variability there is in the scores on the second variable (Y) for those with a given X-score. For example, if the *actual* GPA's received by subjects with test scores of 90 fall close to 1.5, then that would be considered an accurate prediction. However, if the actual grade-point averages received by those with test scores of 90 show a wide spread around 1.5, then this would not be considered such an accurate prediction of GPA based on a test score of 90.

In order to estimate the amount of error in prediction, one must determine the amount of variability observed in the subpopulation of Y-scores corresponding to given X-scores. Since the predicted values of Y all fall on the straight line, then the deviations of those subpopulations from the straight line must represent errors of prediction. That is, the variability of the Y distribution around the line reflects the error. If in each subgroup one could determine how far the actual Y-scores deviate from the predicted Y-scores on the line ($\tilde{Y}$) and calculate the standard deviation, one would have an index of the amount of error involved in the prediction. That is, the error of prediction ($e = Y - \tilde{Y}$) is determined, and the errors are squared and summed. When the square root of the mean of the squared errors is obtained, one has the standard deviation of the errors of prediction. This standard deviation has been given a special name: *the standard error of estimate.* It is a measure of the variation of the Y measure about the regression line. Formula (8.9) shows the definition formula for the standard error of estimate which is defined as the square root of the mean of the squared errors of prediction.

$$\sigma_{yx\ est} = \sqrt{\frac{\Sigma (Y - \tilde{Y})^2}{N}} = \sqrt{\frac{\Sigma e^2}{N}}$$

where: $\sigma_{yx\ est}$ = standard error of estimate of $\tilde{Y}$ predicted from X

Σ = sum of

Y = actual Y-score

$\tilde{Y}$ = Y score predicted on the basis of X

N = number of predicted scores

Formula (8.9)
Standard Error of Estimate
(definition formula)

Formula (8.9) shows that for each X-score, the squared error $(Y - \tilde{Y})^2$ is obtained, and these squared errors are summed across all points. Then the sum is divided by N and the square root found.

To illustrate its calculation let us expand the data from Table 8.1. In Table 8.3 we add a fourth column to show the difference between the predicted (Y) scores and the actual Y-scores. These differences are error scores, e. Then these error scores are squared and shown in a fifth column. The standard deviation of the errors, which is the standard error of estimate, is obtained by summing the squared errors, dividing by N, and taking the square root.

TABLE 8.3 Calculation of Standard Error of Estimate

PAIRED VALUES X	Y	PREDICTED $\tilde{Y}$ Y	ERRORS IN PREDICTION $e = Y - \tilde{Y}$	ERRORS SQUARED $e^2 = (Y - \tilde{Y})^2$
10	8	7.72	.28	.0784
9	8	7.376	.624	.389376
8	4	7.032	− 3.032	9.193024
5	9	6.00	3.00	9.00
6	6	6.344	− .344	.118336
4	6	5.656	.344	.118336
3	7	5.312	1.688	2.849344
2	6	4.968	1.032	1.065024
2	3	4.968	− 1.968	3.873024
1	3	4.624	− 1.624	2.637376
			$\Sigma e = 0$	$\Sigma e^2 = 29.322240$

$$\sigma_{yx\ est} = \sqrt{\frac{\Sigma (Y - \tilde{Y})^2}{N}} = \sqrt{\frac{29.32}{10}} = 1.71$$

218 DESCRIPTIVE STATISTICS

The amount of error in prediction depends on two factors: (a) the strength of the correlation (ρ_{xy}) and (b) the standard deviation of the Y-scores (σ_y). A more convenient formula, which does not require the computation of each of the error scores, has been derived which is based on these two elements. It has been shown that Formula (8.10) is mathematically equivalent to Formula (8.9). Formula (8.10) is the most common formula for estimating the errors of prediction.

$$\sigma_{yx\ est} = \sigma_y \sqrt{1 - \rho_{xy}^2}$$

where $\sigma_{yx\ est}$ = standard error of estimate

σ_y = standard deviation of Y-scores

ρ_{xy}^2 = the square of the correlation between X and Y

Formula (8.10)
Standard Error of Estimate
(computation formula)

Applying this formula to the data in Table 8.1 where the standard deviation of Y-scores is 2 and the correlation between X and Y is .516 we have:

$$\sigma_{xy\ est} = \sigma_y \sqrt{1 - \rho_{xy}^2}$$
$$= 2\sqrt{1 - (.516)^2}$$
$$= 2\sqrt{.7337}$$
$$= 2(.856)$$
$$= 1.71$$

This is the same result we obtained with Formula (8.4). This is to be expected since the two formulas are mathematically equivalent.

What does this $\sigma_{yx\ est}$ of 1.71 indicate? You recall that the σ_{est} represents the standard deviation of the distribution of obtained scores around the predicted scores and as such is an index of the extent of error to be expected in a prediction situation. Thus, the σ_{est} of 1.71 indicates that the standard deviation of the Y-scores predicted from any given X-score (in Table 8.1) is 1.71. The standard deviation of the Y-scores when X is 10 is 1.71 and it is 1.71 when $X = 9$, and so on. For example for the multitude of individuals scoring 10 on the X variable, there is a predicted Y-score of 7.72 with a deviation of 1.71. This measure of variability (1.71) around the predicted Y-score represents the error involved. One needs to know the value of the σ_{est} in a prediction problem in order to estimate the amount of error.

We have seen that the regression line gives us the expected mean of each Y distribution for the corresponding X value and the σ_{est} gives an estimate of the standard deviation of each of those Y distributions. This is a useful way to estimate the error of prediction. However, there are certain assumptions that must be made if these interpretations are to be correct.

1. It must be assumed that the Y distributions for each of the different

values of X have the same variability. This condition of equal variability of the Y-scores for each value of X is known as *homoscedasticity*. In Chapter 7 we referred to homoscedasticity as the condition where the width of the pattern of dots is the same in all parts of the scattergram. Applying this concept to the present context it means that the spread of actual Y-scores around their predicted $\tilde{Y}$ will be the same for each value of $\tilde{Y}$. The Y distributions in Figure 8.6 have equal variability and thus illustrate this property of homoscedasticity. You note that in applying Formula (8.9), one finds the standard deviation of all the $(Y - \tilde{Y})$-scores and assumes that it corresponds to the standard deviation of the subpopulations on Y associated with the given X-scores.
2. It is also assumed that for any single value of X, the associated Y-scores are normally distributed.

If these assumptions are met, the standard error of estimate, $\sigma_{yx\,est}$, and the normal curve table can be used to set limits around a predicted Y-score ($\tilde{Y}$) within which a person's actual score is likely to fall. If it can be assumed that the scores that determined the prediction line $\tilde{Y} = \beta X + \alpha$ came from a normal distribution, then certain statements can be made: In a *large* group of persons to which the equation is applied:

1. Approximately 68 percent will have actual scores that lie within one $\sigma_{yx\,est}$ of their predicted score Y.
2. Approximately 95 percent will have actual scores that lie within two $\sigma_{yx\,est}$ of their predicted score Y.
3. Approximately 99.7 percent will have actual scores that lie within three $\sigma_{yx\,est}$ of their predicted score.

To illustrate, assume that one predicts on the basis of IQ a score of 85 on a proficiency examination and the $\sigma_{yx\,est}$ is 1.5. If we can assume normality in the subpopulation on Y, then we would expect errors in prediction of less than 1.5 approximately 68 percent of the time. Similarly, we would expect errors of less than 3 points (1.5 × 2) approximately 95 percent of the time.

It can be seen from the formula for standard error of estimate: $\sigma_{yx\,est} = \sigma_y \sqrt{1 - \rho^2}$ that the larger the size of ρ, the correlation coefficient, the smaller is the standard error. If ρ should be equal to $+1.0$ or -1.0 and you substitute in the formula, you can see that the σ_{est} will be equal to zero. In other words, no error is involved in the prediction when $\rho = \pm 1.0$. However, if $\rho = 0$, then σ_{est} is equal to the value of the standard deviation of all the Y measures.

Exercises

1. Find the standard error of estimate for the exercise on page 209 involving a qualifying exam for predicting English grade-point average. Use the definition formula.

220 DESCRIPTIVE STATISTICS

2. What is the standard error of estimate if the correlation between intelligence and reading achievement is (a) +.60, (b) −.25, (c) +.45? The standard deviation of the Y-scores is 3.

3. The correlation between an entrance exam and GPA at a midwestern university is .60. The σ of GPA's is 1.00. A student scoring 120 on the exam is predicted to have a GPA of 3.2. What would be the *standard deviation* of the GPA's of all those persons scoring 120 on the entrance exam?

Solutions

1. Recall that $\tilde{Y} = .158X + .77$.

X	Y	$\tilde{Y}$	$e = Y - \tilde{Y}$	$(Y - \tilde{Y})^2$
10	2.0	2.35	−.35	.1225
15	3.0	3.14	−.14	.0196
8	2.0	2.034	−.034	.001156
5	1.5	1.56	−.06	.0036
18	3.5	3.614	−.114	.012996
12	2.5	2.666	−.166	.027556
6	2.0	1.718	.282	.079524
16	3.0	3.298	−.298	.088804
17	4.0	3.456	.544	.295936
12	3.0	2.666	.334	.111556

$$\Sigma = .763228$$

$$\sigma_{est} = \sqrt{\frac{\Sigma(Y-\tilde{Y})^2}{N}} = \sqrt{\frac{.763228}{10}} = \sqrt{.0763228} = .276$$

2. (a) $\sigma_{est} = \sigma_y \sqrt{1 - \rho_{xy}^2} = 3\sqrt{1 - .36} = 3\sqrt{.64} = 3(.8) = 2.4$

(b) $\sigma_{est} = 3\sqrt{1 - .0625} = 3\sqrt{.9375} = 3(.968) = 2.904$

(c) $\sigma_{est} = 3\sqrt{1 - .2025} = 3\sqrt{.7975} = 3(.893) = 2.679$

3. $\sigma_{est} = \sigma_y \sqrt{1 - \rho^2}$
$= 1\sqrt{1 - (.60)^2}$
$= 1\sqrt{1 - .36}$
$= 1\sqrt{.64}$
$= .80$, which means that the standard deviation of all those people scoring 120 on the entrance exam would be .80.

Two Standard Errors of Estimate Just as there were two different regression lines in a bivariate situation, there are also two different standard errors of estimate. We have discussed the standard error of estimate for predicting Y from X ($\sigma_{yx\ est}$). There is a standard error of estimate for X values predicted from Y. It is symbolized $\sigma_{xy\ est}$ and represents the standard deviation of the $(X - \tilde{X})$-scores. The formula for $\sigma_{xy\ est}$ is:

$$\sigma_{xy\ est} = \sigma_x \sqrt{1 - \rho_{xy}^2} \quad \text{where:} \quad \sigma_x = \text{standard deviation of } X\text{-scores}$$
$$\rho_{xy}^2 = \text{square of the correlation coefficient}$$

Formula (8.11)
Standard Error of Estimate
for $\tilde{X}$ from Y

Of course, the two standard errors of estimate may not be equal in value.

SUMMARY

In this chapter, we have looked at the process of predicting from scores on one variable (X) to scores on another variable (Y). Prediction is possible in situations where prior studies have established a systematic relationship between X and Y for a particular sample. Then the findings are later applied to predict $\tilde{Y}$ for another sample when only the X scores are known. It is assumed that the later samples come from the same population as the original sample studied.

The relationship between the measured variable X and the one to be predicted Y is expressed as a mathematical equation. A linear prediction line of the form $\tilde{Y} = \beta X + \alpha$ is drawn where β and α are constants and X and $\tilde{Y}$ are the variables. We saw that it was possible to fit two straight lines to bivariate data, one for predicting Y-scores from known X values and one for predicting X-scores from known Y-scores. The best possible prediction line is defined as the line that minimizes the sum of the squared errors of prediction.

The predicted Y-score will be the mean of all the Y-scores corresponding to a particular value of X. Not all observed Y-scores will exactly equal the predicted Y-score. The difference between Y and $\tilde{Y}$ is e, the error of prediction. The standard deviation of the errors of prediction is known as the standard error of estimate and it is used as an index of the accuracy of prediction.

The accuracy of prediction is related to the magnitude of the correlation between the variables. When the correlation is perfect, prediction is perfect. When the correlation is less than perfect, some error is involved.

We also examined the phenomenon of regression toward the mean which occurs any time that the correlation is less than perfect.

FLOWCHART 8

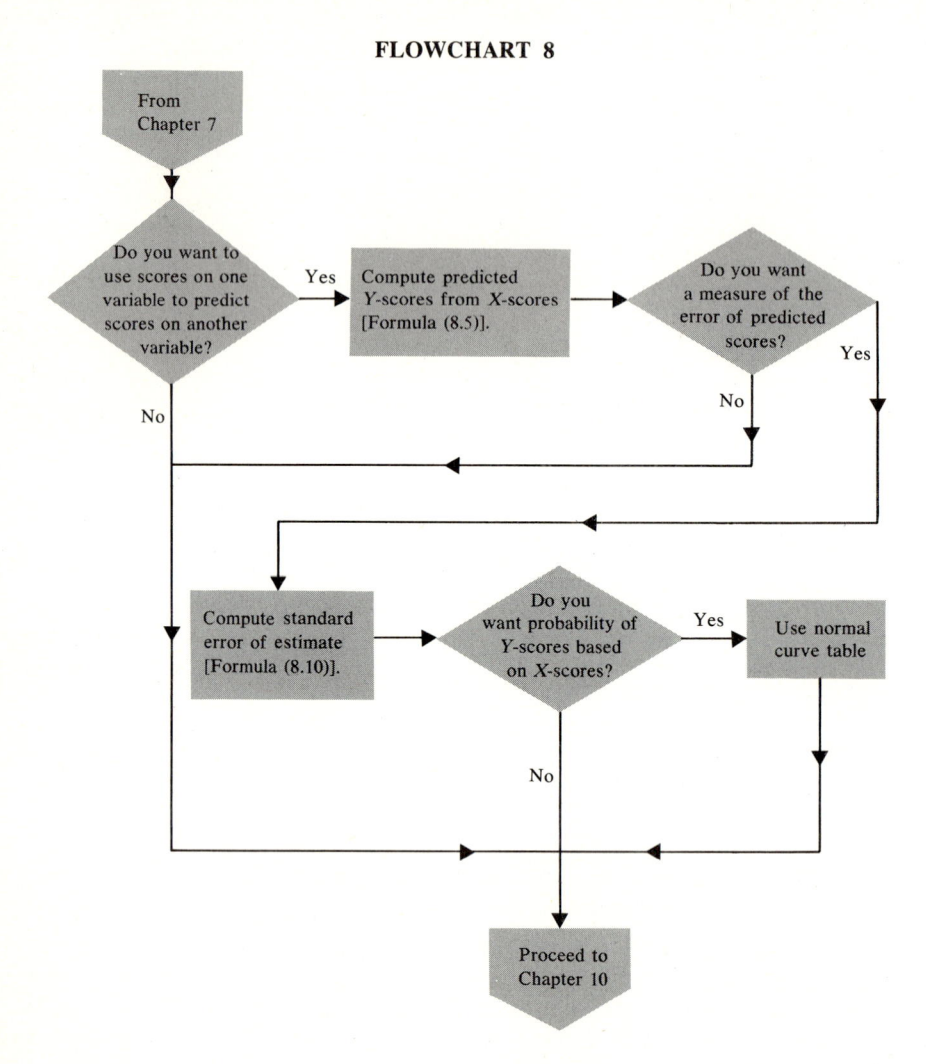

```
From Chapter 7
    │
    ▼
Do you want to use scores on one variable to predict scores on another variable?
    │ Yes → Compute predicted Y-scores from X-scores [Formula (8.5)]. → Do you want a measure of the error of predicted scores?
    │                                                                      │ Yes ──┐
    │ No                                                           No ─────┘       │
    │                                                                              │
    ▼                                                                              ▼
Compute standard error of estimate [Formula (8.10)]. → Do you want probability of Y-scores based on X-scores?
                                                           │ Yes → Use normal curve table
                                                           │ No
                                                           ▼
                                                    Proceed to Chapter 10
```

EXERCISES

1. A college math department is interested in predicting performance on a math proficiency test using the number of years of math studied as the predictor. The following data were gathered on a sample. (a) Calculate the constants β_{yx} and α and write the regression equation for these data. (b) Use the equation to predict the score on the proficiency test of an individual who had studied math 5 years and also the score for one who had studied math 2 years.

NO. YEARS OF MATH X	PROFICIENCY SCORE Y
3	60
5	86
2	55
4	69
4	75
3	54
5	81
3	72
2	45
4	70

2. The following represent hours of practice (X) and scores on a physical fitness proficiency performance test involving a physical skill (Y). (a) Compute the regression equation for predicting test scores from hours of practice; (b) compute the predicted test scores for students who practice 10 hours and for those who practice 15 hours.

HOURS OF PRACTICE X	PROFICIENCY SCORE Y
2	33
8	65
15	85
18	94
5	54
8	56
10	70
5	44
13	72
11	79

3. A high school has determined a prediction equation for predicting college GPA at the state university from high school GPA's. The equation is $\hat{Y} = .75X + .62$. Predict the college GPA which could be expected for students with high school GPA of (a) 3.25; (b) 2.50.

4. Given IQ scores (X) and the grades received in a math course (Y):

X	Y	X^2	Y^2	XY
120	4	14,400	16	480
110	3	12,100	9	330
100	2	10,000	4	200
95	0	9,025	0	0
90	1	8,100	1	90
515	10	53,625	30	1,100

(a) Plot a scattergram of the X- and Y-scores above.
(b) Calculate the correlation coefficient between X and Y.

224 DESCRIPTIVE STATISTICS

(c) Determine the slope of the regression line of Y on X.
(d) Determine the Y-intercept of the regression line of Y on X.
(e) Write the regression equation.
(f) Calculate the predicted Y-score for each X-score recorded above.
(g) Draw the regression line on the scattergram.
(h) Calculate the standard error of estimate.

5. Given: $\mu_x = 100$, $\sigma_x = 10$, $\rho = .85$, $\mu_y = 50$, $\sigma_y = 5$.
(a) Calculate the standard error of estimate of Y on X.
(b) If $\rho = 1.00$, calculate the standard error of estimate.
(c) If $\rho = 0.0$, calculate the standard error of estimate.

ANSWERS

1. $\Sigma X = 35$; $\Sigma Y = 667$; $\Sigma XY = 2{,}449$; $\mu_x = 3.5$; $\mu_y = 66.7$; $\Sigma X^2 = 133$

(a) $\beta_{yx} = \dfrac{N \Sigma XY - \Sigma X \Sigma Y}{N \Sigma X^2 - (\Sigma X)^2}$

$= \dfrac{10(2{,}449) - (35)(667)}{10(133) - 1{,}225}$

$= \dfrac{24{,}490 - 23{,}345}{105}$

$= 10.9$.

$\alpha = \mu_y - \beta_{yx}\mu_x$
$= 66.7 - 10.9(3.5)$
$= 66.7 - 38.15$
$= 28.55$
$\tilde{Y} = 10.9X + 28.55$

(b) $\tilde{Y} = 10.9X + 28.55$
$= 10.9(5) + 28.55$
$= 83.05$ (predicted for one with X of 5)
$= 10.9(2) + 28.55$
$= 50.35$ (predicted for one with X of 2)

2. $\Sigma X = 95$; $\Sigma Y = 652$; $\Sigma XY = 6{,}996$; $\Sigma X^2 = 1{,}121$

$\beta_{yx} = \dfrac{10(6{,}996) - (95)(652)}{10(1{,}121) - 9{,}025}$

$= \dfrac{69{,}960 - 61{,}940}{2{,}185}$

$= 3.67$

$\alpha = \mu_y - \beta\mu_x$
$= 65.2 - 3.67(9.5)$
$= 30.34$

(a) $\tilde{Y} = 3.67X + 30.34$
(b) If $X = 10$,
$\tilde{Y} = 3.67(10) + 30.34$
$= 67.04$
If $X = 15$,
$\tilde{Y} = 3.67(15) + 30.34$
$= 85.39$

3. (a) $\tilde{Y} = .75(3.25) + .62$
$= 3.06$

(b) $\tilde{Y} = .75(2.50) + .62$
$= 2.495$

4. (a) and (g)

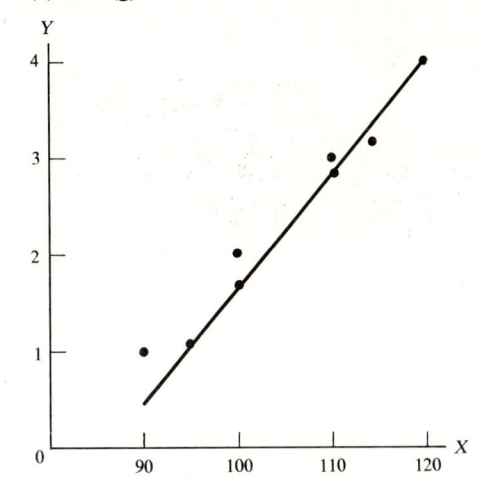

[Scatter plot with Y axis 0-4 and X axis 90-120, showing data points and a fitted regression line]

(b) $\rho = \dfrac{N \Sigma XY - \Sigma X \Sigma Y}{\sqrt{[N \Sigma X^2 - (\Sigma X)^2][N \Sigma Y^2 - (\Sigma Y)^2]}}$

$= \dfrac{5(1,100) - (515)(10)}{\sqrt{[5(53,625) - (515)^2][5(30) - (10)^2]}}$

$= \dfrac{5,500 - 5,150}{\sqrt{(2,900)(50)}}$

$= \dfrac{350}{\sqrt{145,000}} = \dfrac{350}{380.7} = .919$ or $.92$

(c) $\beta_{yx} = \dfrac{N \Sigma XY - \Sigma X \Sigma Y}{N \Sigma X^2 - (\Sigma X)^2}$

$= \dfrac{5(1,100) - (515)(10)}{5(53,625) - (515)^2}$

$= \dfrac{350}{2,900} = .121$

(d) $\alpha = \mu_y - \beta \mu_x$
$= 2 - (.121)(103)$
$= 2 - 12.463$
$= -10.46$

(e) $\tilde{Y} = .121X - 10.46$

(f) $\tilde{Y} = .121(120) - 10.46$
$= 4.06$
$\tilde{Y} = .121(110) - 10.46$
$= 2.85$
$\tilde{Y} = .121(100) - 10.46$
$= 1.64$
$\tilde{Y} = .121(95) - 10.46$
$= 1.04$
$\tilde{Y} = .121(90) - 10.46$
$= .43$

(h) $\sigma_{est} = \sigma_y \sqrt{1 - \rho^2}$
$= 1.414\sqrt{1 - (.92)^2}$
$= 1.414\sqrt{.1536}$
$= 1.414(.392)$
$= .554$

5. (a) $\sigma_{est} = \sigma_y \sqrt{1 - \rho^2}$
$= 5\sqrt{1 - (.85)^2}$
$= 5\sqrt{1 - .7225}$
$= 5\sqrt{.2775}$
$= 5(.526)$
$= 2.63$

(b) $\sigma_{est} = 5\sqrt{1 - 1}$
$= 5(0)$
$= 0$
$\sigma_{est} = 5\sqrt{1 - 0}$
$= 5\sqrt{1}$
$= 5(1)$
$= 5$

CHAPTER NINE
OTHER MEASURES OF RELATIONSHIP

OBJECTIVES

In Chapter Seven we discussed the Pearson product-moment coefficient of correlation, which we described as the best known and the most frequently used index of relationship. However, there are some situations where other indexes are appropriate. Where there are bivariate data X and Y for each of a number of individuals, and either X or Y can be measured in any one of several ways, there are several possibilities for different types of correlation coefficients. For example, X (or Y) may be measured:

1. On an interval or ratio scale, like intelligence or height;
2. On an ordinal scale, like rank in class or rank on leadership abilities;
3. Dichotomously on a nominal scale, like male-female or graduate-undergraduate (a genuine dichotomy);
4. Or on a dichotomy where an underlying normal distribution is assumed (an artificial dichotomy) like pass-fail on a test or above-average/below-average intelligence.

These four types of measures in various combinations result in a number of different types of correlation coefficients. In this chapter we will discuss several types of coefficients of correlation which are used with these particular types of data.

Many of the other coefficients of correlation are algebraic simplifications of the Pearson product-moment coefficient applied to a particular type of data. Others have been developed independently of the product-moment correlation. After reading this chapter, the student should be able to:

1. Recognize situations in which correlation coefficients other than the Pearson should be used and know which coefficient is the appropriate one to use with the data.
2. Define a Spearman ρ correlation coefficient.
3. Calculate a Spearman ρ from the given data and interpret the coefficient.
4. Define a point-biserial correlation coefficient.
5. Calculate a point-biserial coefficient from the given data.
6. Define a biserial correlation coefficient.
7. Calculate a biserial coefficient from the given data.
8. Define a tetrachoric coefficient of correlation.
9. Summarize the given data in a 2×2 table and calculate the tetrachoric coefficient.
10. Define a ϕ coefficient of correlation and arrange the data in a 2×2 table for correlation.
11. Calculate ϕ from the given data.

Spearman Rank Correlation Coefficient: ρ_s

The Spearman rank correlation coefficient is an adaptation of the Pearson product-moment coefficient designed for use with data that are in the form of ranks. If both of the two variables to be correlated are measured on an ordinal (rank-order) scale, the Spearman rank coefficient is the technique generally applied. This coefficient is usually referred to as the Spearman rho since it is symbolized by the Greek letter rho$_s$ (ρ_s).[1] The subscript s is added in order to distinguish the Spearman ρ from the Pearson population coefficient, which is also symbolized by ρ.

There are many situations in which one might have ordinal data for correlation and use the Spearman ρ. Teachers often ask questions concerning the relationship between variables on which school children can be ranked, such as sociability, cooperativeness, socioeconomic status, ability, interest, attitudes toward certain issues, adjustment to the school, and performance in art class. The students could be ranked according to their interest in art as indicated by an interest inventory, and also ranked by judges according to the quality of their art work. Then a Spearman rank coefficient could be computed to indicate the extent of relationship between these two variables. Similarly, one might correlate attitudes toward busing with socioeconomic status, intelligence test scores with social adjustment in the classroom, or age of first graders at entrance to school with their social

[1] *This coefficient is named for the British psychologist, Charles Spearman, who was the first to make wide use of it.*

adjustment. Members of a high school graduating class are ranked and their rank in class is correlated with their academic performance in college.

Although the Spearman coefficient is designed for use with ranked data, it can be used with interval data that have been expressed as ranks. As such it is an alternative to the Pearson ρ. For example, instead of using the raw scores 125, 120, 100, and so on, one would use the ranks 1, 2, 3, and so on. A classroom teacher who wishes to know the relationship between mathematics scores and science scores could rank the students on both sets of test scores and use the Spearman rather than the Pearson coefficient, which uses the raw scores and involves considerably more computation. Since it is relatively simple to compute, the Spearman rank correlation coefficient is frequently chosen instead of the Pearson ρ. It closely approximates the Pearson ρ and is satisfactory for use by the teacher to investigate the relationship between sets of scores in the classroom.

The use of ranks instead of the original raw scores results in a marked simplification in the formula for ρ. The formula for ρ_s is:

$$\rho_s = 1 - \frac{6 \Sigma d^2}{N(N^2 - 1)}$$

where: ρ_s = Spearman rho

d^2 = difference in subject's rank on the two measures squared

N = number of subjects in the sample

6 = constant always employed in this calculation

Formula (9.1)
Spearman Rho Correlation Coefficient

All the subjects are ranked on both variables and then one determines the difference between the paired ranks of each individual. The differences are squared over all cases, the sum of the squared differences found, and then substituted in the formula. If there are individual subjects with the same score, they are assigned the average of the ranks that they occupy. For example, the scores 16, 14, 14, 14, 12, 12, 10, 9, 9, 8 would be ranked as follows: 1, 3, 3, 3, 5.5, 5.5, 7, 8.5, 8.5, 10.

The calculation of ρ_s is illustrated in Table 9.1. A group of subjects was given instruction in hitting an underwater target and a correlation determined between hours spent in practice and proficiency of performance.

It is suggested that the student work through the calculations in Table 9.1 in order to check his understanding of ranking and substituting in the formula.

We see from Table 9.1 that ρ_s can be found much more quickly than the Pearson ρ. Although the simpler formula does not look much like the computational formula we used for the Pearson, it is algebraically equivalent to the Pearson when it is used with ranks. Table 9.2 shows the calculation of the Spearman ρ_s on data that are ranked and also the Pearson coefficient of correlation calculated on the same ranks.

TABLE 9.1 Calculation of Spearman Rank Coefficient of Correlation

INDIVIDUAL	HOURS OF PRACTICE (X)	PROFICIENCY SCORE (Y)	RANK X	RANK Y	DIFFERENCE d	d^2
A	12	73	2	1	1	1
B	14	65	1	2	−1	1
C	8	55	3.5	4	−.5	.25
D	8	60	3.5	3	.5	.25
E	7	54	5	5	0	0
F	6	49	6	8	−2	4
G	4	50	9	7	2	4
H	5	48	7.5	9	−1.5	2.25
I	5	51	7.5	6	1.5	2.25
J	2	30	10	10	0	0
						$\Sigma d^2 = 15$

$$\rho_s = 1 - \frac{6\Sigma d^2}{N(N^2 - 1)} \quad \rho_s = 1 - \frac{6(15)}{10(99)} \quad \rho_s = 1 - \frac{90}{990} \quad \rho_s = .91$$

TABLE 9.2 Spearman ρ and the Pearson Product-Moment Coefficients Calculated on the Same Data

ORIGINAL DATA		SPEARMAN ρ_s				PEARSON ρ FROM RANKS				
X	Y	RANK X	RANK Y	d	d^2	RANK X	RANK Y	X^2	Y^2	XY
0	4	5	3	2	4	5	3	25	9	15
12	5	1	2	−1	1	1	2	1	4	2
4	1	4	5	−1	1	4	5	16	25	20
8	7	2	1	1	1	2	1	4	1	2
6	3	3	4	−1	1	3	4	9	16	12
					$\Sigma = 8$	$\Sigma = 15$	$\overline{15}$	$\overline{55}$	$\overline{55}$	$\overline{51}$

$\rho_s = 1 - \dfrac{6 \Sigma d^2}{N(N^2 - 1)}$

$\rho_s = 1 - \dfrac{48}{120}$

$\rho_s = 1 - .40$

$\rho_s = .60$

$\rho = \dfrac{N \Sigma XY - (\Sigma X)(\Sigma Y)}{\sqrt{[N \Sigma X^2 - (\Sigma X)^2][N \Sigma Y^2 - (\Sigma Y)^2]}}$

$\rho = \dfrac{5(51) - (15)(15)}{\sqrt{[5(55) - 15^2][5(55) - 15^2]}}$

$\rho = \dfrac{255 - 225}{\sqrt{(50)(50)}} = \dfrac{30}{50} = .60$

Since the Spearman rho (ρ_s) is essentially a Pearson coefficient of correlation computed on ranks, no new interpretation is made of ρ_s. It too represents the extent to which individuals occupy the same relative position on two variables. In Table 9.1 we see that the differences in the individuals' paired ranks are not large. This means that individuals have nearly the same relative position on X and Y and the size of the coefficient (.91) reflects this relationship. The Spearman formula, like the Pearson, yields a value of +1.0 for perfect positive correlation, -1.0 for perfect negative correlation, and 0.0 for an absence of relationship between the variables. If all the individuals have exactly the same rank on the two variables (perfect positive relationship) the differences in rank will all be zero, the fraction will disappear from the formula, and $\rho_s = 1$. If the individuals' ranks on the two variables are exactly reverse, $\rho_s = -1.00$. If there is no relationship between the paired rankings, $\rho_s = 0.0$.

An assumption of the Spearman coefficient is that the measures are continuous variables and, of course, with continuously distributed variables no two persons have precisely the same quantity of the variable. If there are many duplicate scores, the use of the Spearman ρ will underestimate the actual correlation. However, if there are only a few ties, the Spearman can be safely used and will be a good estimate of the Pearson product-moment coefficient.

Exercise

1. Two teachers were asked to rank five student essays according to their quality. The results are listed below. Calculate the Spearman rho correlation coefficient for these data.

		TEACHER 1 RANKINGS	TEACHER 2 RANKINGS
ESSAY	1	1	1
	2	2	4
	3	3	2
	4	4	3
	5	5	5

Solution

$\rho_s = 1 - \dfrac{6 \Sigma d^2}{N(N^2 - 1)}$

RANKS	RANKS	d	d^2
1	1	0	0
2	4	2	4
3	2	1	1
4	3	1	1
5	5	0	0

$\rho_s = 1 - \dfrac{6(6)}{5(24)}$

$= 1 - \dfrac{36}{120}$

$= .70$

Point-Biserial Correlation Coefficient: ρ_{pbi}

One may wish to correlate data where one variable is *continuous* and measured on an interval or ratio scale and the other variable is dichotomous, that is, has one of only two possible values. For example, a high school counselor may be interested in the relationship between the sex of the student and performance on a comprehensive achievement test given to high school seniors. In this case, the sex variable is measured dichotomously (M-F) and the achievement variable is measured on an interval scale. The correlation coefficient used for measuring the relationship between such variables is a variation of the Pearson coefficient known as the *point-biserial correlation coefficient*; it is designated ρ_{pbi}. The use of the point-biserial correlation coefficient is based on the assumption that the continuous variable is represented by a normal distribution and that the dichotomous variable represents a *genuine* dichotomy (on a nominal scale). For example, male-female, graduate-undergraduate, married-unmarried, urban-rural, smoker-nonsmoker, alive-dead, college graduate-noncollege graduate all represent genuine dichotomies.

The point-biserial is simply a Pearson product-moment coefficient of correlation computed from data where one variable is dichotomized and the other is normally distributed. If one used the Pearson formula (7.2) for ρ on these data, one would compute a correlation coefficient equal to the point-biserial coefficient, ρ_{pbi}. However, a formula has been derived for ρ_{pbi} which, although mathematically equivalent to the Pearson formula, is somewhat simpler.

For the computation, one has measurements on two variables X (the interval measure) and Y (the dichotomy) for each individual. The dichotomous variable Y is assigned two numerical values. Any values would do since the coefficient is independent of the values used; however, most often the variables are assigned values of 1 and 0 simply to facilitate computation. One arbitrarily assigns a 1 to individuals in one category of the variable Y and a 0 to individuals in the other category and then proceeds to calculate the correlation of these values with the values on the continuous variable (X). In fact, the word "biserial" indicates that there are two series of individuals being measured on X: (1) those who score 0 on Y and (2) those who score 1 on Y.

The computation of a point-biserial involves the *proportion* of observations falling in the two categories of the dichotomous variable. The two proportions are designated p and q and since there are only two groups the total of $p + q$ will always equal 1. The continuous variable is measured on an interval scale and the first step is to compute the mean and standard deviation for this variable. These values are then used for the calculation of

Other Measures of Relationship 233

the point-biserial correlation coefficient. The formula is:

$$\rho_{pbi} = \frac{\mu_p - \mu_q}{\sigma_x}\sqrt{pq}$$

where: σ_x = standard deviation of all the scores on the continuous variable X

p = proportion of the total group of individuals scoring 1 on the dichotomous variable Y

q = proportion of the total group scoring 0 on Y

pq = the product of the two proportions

μ_p = the mean score on the continuous variable X for the group scoring 1 on the dichotomy Y

μ_q = the mean score on the continuous variable X for the group scoring 0 on the dichotomy Y

Formula (9.2)
Point-Biserial Correlation Coefficient

From Formula (9.2) we see that the ρ_{pbi} is a measure of the difference between the mean score on X of the individuals scoring 1 on Y and the individuals scoring 0 on Y. Let us apply Formula (9.2) to a correlation situation. Assume that one wants to establish the relationship between the scores on a comprehensive achievement test given to high school seniors and the sex of the student. Although more students would be included in an actual study, for this example, ten students are picked at random and data are gathered on the X and Y variable for each student. The continuous variable X in this example is the achievement test scores. The dichotomous variable is the sex of the student. The number 1 represents a male student; 0 represents a female student; μ_p is the mean achievement test score for males; and μ_q is the mean achievement test score for females.

The data and computation of the point-biserial coefficient on the sex and achievement data are shown in Table 9.3. The ρ_{pbi} of .58 for the data in Table 9.3 indicates that there is some relationship between the variables sex and performance on the achievement test. The achievement test does differentiate between the sexes. Examination of the data in Table 9.3 reveals that males tended to receive the higher grades and females received the lower grades. Only if the six individuals making the highest scores were males and the four making the lowest scores were females would the ρ_{pbi} have its highest positive value. If the females (those coded 0 on the dichotomous variable) had the highest scores and the males ($Y = 0$) had the lowest scores, then ρ_{pbi} would have the highest negative value. Considerable error would be involved, however, if one tried to predict achievement from a knowledge of the sex of the student.

Since the point-biserial correlation is actually a Pearson product-moment coefficient for particular data, it too has a theoretical range of +1 to −1, inclusive. If those scoring 1 on Y have the same mean on X as those scoring

234 DESCRIPTIVE STATISTICS

TABLE 9.3 Calculation of Point-Biserial Correlation Coefficient from a Continuous and a Genuine Dichotomous Variable

INDIVIDUAL	ACHIEVEMENT TEST SCORE (X)	SEX (Y) 1, MALE; 0, FEMALE
A	60	1
B	56	1
C	51	1
D	58	1
E	49	0
F	48	1
G	55	0
H	45	0
I	47	0
J	55	1

$N_1 = 6 \quad N_0 = 4 \quad N = 10$

Mean X-Score for Males $\mu_p = \dfrac{328}{6} = 54.67$

Mean X-Scores for Females $\mu_q = \dfrac{196}{4} = 49$

$\sigma_x = 4.82$

$p = 6/10 = .60$

$q = 4/10 = .40$

$\rho_{pbi} = \dfrac{\mu_p - \mu_q}{\sigma_x} \sqrt{pq}$

$= \dfrac{54.67 - 49}{4.82} \sqrt{.60 \times .40}$

$= \dfrac{5.67}{4.82} \sqrt{.2400}$

$= 1.18(.4899)$

$= .58$

---

0 on Y, the $\rho_{pbi} = 0$. The size of the coefficient is also dependent upon the proportions (p and q) in the two categories of the dichotomous variable. A ρ_{pbi} can only equal +1 or −1 when p and q are .50, that is, $p = q$. If the proportions differ from p and q = .50 it is mathematically impossible for the ρ_{pbi} to reach + or − 1. In the example above, the maximum possible value would be $\rho_{pbi} = .87$.

Table 9.4 shows the calculation of the Pearson ρ on these same data.

TABLE 9.4 Calculation of the Pearson Product-Moment Correlation Coefficient from Data in Table 9.3

	X	Y	X^2	Y^2	XY
	60	1	3,600	1	60
	56	1	3,136	1	56
	51	1	2,601	1	51
	58	1	3,364	1	58
	49	0	2,401	0	0
	48	1	2,304	1	48
	55	0	3,025	0	0
	45	0	2,025	0	0
	47	0	2,209	0	0
	55	1	3,025	1	55
$\Sigma X =$	524	6	27,690	6	328

$$\rho = \frac{N \Sigma XY - (\Sigma X)(\Sigma Y)}{\sqrt{[N \Sigma X^2 - (\Sigma X)^2][N \Sigma Y^2 - (\Sigma Y)^2]}}$$

$$= \frac{10(328) - (524)(6)}{\sqrt{[10(27,690) - 524^2][10(6) - (6)^2]}}$$

$$= \frac{3,280 - 3,144}{\sqrt{(2,324)(24)}} = \frac{136}{\sqrt{55,776}}$$

$$= \frac{136}{236} = .576 = .58$$

Notice that one arrives at the same coefficient of correlation obtained with Formula (9.2), but the calculations are much more extensive. This illustrates that the ρ_{pbi} is a Pearson correlation coefficient and that Formula (9.2) is an algebraic simplification of the Pearson ρ formula.

Exercises

1. Below are listed data on X, a continuous variable, and Y, a dichotomous-nominal variable. Calculate the point-biserial coefficient of correlation for these data:

X	Y		X	Y
0	1		7	0
1	1		8	0
2	0		8	1
4	1		9	0
5	1			
6	0			

$\mu_x = 5$
$\sigma_x = 3$

2. Calculate the Pearson product-moment coefficient of correlation for the above data. How does the Pearson compare in value with the point-biserial?

Solutions

1. $\mu_p = 3.6 \quad \left(\frac{18}{5}\right)$

   $\mu_q = 6.4 \quad \left(\frac{32}{5}\right)$

   $p = .5; \quad q = .5$

   $\rho_{pbi} = \dfrac{3.6 - 6.4}{3} \sqrt{(.5)(.5)}$

   $= \dfrac{-2.8}{3} \sqrt{.25}$

   $= -.933(.5)$

   $= -.4665 = -.47$

2.

X	Y	X^2	Y^2	XY
0	1	0	1	0
1	1	1	1	1
2	0	4	0	0
4	1	16	1	4
5	1	25	1	5
6	0	36	0	0
7	0	49	0	0
8	0	64	0	0
8	1	64	1	8
9	0	81	0	0
50	5	340	5	18

$\rho = \dfrac{N \Sigma XY - (\Sigma X)(\Sigma Y)}{\sqrt{[N \Sigma X^2 - (\Sigma X)^2][N \Sigma Y^2 - (\Sigma Y)^2]}}$

$= \dfrac{10(18) - (50)(5)}{\sqrt{[10(340) - 50^2][10(5) - 5^2]}} = \dfrac{180 - 250}{\sqrt{(900)(25)}}$

$= \dfrac{-70}{\sqrt{22,500}} = \dfrac{-70}{150}$

$= -.4666 = -.47 \quad (\rho_{pbi} = \rho)$

Biserial Correlation Coefficient: ρ_{bi}

Closely related to the point-biserial correlation coefficient is the biserial coefficient. It is symbolized ρ_{bi}. The biserial correlation coefficient is an estimate of the product-moment correlation between *one continuous* variable X and another continuous variable Y which is being treated as a dichotomy, that is, *an artificial dichotomy*. For example, the X-scores might be the scores on a mathematics achievement test and the Y-scores might be 0's (favorable) and 1's (unfavorable) on an inventory measuring attitudes toward mathematics. Although one treats the Y variable as a dichotomy, it is assumed that the scores on Y which underlie the dichotomy are *normally distributed*. In other words, the variable is only measured dichotomously, but there is a normal distribution underlying the dichotomy.

For example, one may wish to determine the relationship between intelligence and whether one passes or fails a mathematics examination. The original distribution of the mathematics scores is continuous, but one could dichotomize performance in mathematics simply as pass or fail and then correlate with intelligence test scores. As in the example given above, performance on an attitude scale is often dichotomized as simply favorable-unfavorable and a biserial correlation established with some continuous variable. Age is a continuous variable which is sometimes dichotomized as young-old for correlation purposes.

The question arises why one sets up an artificial dichotomy when the potential for more precise measurement is present. That is, the Y variable could be a wide range of scores normally or near normally distributed instead of a dichotomy. Then the two distributions of X- and Y-scores could be correlated and the Pearson coefficient obtained. This would be a far more precise procedure to use. However, the reason for the dichotomy may be that a specific research question is answered more appropriately in terms of the dichotomy. For example, the researcher may want to know only the relationship between passing or failing on a qualifying examination and intelligence. Thus, he dichotomizes performance on the qualifying examination as pass-fail even though the actual scores are available.

There may be situations in which, even though the variable itself is continuous, the measuring technique may be appropriate only for placing subjects into dichotomous categories. For example, an adjustment inventory may be valid only for placing subjects into the two categories: normal and neurotic.

The data gathered for the computation of ρ_{bi} consist of an X-score which can assume any one of several different values, and a Y-score, which is either 0 or 1 for each of the persons in the study. For example, assume one wishes to know the relationship between scores on a law school comprehensive examination (X) and whether one passes or fails the bar examination (Y). The Y variable scores are recorded as 1 (pass) or 0 (fail), although a normal distribution of ability is assumed to underlie this dichotomy.

The computational formula for the biserial coefficient is similar to the point-biserial. One first finds the difference between the mean score on X of those scoring 1 on Y and the mean X score of those scoring 0 on Y. The proportion of the total number who score 1 on Y (p) and the proportion scoring 0 on Y are also determined (q). Then y, which represents the height of the ordinate of the normal curve at the point of division for the proportions p and q, is found. These components are combined into the following computational formula for ρ_{bi}:

$$\rho_{bi} = \frac{\mu_p - \mu_q}{\sigma_x} \cdot \frac{pq}{y}$$

where: μ_p and μ_q = mean scores on the continuous variable X of individuals in the two categories on Y

p and q = proportions in the two categories on the Y variable

σ_x = standard deviation of all scores on the continuous variable X

y = height of ordinate of normal curve at the point of division for proportions p and q

Formula (9.3)
Biserial Correlation Coefficient

The value of y may be determined from the area table of the normal curve that appears in Table A2 of the Appendix. Thus, if $p = .70$ and $q = .30$, we know that for p we have all the area in one-half of the distribution plus an additional 20 percent located between the mean and the split of the dichotomy. In the *area* column of Table A2, we see that for area .20, the height of the ordinate y is .348. If $p = .80$ and $q = .20$, we know that we have all the area in one-half of the distribution (.50) plus an additional 30 percent located between the mean and the split in the dichotomy. Checking the area column of Table A2, we see that for area .80, the height of the ordinate y is .280.

Using the data from Table 9.5 let us determine the correlation between pass-fail on a bar examination and grades on a law school comprehensive examination.

This calculation shows us that a biserial correlation coefficient of .60 exists between scores on a law school comprehensive examination and pass-fail on a bar examination.

Theoretically, the value of ρ_{bi} ranges from +1.00 to −1.00. However, unlike any other correlation coefficient that we have discussed, ρ_{bi} can sometimes assume values below −1.00 and above +1.00. However, such extraordinary values of ρ_{bi} can only occur as a result of some gross departures from normality in the data used for correlation. You recall than an assumption underlying use of the ρ_{bi} is that the continuous variable X is normally distributed, as well as the variable Y underlying the dichotomy.

As you probably noticed, the data on which both ρ_{pbi} and ρ_{bi} are calculated are very similar. However, the two coefficients are quite different in terms of the assumptions made about the data. In the case of ρ_{bi}, one assumes that the distribution underlying the dichotomy on Y is normal. The dichotomy is an

Other Measures of Relationship 239

TABLE 9.5 Calculation of Biserial Correlation from a Continuous Variable and an Artificial Dichotomy

SUBJECT	X GRADES ON LAW SCHOOL COMPREHENSIVE EXAM	Y BAR EXAMINATION PASS—1 FAIL—0
A	40	1
B	45	1
C	35	0
D	30	1
E	40	1
F	25	0
G	20	1
H	5	0
I	0	0
J	10	1
		$N_1 = 6$ $N_0 = 4$ $N = 10$

Mean Score on X for Pass Group: $\mu_p = \dfrac{185}{6} = 30.83$

Mean Score on X for Fail Group: $\mu_q = \dfrac{65}{4} = 16.25$

$\sigma_x = 15$; $p = 6/10 = .60$; $q = 4/10 = .40$; $y = .386$

$$\rho_{bi} = \frac{\mu_p - \mu_q}{\sigma_x} \times \frac{pq}{y}$$

$$= \frac{30.83 - 16.25}{15} \times \frac{(.60)(.40)}{.386}$$

$$= \frac{14.58}{15} \times \frac{.2400}{.386}$$

$$= .972 \times .622$$

$$= .60$$

artificial one made for research purposes. In the case of ρ_{pbi}, the Y variable is a genuine dichotomy on a nominal scale. Thus, each of these coefficients is appropriate for a certain type of data and for a particular purpose. They do not compete for use.

There is a formula which relates ρ_{bi} and ρ_{pbi} and which may be used to convert from one to the other if at a later time one decides that the data do have an unlying normal distribution and, therefore, ρ_{bi} should be used. The conversion formula is:

$$\rho_{bi} = \rho_{pbi} \frac{\sqrt{pq}}{y}$$

240 DESCRIPTIVE STATISTICS

From the formula we see that (for the same set of data): if ρ_{pbi} is positive, then ρ_{bi} will be positive but larger than ρ_{pbi} since the latter is multiplied by the factor $\sqrt{pq}/y$ which is greater than 1. If ρ_{pbi} is negative, ρ_{bi} is negative and farther from zero. If $\rho_{pbi} = 0$, then $\rho_{bi} = 0$. Thus, the same data will give evidence for a higher degree of relationship between X and Y variables when the ρ_{bi} is calculated. On the other hand, ρ_{pbi} will yield a noticeably smaller coefficient if applied to a situation where the biserial should have been used. One must know what can be assumed about the distribution of the Y variable (true dichotomy or a dichotomy with an underlying normal distribution) in order to know whether ρ_{bi} or ρ_{pbi} is the appropriate one to use.

One of the main applications of the ρ_{pbi} is in the item analysis of educational and psychological tests. The point biserial correlation of a test item with the total test score is used as an index of the discriminating power of the item (that is, whether the item discriminates between the high and low scorers on the total test). The total test scores are the continuous variable X and the right (1) or wrong (0) scores on a specific item represent the dichotomous variable Y. An item that has a positive correlation with the total test score is assumed to be discriminating and is retained in the test. An item that is not correlated with total test is not discriminating and therefore is not playing a useful role in the test. An item that has a negative correlation with the total test score is discriminating in the wrong direction and is also a candidate for elimination from the test.

Exercise

In order to continue in its program, a college mathematics department requires all its majors to pass a qualifying examination at the end of their sophomore year. The examination consists of one comprehensive question and is graded either pass (1) or fail (0). For the given data find the relationship between mathematics GPA and score on the comprehensive examination.

STUDENT	X MATH GPA	Y COMPREHENSIVE EXAM
Bob	2.50	0
Carol	3.20	1
Ted	3.35	1
Alice	3.25	0
Judy	2.75	1
Sally	2.80	0
John	3.40	0
Linda	2.50	0
Tim	3.75	1
Mike	3.80	1

Mean Score on X for Pass Group: $\mu_p = \dfrac{16.85}{5} = 3.37$

Mean Score on X for Fail Group: $\mu_q = \dfrac{14.45}{5} = 2.89$

$p = \dfrac{5}{10} = .50$

$q = \dfrac{5}{10} = .50$

$y = .3989$

$\sigma_x = .45$

Solution

$$\rho_{bi} = \dfrac{\mu_p - \mu_q}{\sigma_x} \times \dfrac{pq}{y} = \dfrac{3.37 - 2.89}{.45} \times \dfrac{(.50)(.50)}{.3989}$$

$$= \dfrac{.48}{.45} \times \dfrac{.25}{.3989}$$

$$= (1.0667)(.6267)$$

$$= .6685$$

Tetrachoric Correlation: r_t

The tetrachoric correlation is an extension of the biserial coefficient to a correlation situation in which *both* variables have been *artificially dichotomized*. That is, both X and Y are measured dichotomously for a group, although it is assumed that the distributions underlying both variables are normal. For example, one may wish to determine the relationship between attitudes toward mathematics (X) which had been dichotomized as favorable (1) or unfavorable (0) and performance on a mathematics achievement test (Y) which had been dichotomized as pass (1) or fail (0). Thus, only 0 and 1 data are available for the calculation, but the researcher's interest is in the correlation between X and Y that he would have obtained had he used the normally distributed measures.

The observable data for a tetrachoric correlation are dichotomous scores (0 or 1) for each individual on both X and Y. Since the data are reduced to two categories on the two variables, they are generally presented in a 2×2 table. The table contains two columns of 0's and 1's for the X variable and two rows of 0's and 1's for the Y variable. Then each individual is placed in one of the four categories of the table corresponding to his two scores (X and Y). The cells of the table are labeled *a*, *b*, *c*, and *d*. The number of persons in each of the four cells is determined and that number is placed in

242 DESCRIPTIVE STATISTICS

TABLE 9.6 Dichotomized Data on Two Variables X and Y for a Tetrachoric Correlation

STUDENT	ATTITUDE (X) FAVORABLE, 1 UNFAVORABLE, 0	MATH PERFORMANCE (Y) PASS, 1 FAIL, 0
A	1	1
B	1	1
C	1	1
D	1	1
E	1	0
F	0	1
G	0	1
H	0	0
I	0	0
J	0	0

		ATTITUDE (X) 0	1	
ACHIEVEMENT (Y)	1	2(a)	4(b)	6
	0	3(c)	1(d)	4
		5	5	10

the cell. Table 9.6 shows dichotomized data on X (attitudes toward mathematics) with 1 being favorable and 0 unfavorable and Y (mathematics achievement scores) with 1 being pass and 0 fail. The 2×2 frequency summary table is also shown.

The calculation of r_t uses the frequencies of a, b, c, and d in the 2×2 table in order to obtain an approximation of the Pearson coefficient ρ_{xy} that would be calculated if more refined measurements of X and Y were made. In order for a positive relationship to produce a positive tetrachoric coefficient, cell b in the table should contain the count of individuals who score 1 on both variables and cell c the count of individuals who score 0 on both variables. In the example, cell b represents those with a favorable attitude toward mathematics (1) and who passed the mathematics achievement test (1). There are four persons in this cell. Cell c contains those with unfavorable attitudes (0) who failed the test (0). There are three persons in this cell.

With the cells appropriately labeled and the frequencies determined, one can calculate the tetrachoric ratio which is the basis for determining the tetrachoric correlation coefficient, r_t. The steps are:

1. Multiply a by d to get ad; multiply b by c to get bc. In Table 9.6 $ad = 2$ and $bc = 12$. When bc is greater than ad the correlation is positive; when ad is greater than bc, the correlation is negative.
2. Divide the larger of the two products above by the smaller to obtain the ratio. For the data in Table 9.6, we divide bc by ad or 12 by 2 and obtain a ratio of 6.
3. Next, enter Appendix Table A8 with bc/ad or ad/bc, whichever is larger, and read the corresponding value of r_t to the left of it. In the example in Table 9.6, we enter the table with a bc/ad ratio of 6 and find the tetrachoric coefficient to be .61. This coefficient, .61, represents an approximation of the value of the Pearson product-moment correlation coefficient between the normally distributed variables attitude and achievement for which only dichotomous measures were used.

As an additional example, let us consider the data in the 2 × 2 table, Table 9.7.

TABLE 9.7 Calculation of the Tetrachoric Correlation

		X = 0	X = 1
Y	1	30(a)	10(b)
	0	4(c)	16(d)

$bc = 40$
$ad = 480$

$\dfrac{ad}{bc} = \dfrac{480}{40} = 12$

The correlation is negative since $ad > bc$.

Entering Table A8 with $ad/bc = 12$, we see that the value of r_t is $-.76$.

The above procedure for calculating r_t is most accurate when the variables are dichotomized close to the median. That is, the proportion of 1's on the Y variable, should be close to .50 of the total N. Likewise the proportion of 1's on the X variable should be close to .50. If the proportions depart greatly from .50, the tabled values will be overestimates of the true value of r_t when it is positive.

The value of r_t will be within the limits of -1.0 to $+1.0$.

The tetrachoric coefficient is not a widely used correlation index. It is a rather crude correlational procedure since both variables have been reduced to dichotomies. Consequently, there is a great deal of information lost. Instead of using the raw scores on a test, one simply categorizes the scores as pass or fail. Individuals who are very close in test scores may be placed in different categories. One person may be one point above the cutoff and another one point below, but this closeness in performance is not apparent in the computational procedure. The former person is placed in the pass category and the latter in the fail group. This reduces the accuracy of the procedure. The advantage of the tetrachoric is the ease of computation.

244 DESCRIPTIVE STATISTICS

In summary, the tetrachoric coefficient may be used as a measure of relationship between two dichotomous variables when normal distributions underlying the dichotomies can be assumed. The value of r_t is an approximation of the product-moment correlation that would be obtained if the normally distributed measures were used.

Exercise

The following are dichotomous data on anxiety X and intelligence Y. Anxiety was measured as low (0) or high (1) and intelligence was measured as below average (0) or above average (1).

INTELLIGENCE Y

	ANXIETY X LOW (0)	HIGH (1)	
ABOVE AVERAGE (1)	35(a)	58(b)	93
BELOW AVERAGE (0)	65(c)	42(d)	107

Assume that normally distributed variables underlie both dichotomies and calculate the value of r_t for the above data.

Solution

$ad = 1470$
$bc = 3770$

$\dfrac{bc}{ad} = \dfrac{3770}{1470} = 2.565$

Entering the table with the ratio 2.565, one finds the r_t to be .36.

The Phi Coefficient: r_ϕ Or Simply ϕ

The phi coefficient is used as a measure of correlation when *both* variables represent nominal-dichotomous measures. That is, both X and Y are *genuine* dichotomies—only two classes of observations for each variable. For example, one may be interested in the relationship between the sex (X) of a group of high school seniors and whether or not one wins a scholarship (Y). Sex is dichotomized as male-female and "winning a scholarship" is dichotomized as yes-no. It is decided arbitrarily that 1 means male and 0 means female on the X variable; and 1 means yes on winning a scholarship and 0 means no. Each student will have two scores—an X and a Y. Table 9.8 shows data on these two variables for 12 students.

One could calculate the Pearson product-moment correlation coefficient for the data in Table 9.8. If this is done, one would have the *phi coefficient*, ϕ. That is, the phi coefficient is actually the Pearson product-moment coefficient for nominal-dichotomous data.

TABLE 9.8 Data for the Correlation of Two Dichotomous Variables: Sex and Winning a Scholarship

STUDENT	X SEX (MALE, 1; FEMALE, 0)	Y SCHOLARSHIP (YES, 1; NO, 0)
A	1	1
B	0	0
C	0	1
D	0	0
E	1	1
F	0	0
G	1	0
H	1	1
I	0	0
J	0	1
K	1	1
L	0	0

However, the phi coefficient can be calculated by a much simpler procedure. The dichotomous data on the two variables are tabulated in a 2 × 2 table similar to the one used to calculate r_t. Table 9.9 presents the general form for arranging the data in the 2 × 2 table with the cells labeled a, b, c, and d as before.

TABLE 9.9 General Form for the 2 × 2 Arrangement of Data for Calculation of the Phi Coefficient

		VARIABLE X		TOTALS
		0	1	
VARIABLE Y	1	a	b	$a + b$
	0	c	d	$c + d$
Totals:		$a + c$	$b + d$	N

For example, the number of persons scoring 0 on X and 1 on Y is designated a. The number of persons scoring 0 on X and 0 on Y is designated c. The total number scoring 0 on the X variable is $a + c$. Then the number scoring 1 on X and 1 on Y is denoted by b, and so on. Formula (9.4) is the formula for the calculation of the phi coefficient from the data in a 2 × 2 table.

$$\phi = \frac{bc - ad}{\sqrt{(a+b)(c+d)(a+c)(b+d)}}$$ where: a, b, c, d represent the frequency of observations in the four cells of the 2×2 table

Formula (9.4)
Phi Coefficient

The summary table and the calculation of ϕ for the data in Table 9.8 are shown in Table 9.10.

TABLE 9.10 Calculation of the Phi Coefficient of Correlation between Two Dichotomous Variables (Based on Data in Table 9.8)

		SEX		TOTALS
		FEMALE, 0	MALE, 1	
SCHOLAR- SHIP	YES(1)	2(a)	4(b)	6
	NO(0)	5(c)	1(d)	6
		7	5	12

$$\phi = \frac{bc - ad}{\sqrt{(a+b)(c+d)(a+c)(b+d)}}$$

$$= \frac{20 - 2}{\sqrt{(6)(6)(7)(5)}} = \frac{18}{\sqrt{1,260}}$$

$$= \frac{18}{35.5} = .507 = .51$$

Theoretically the range of possible values of ϕ is -1.0 to $+1.0$. However, these limits can be attained only when $(a+b)/N$ and $(b+d)/N$ are equal in the 2×2 table; that is, when there are the same proportions of 1's on both X and Y.[2] For example, consider the data in Table 9.11 where the proportion of 1's on $X = .80$ $((b+d)/N)$ and the proportion of 1's on $Y = .40$ $((a+b)/N)$. No higher correlation could be obtained for the data in Table 9.11 than the .41 obtained. Notice that all of the 20 1's on Y are paired with 1's on X, but there are 20 other 1's on X that are paired with 0's on Y. Thus, there could not be a perfect correlation between X and Y because not all of the 1's on X are also 1's on Y. And, of course, one could not predict perfectly either from X to Y because knowing that a person's score on X is 1 does not tell us that his score on Y is also 1; it could be 0.

[2] Or, what amounts to the same thing, when $(a+c)/N = (c+d)/N$.

TABLE 9.11 Phi Correlation when There Are Unequal Proportions of 1's on the Two Variables X and Y

		X 0	X 1	TOTALS
Y	1	0(a)	20(b)	20
	0	10(c)	20(d)	30
		10	40	50

$$\phi = \frac{bc - ad}{\sqrt{(a+b)(c+d)(a+c)(b+d)}}$$

$$= \frac{200 - 0}{\sqrt{(20)(30)(10)(40)}}$$

$$= \frac{200}{\sqrt{240{,}000}} = \frac{200}{489} = .409 = .41$$

Consider the following 2 × 2 tables where the proportions are equal. In (a) a phi coefficient of +1.0 is possible and in (b) a ϕ of −1.0 is possible.

(a)

		X 0	X 1	TOTALS
Y	1	0	20	20
	0	20	0	20
		20	20	40

(b)

		X 0	X 1	TOTALS
Y	1	20	0	20
	0	0	20	20
		20	20	40

248 DESCRIPTIVE STATISTICS

In the above illustrations we see that the value of phi is influenced by the marginal totals in the 2×2 table. This influence of the marginal totals on the range of values of phi is one of the disadvantages of the phi coefficient. The tetrachoric has an advantage in that its maximum and minimum values are +1.0 and −1.0 regardless of how far the proportions $((a+b)/N)$ or $((b+d)/N)$ depart from equality.

Exercise

Given the following data in a 2×2 table, calculate the phi coefficient between X and Y.

		X 0	X 1	TOTALS
Y	1	20(a)	60(b)	80
Y	0	20(c)	0(d)	20
		40	60	100

Solution

$$\phi = \frac{bc - ad}{\sqrt{(a+b)(c+d)(a+c)(b+d)}}$$

$$= \frac{1{,}200 - 0}{\sqrt{(80)(20)(40)(60)}}$$

$$= \frac{1{,}200}{\sqrt{3{,}840{,}000}}$$

$$= \frac{1{,}200}{1{,}959.6}$$

$$= .61$$

Coefficient of Contingency

The phi coefficient can only be used when the data fit a 2×2 bivariate frequency table. The coefficient of contingency can be used with a bivariate frequency table with any number of categories on either of the variables.

For example, the contingency coefficient would be used to relate the variable socioeconomic status categorized as high, middle, and low to the extent of preschool experience children might have had. The latter variable could be categorized as kindergarten only, kindergarten + one year of nursery school, or kindergarten + two years of nursery school.

The computation of this coefficient requires a knowledge of chi-square statistics which has not yet been explained in this text. Thus, it will be necessary to postpone a discussion of its computation until Chapter 13.

SUMMARY

In this chapter we have examined a number of correlation coefficients which have been developed for use with particular types of data. The Spearman rank correlation coefficient is designed for use with naturally ordinal data or with interval data that have been expressed as ranks. The point-biserial and biserial coefficients are used to correlate one continuous with one dichotomous variable. Point-biserial assumes that the dichotomous variable is genuine; biserial assumes that the dichotomous variable is in fact continuous and normally distributed and is artificially reduced to a dichotomy.

Tetrachoric correlation is used when both variables are artificial dichotomies with data presented in a 2 × 2 table. Both variables are assumed to be normally distributed. The phi coefficient is used with 2 × 2 tables when the dichotomous variables are genuine.

The contingency coefficient is an index of the relationship between nominal variables when each is expressed in several categories.

Table 9.12 presents a summary of the types of correlation coefficients and the particular type of data for which they are appropriate.

TABLE 9.12 Summary of Types of Correlation Coefficients and the Corresponding Types of Data

CORRELATION COEFFICIENT	TYPE OF DATA
1. Pearson product-moment	1. Interval scales characteristic of both variables
2. Spearman rank	2. Ordinal scales characteristic of both variables
3. Point-biserial	3. One variable on interval scale; the other a genuine dichotomy
4. Biserial	4. One variable on interval scale; the other an artificially dichotomized normal variable
5. Tetrachoric	5. Both variables are artificial dichotomies with underlying continuous normal distributions
6. Phi	6. Genuine dichotomy characteristic of both variables
7. Coefficient of contingency	7. Nominal scale characteristic of both variables

FLOWCHART 9

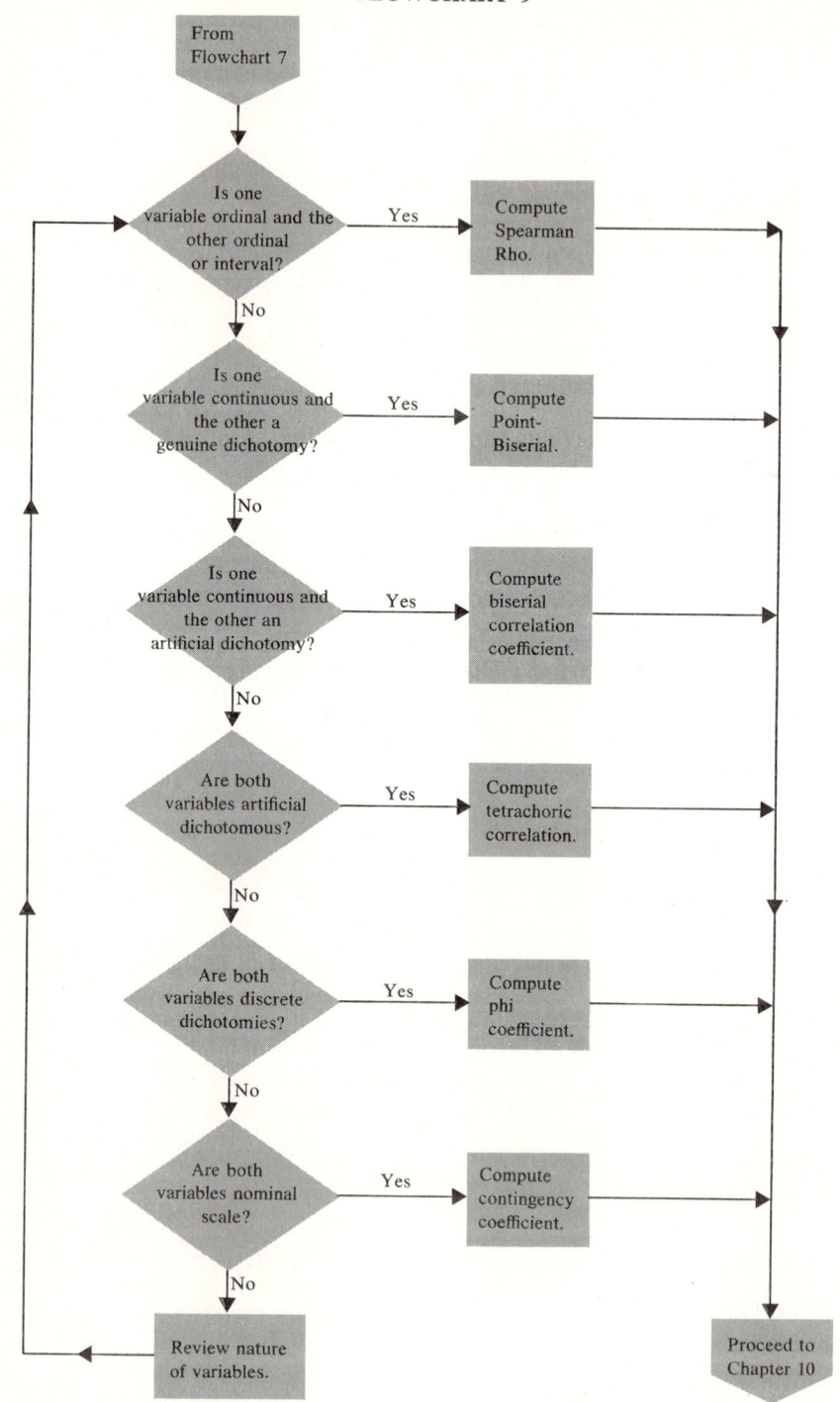

```
From Flowchart 7
      │
      ▼
┌─────────────────┐
│ Is one variable │  Yes  ┌──────────┐
│ ordinal and the │──────▶│ Compute  │──────▶
│ other ordinal   │       │ Spearman │
│ or interval?    │       │ Rho.     │
└─────────────────┘       └──────────┘
      │ No
      ▼
┌─────────────────┐
│ Is one variable │  Yes  ┌──────────┐
│ continuous and  │──────▶│ Compute  │──────▶
│ the other a     │       │ Point-   │
│ genuine         │       │ Biserial.│
│ dichotomy?      │       └──────────┘
└─────────────────┘
      │ No
      ▼
┌─────────────────┐
│ Is one variable │  Yes  ┌────────────┐
│ continuous and  │──────▶│ Compute    │──────▶
│ the other an    │       │ biserial   │
│ artificial      │       │ correlation│
│ dichotomy?      │       │ coefficient│
└─────────────────┘       └────────────┘
      │ No
      ▼
┌─────────────────┐
│ Are both        │  Yes  ┌────────────┐
│ variables       │──────▶│ Compute    │──────▶
│ artificial      │       │ tetrachoric│
│ dichotomous?    │       │ correlation│
└─────────────────┘       └────────────┘
      │ No
      ▼
┌─────────────────┐
│ Are both        │  Yes  ┌──────────┐
│ variables       │──────▶│ Compute  │──────▶
│ discrete        │       │ phi      │
│ dichotomies?    │       │ coeff.   │
└─────────────────┘       └──────────┘
      │ No
      ▼
┌─────────────────┐
│ Are both        │  Yes  ┌────────────┐
│ variables       │──────▶│ Compute    │──────▶
│ nominal scale?  │       │ contingency│
│                 │       │ coefficient│
└─────────────────┘       └────────────┘
      │ No
      ▼
┌─────────────────┐
│ Review nature   │
│ of variables.   │
└─────────────────┘

                          Proceed to Chapter 10
```

250

EXERCISES

1. In each of the following examples, identify the correlation coefficient appropriate for expressing the relationship between the two variables listed:
   (a) time spent in study and performance on a test dichotomized as pass-fail
   (b) age in months and height in inches
   (c) age in years and political preference dichotomized as Republican-Democrat
   (d) intelligence measured as above-average/below-average and performance on an examination measured as pass-fail
   (e) sex and intelligence test scores
   (f) marital status (married-unmarried) and attrition (drop-out or remain in school)
   (g) raw scores on a mathematics test and raw scores on a chemistry test
   (h) social maturity measured by ranks and reading readiness ranks for a group of kindergarten students
2. At a piano concerto competition two judges, a professional musician and an amateur musician, each ranked the contestants in order of best performance. How closely did the judges tend to agree with each other?

CONTESTANT	PROFESSIONAL	AMATEUR
A	4	1
B	1	6
C	3	7
D	6	4
E	2	5
F	7	10
G	9	8
H	10	2
I	5	3
J	8	9

3. As the director of admissions at an engineering school, you wish to investigate the relationship between sex and performance on a test of Spatial Relations. You have data for the following high school juniors. The number 1 represents a male student; 0 represents a female student. Calculate the appropriate coefficient.

INDIVIDUAL	SPATIAL RELATIONS SCORE	SEX
A	22	0
B	38	1
C	35	1
D	20	0
E	18	0
F	34	1
G	30	1
H	28	1
I	12	0
J	33	1

252 DESCRIPTIVE STATISTICS

4. A high school principal wants to determine whether students with high Verbal Reasoning scores tend to have a more favorable attitude in English courses than do students with lower Verbal Reasoning scores. An attitude scale given to the following ten students was scored as either favorable (1) or unfavorable (0). It is assumed that the distribution underlying the dichotomy is normal. Calculate the appropriate coefficient.

STUDENT	VERBAL REASONING SCORE	ATTITUDE
A	18	1
B	10	0
C	15	1
D	10	1
E	14	1
F	8	0
G	9	0
H	16	1
I	17	1
J	13	0

5. A basketball coach gathered dichotomous data on game performances (X) and a measure of peripheral vision (Y). He gathered data in the form of poor (0) and good (1) game performance and narrow (0) and wide (1) field of peripheral vision. It is assumed that there are normal distributions underlying the dichotomies. Calculate the appropriate coefficient.

		GAME PERFORMANCE POOR (0)	GAME PERFORMANCE GOOD (1)
VISUAL FIELD	WIDE (1)	15(a)	32(b)
	NARROW (0)	35(c)	18(d)

6. You wish to correlate the variables sex and whether or not one is a member of the National Honor Society. You have gathered the following data from your high school. Calculate the appropriate coefficient.

		SEX MALE (0)	SEX FEMALE (1)
MEMBER	YES (1)	28(a)	37(b)
	NO (0)	38(c)	32(d)

7. A high school principal would like to examine the relationship between teacher effectiveness and teacher grade-point average in college. He rated the 20

teachers in his school from most effective (1) to least effective (20). Calculate the appropriate coefficient.

TEACHER	EFFECTIVENESS	GPA
A	12	3.50
B	19	3.00
C	2	3.50
D	17	2.50
E	11	2.25
F	18	2.20
G	6	3.98
H	15	2.45
I	10	3.73
J	3	3.30
K	14	3.00
L	1	4.00
M	9	2.65
N	16	3.25
O	13	3.67
P	4	3.15
Q	8	3.25
R	20	2.80
S	7	3.05
T	5	2.90

8. A researcher is interested in anxiety and how it might affect performance on intelligence tests. He has a clinical psychologist assess the anxiety of subjects by ranking them from 1 through 20. He then administers a standardized intelligence test to each of the 20 subjects and converts their IQ scores to ranks. Which correlation coefficient should the researcher calculate for these data? Explain your answer.

9. An English teacher judged 12 essays (X) as either "below average" (0) or "above average" (1), and then gave a 100-item vocabulary test (Y) to each of the 12 students. Raw scores, the number of correct items, were used as a measure of Y. Which correlation coefficient is appropriate for this situation?

ANSWERS

1. (a) biserial
   (b) Pearson product-moment
   (c) point-biserial
   (d) tetrachoric
   (e) point-biserial
   (f) phi
   (g) Pearson product-moment
   (h) Spearman rho

254 DESCRIPTIVE STATISTICS

2.

CONTESTANT	PROFESSIONAL	AMATEUR	d	d^2
A	4	1	3	9
B	1	6	−5	25
C	3	7	−4	16
D	6	4	2	4
E	2	5	−3	9
F	7	10	−3	9
G	9	8	1	1
H	10	2	8	64
I	5	3	2	4
J	8	9	−1	1
$N = 10$				$\Sigma d^2 = 142$

$$\rho = 1 - \frac{6\Sigma d^2}{N(N^2-1)} = 1 - \frac{6(142)}{10(99)} = 1 - \frac{852}{990} = 1 - .86 = .14$$

There was not much agreement between the two judges.

3. Mean Test Score for Males: $\mu_p = \frac{198}{6} = 33$

Mean Test Score for Females: $\mu_q = \frac{72}{4} = 18$

$p = 6/10 = .60 \quad \sigma_x = 8.124$
$q = 4/10 = .40$

$$\rho_{pbi} = \frac{\mu_p - \mu_q}{\sigma_x} \sqrt{pq}$$

$$= \frac{33 - 18}{8.124} \sqrt{.24}$$

$$= \frac{15}{8.124}(.4899)$$

$$= (1.846)(.4899)$$

$$= .90$$

4. Mean Score for Favorable Attitude Group: $\mu_p = \frac{90}{6} = 15.0$

Mean Score for Unfavorable Attitude Group: $\mu_q = \frac{40}{4} = 10.0$

$p = \frac{6}{10} = .60 \qquad y = .386$

$q = \frac{4}{10} = .40 \qquad \sigma_x = 3.376$

$$\rho_{bi} = \frac{\mu_p - \mu_q}{\sigma_x} \times \frac{pq}{y}$$

$$= \frac{15-10}{3.376} \times \frac{(.6)(.4)}{.386}$$

$$= \frac{5}{3.376} \times \frac{.24}{.386}$$

$$= (1.48)(.62)$$

$$= .92$$

5. $bc = 1{,}120$
   $ad = 270$
   $\dfrac{bc}{ad} = \dfrac{1{,}120}{270} = 4.148$

   Since $bc > ad$, r_t will be $+$.

   We enter 4.148 in the table and find the r_t to be positive .51.

6. $\phi = \dfrac{bc - ad}{\sqrt{(a+b)(c+d)(a+c)(b+d)}}$

   $= \dfrac{(37)(38) - (28)(32)}{\sqrt{(65)(70)(66)(69)}}$

   $= \dfrac{1{,}406 - 896}{\sqrt{20{,}720{,}700}}$

   $= \dfrac{510}{4{,}551.9995}$

   $= .112$

The phi of .112 indicates that the relationship between the variables sex and membership in the National Honor Society is slight.

7.

TEACHER	EFFECTIVENESS	GPA	GPA RANK	d	d²
A	12	3.50	5.5	6.5	42.25
B	19	3.00	12.5	6.5	42.25
C	2	3.50	5.5	−3.5	12.25
D	17	2.50	17	0	0
E	11	2.25	19	−8	64
F	18	2.20	20	−2	4
G	6	3.98	2	4	16
H	15	2.45	18	−3	9
I	10	3.73	3	7	49
J	3	3.30	7	−4	16
K	14	3.00	12.5	1.5	2.25
L	1	4.00	1	0	0
M	9	2.65	16	−7	49
N	16	3.25	8.5	7.5	56.25

256 DESCRIPTIVE STATISTICS

TEACHER	EFFECTIVENESS	GPA	GPA RANK	d	d^2
O	13	3.67	4	9	81
P	4	3.15	10	−6	36
Q	8	3.25	8.5	−.5	.25
R	20	2.80	15	5	25
S	7	3.05	11	−4	16
T	5	2.90	14	−9	81
					$\Sigma d^2 = 601.5$

$$\rho = 1 - \frac{6 \Sigma d^2}{N(N^2-1)}$$

$$= 1 - \frac{6(601.5)}{20(399)}$$

$$= 1 - \frac{3,609}{7,980}$$

$$= 1 - .45$$

$$= .55$$

8. Since we have ordinal or rank-order data, the Spearman rank correlation coefficient should be calculated.
9. The biserial correlation coefficient, ρ_{bi}.

PART TWO
INFERENTIAL STATISTICS

CHAPTER TEN
THE LOGIC OF INFERENTIAL STATISTICS

OBJECTIVES

After studying this chapter the reader should be able to:

1. Describe the general strategy of inferential statistics.
2. Define null hypotheses.
3. Distinguish between Type I and Type II errors.
4. Define alpha.
5. Describe the process involved in selecting alpha.
6. Distinguish between situations leading to the rejection of a null hypothesis and situations leading to the retention of a null hypothesis.
7. Describe the general nature of the binomial expansion.
8. Employ Table A3 to identify probabilities for small samples when $P = Q = .5$.
9. Describe the use of the normal curve to approximate probabilities.
10. Calculate approximate probabilities using the normal curve.

Introduction

We shall begin this chapter with a fable intended to introduce the general nature of inferential statistics.

The Potion and the Potentate

Once upon a time an oriental potentate was sorely troubled because his wife had produced for him many daughters but no sons. The law of the land

260 INFERENTIAL STATISTICS

declared that only sons could inherit the throne. Therefore, it troubled the potentate to ponder the fate awaiting his people if he should have no sons and the throne should pass to his wicked cousin.

About this time a Yankee peddler appeared in court saying, "I have in my hand a potion which, when taken diligently daily by a man and his wife, greatly increases the probability that the wife will bring forth sons instead of daughters. I am willing to sell you this potion for the paltry sum of $100,000."

"By what signs can I know that this potion really does increase the likelihood of male heirs, as you claim?" asked the potentate.

"My wife and I took the potion daily and we now have a son," replied the peddler.

"This is not reason enough to pay $100,000 for your potion when I know that 50 percent of all children born are males. Therefore you and your wife had a 50 percent chance of having a son through chance alone," responded the potentate. "Is that all the evidence you can give me?"

The peddler replied, "My brother and sister-in-law also took the potion and they too have a son. No one else has yet taken the potion, so I can say to you that all the couples who have taken the potion have had male heirs."

At this point the potentate called for his royal statistician and asked him, "How likely is it that two men shall each beget a male heir through chance alone, since it has long been known that we can assume that there is a 50-50 chance of either son or daughter in each case?"

The royal statistician replied, "Sire, we learned statisticians would say, 'Given the a priori (assumed) probability of .5 for a son and .5 for a daughter we can determine the probability of sons or daughters in any number of cases through the use of deductive logic.'"

If a couple has .5 probability of begetting a son, the probability of two couples begetting a son is $.5^2 = .25$, the probability of the first begetting a son and the second a daughter is $.5^2 = .25$. The probability of the first begetting a daughter and the second a son is $.5^2 = .25$, the probability of the first a daughter and the second a daughter is $.5^2 = .25$. Summing these probabilities we have 25 percent chance of two sons, 50 percent chance of one son and one daughter, and 25 percent chance of two daughters.

The potentate said, "My royal treasurer would not permit me to give this Yankee peddler $100,000 for a potion when he had a 25 percent chance of getting his result through chance alone. Is there some way we can gather more evidence concerning the worth of this potion?"

"Verily, sire," responded the royal statistician, "we have many young officers among the palace guard who would be delighted to beget sons. We could ask for volunteers from among them to test the potion."

"Then would we know for a certainty whether the potion works or not?" asked the potentate.

"I regret to say, my lord, that in this world we can never certainly know whether observed events such as the births of sons or daughters are truly caused by some variable such as use or nonuse of a potion or whether such events are chance and chance alone. None of us will know the ultimate truth concerning whether the potion works or not until we join our revered ancestors in heaven."

"Then why," demanded the potentate, "are you suggesting that I ask my officers and their wives to test the potion? You tell me that no matter what I do I must remain in ignorance concerning the efficacy or lack thereof of the potion until I join my ancestors in heaven. By then it will be too late to beget sons and my wicked cousin will ascend the throne."

"Wise men have taught us that all men must live with their ignorance and no mortal can know ultimate truth. But although we cannot know ultimate truth, our senses can give us knowledge and that knowledge can lead us in making reasonable decisions."

"We statisticians are fond of saying, 'inferential statistics is the science of making reasonable decisions with limited information.' Let me illustrate this with the present problem. If 20 couples took the Yankee's potion and all 20 had male heirs is it possible that such a thing could occur by chance alone?"

"It is possible," answered the potentate, "but I think it unlikely that even a Yankee peddler would be favored by fortune 20 times in a row."

"Truly said, oh wise ruler, we statisticians have computed such possibilities and have determined that if we assume that a male heir and a female heir are equally likely in any birth, then the probability of 20 males in 20 births is $(\frac{1}{2})^{20}$, or one chance in 1,048,576. We know that the chances of obtaining 19 sons and 1 daughter is one chance in 52, 429, still a possible but highly unlikely outcome with pure luck. Recall, the Yankee peddler did not say his potion guaranteed male heirs, he said only that it increased the probability of male heirs."

"I see," said the potentate, "being mortal, neither I nor any other man can truly say whether the Yankee's potion really does what he claims. However, I *can* determine how likely various outcomes would be, given chance alone. Then I can use these probabilities in order to reach a reasonable decision on this matter. I have already said that two sons in two births is not sufficient evidence to lead me to pay this Yankee a great sum of money for his potion, since there is one chance in four that this could occur by chance. But if 20 couples take the potion and have 19 sons and 1 daughter among them, I know that this outcome is so unlikely to be due to mere chance that it would be reasonable to buy the potion, even though I can never be absolutely certain that the potion really works."

"That is well said, Enlightened Ruler," exclaimed the royal statistician. "Now if the Yankee peddler will agree to let us test his potion you can make a decision that, though fallible, shall be reasonable."

262   INFERENTIAL STATISTICS

The peddler agreed to the test but protested that he could only spare enough potion for ten couples. Ten officers and their wives began taking the potion and the test began.

Type I and Type II Errors   During the months he was awaiting the outcome of the test the potentate again summoned his royal statistician saying, "I have been pondering the consequences of the decisions I may make concerning the Yankee peddler's potion. Each decision could be either right or wrong. I have prepared a chart to illustrate this:

	IN TRUTH THE POTION:	
	(a) WORKS	(b) DOES NOT WORK
1. I BUY THE POTION	right	wrong
2. I DO NOT BUY THE POTION	wrong	right

Now there are two ways I may be right and two ways I may be wrong. If I pay the Yankee peddler $100,000 for the potion I am right if it works but wrong if it doesn't. If I do not buy the potion I am right if it doesn't work but wrong if it does. If I make the one error I will have spent money foolishly and will lose face with my people. If I make the other I will fail to buy the potion when it would have been to my benefit to do so."

"Oh, splendid ruler!" exclaimed the royal statistician, "You in your wisdom have devised in your own mind in a short span of days what it took learned statisticians many years to devise, the distinction between Type I and Type II errors. A Type I error would occur in this situation if you conclude that a relationship exists between variables when no genuine relationship exists. In your case you would make a Type I error if you concluded there was a relationship between the variable 'taking or not taking the potion' and the variable 'proportion of male heirs.' A Type II error would occur if you conclude there is no relationship between variables when there genuinely is such a relationship."

"The chart that learned statisticians have devised is like your own chart but is designed to illustrate any case rather than a specific case. It is based on the truth or falsity of any null hypothesis."

"Explain to me," said the potentate, "this strange thing you call a null hypothesis."

The statistician then expounded, "The null hypothesis is the statement that there are no differences among population parameters. In your case the null hypothesis states: the parameter 'proportion of sons' is no different for a population of people who take the potion than for a population of people who do not take the potion. Since we know that in the population at large the proportion of sons is .50, the null hypothesis states that the proportion of sons among a population taking the potion is also .50."

"Although there are various methods of formulating a null hypothesis, the most common method is to determine what population value would be expected if there is no relationship between variables. In the present case if there is no relationship between the variable 'taking or not taking the potion' and the variable 'proportion of sons' we would expect the proportion of sons among a population of potion takers to be the same as that proportion in the general population (that is, those who have not taken the potion)."

"Next we consider your officers and their wives as a sample of the population, all people who might take the potion. Then we determine the probability of various outcomes in our sample if the null hypothesis is true. In the present case we would ask 'How likely are various proportions of sons in a sample of ten when the true population proportion is .50?'"

"The use of the null hypothesis enables us to determine whether there is sufficient evidence for tentatively concluding that there is a relationship between variables or not. If the proportion of sons in your sample is .50 or is near enough to .50 that the difference between the sample proportion and .50 could easily be due to chance you will decide that there is not sufficient evidence to conclude that in a population of couples taking the potion the proportion is other than .50. We statisticians call this 'retaining the null hypothesis.'"

"If the null hypothesis, although possible, seems a sufficiently unlikely explanation of the observations, you would be wise to reject it and to declare with cautious confidence that an actual relationship exists between variables. Your highness has already said that if 20 couples should take the potion and have 20 sons, since this would occur by chance alone only once in 1,048,576 cases, you would buy the peddler's potion. In other words, you would reject the null hypothesis and decide that there probably is a relation between the one variable, taking or not taking the potion, and the other variable, the proportion of sons and daughters. We call this rejecting the null hypothesis."

"I see," replied the potentate, "I have also declared that the fact that the peddler and his brother-in-law and their wives took the potion and begot sons is not sufficient evidence to convince me to buy the potion since there was one chance in four that this could happen through chance alone."

Alpha "Verily, Sire, you have determined that you will retain the null hypothesis when an event has a .25 probability of occurring by chance alone but reject the null hypothesis if it has only a .0000001 probability of occurring by chance alone."

The potentate interrupted impatiently, "Your words sound wise, but I perceive not sufficient wisdom to tell me at what level of probability I should change from retention to rejection of my null hypothesis."

The statistician responded soothingly, "O Great One, this dividing point between probabilities leading to the rejection of the null hypothesis and probabilities leading to its retention is what we statisticians call alpha (α).[1] Alpha is the probability of a Type I error that one is willing to accept. It is set according to the relative seriousness of Type I and Type II errors."

"Let us look, Sire, at your chart and the chart we statisticians use and ponder the relative seriousness of Type I and Type II errors in your case," continued the royal statistician. THE NULL HYPOTHESIS IS IN FACT:

	FALSE	TRUE
THE INVESTIGATOR REJECTS THE NULL HYPOTHESIS	Correct	Type I error
THE INVESTIGATOR RETAINS THE NULL HYPOTHESIS	Type II error	Correct

"If you reject the null hypothesis your action is based on the assumption that there is a genuine relationship between taking or not taking the potion and the proportion of male children. If there really is such a relationship you are right. If you reject the null hypothesis when there is really no relationship you will make a Type I error. A Type I error in your case would mean buying the potion when it does not work.

"If you retain the null hypothesis your action is based on the assumption that there is no relationship between variables, that is, the potion does not work. If the potion in truth does not work you are right, but if in truth it does work, you will commit a Type II error. A Type II error would mean you fail to buy the potion when taking the potion would increase the probability of having a son. A Type II error occurs when one retains a false null hypothesis. Now you must determine the relative seriousness of Type I and Type II errors."

"Verily," responded the potentate, "with a Type I error I pay $100,000 for a worthless potion. With a Type II error I fail to buy a remedy for my dilemma. Methinks for a man in my position a Type I error would be more serious. Not only would I deplete the treasury, I would lose face with my people for spending money foolishly. Unstable is the throne of a ruler whose people think he is foolish."

"A Type II error would be sad indeed. I would miss an opportunity for the queen and myself to produce a male child. However, many wise rulers

[1] *This is an entirely different use of alpha than the use we encountered in Chapter 8 where it represented the Y-intercept constant. There just are not enough Greek letters to represent all the indexes statisticians need; so frequently the same letter will have entirely different meanings in various contexts.*

before me have employed a different solution. They have divorced wives who bear no sons and have taken new wives. My people would not judge me foolish for doing this."

"A queen's alimony would be much more than the $100,000 the peddler asks but I deem a Type II error to be less serious than a Type I error in this case. With a Type I error I lose face with my people, with a Type II error I do not."

"Therefore, I would be willing to take only 1 chance in 100 of a Type I error."

"Very good, Sire," responded the royal statistician. "One must first consider the relative seriousness of a Type I or Type II error in the particular circumstances of each case. You have decided that for you a Type I error would be much more serious, and, as we would say in our profession, you have set your alpha at the .01 level (1 chance in 100)."

"Let me now show you a chart which shows the probabilities of sons and daughters in ten trials under a true null hypothesis, assuming the probability of a son is .5 and the probability of a daughter is .5."

NUMBER OF SONS	NUMBER OF DAUGHTERS	PROBABILITY
10	0	.001
9	1	.010
8	2	.044
7	3	.117
6	4	.205
5	5	.245
4	6	.205
3	7	.117
2	8	.044
1	9	.010
0	10	.001

"Since you have chosen an alpha of .01 you are committed to rejection of the null hypothesis if your officers and their wives have ten sons and no daughters because the probability of this occurring by chance alone is less than the .01 alpha you have chosen. With any other outcome you are committed to retention of the null hypothesis."

"Shouldn't I also reject the null hypothesis if we see nine sons and one daughter?" asked the potentate. "The Yankee peddler did not claim his potion guaranteed a male child; he merely said it increased the likelihood of a male child. Your chart shows that the probability of nine sons and one daughter is .010, exactly equal to my alpha."

"Sire," replied the royal statistician, "when deciding whether to retain or reject a null hypothesis one must consider not only the probability of a particular event, one must combine this probability with the probabilities of all other less likely events and use this total sum as the basis for making one's decision. Thus, although the probability of nine sons is .01, exactly equal to your alpha, you must combine this with the probability of ten sons, .001, giving a total probability of .011. Since this exceeds your predetermined alpha of .01 you cannot reject the null hypothesis if nine sons and one daughter are born."

"If you had decided to take two chances in 100 of a Type I error you would have set your alpha at .02. In this case you could reject the null hypothesis if either nine or ten sons are born."

"Let me ponder that," responded the potentate.

The distraught queen, hearing of the potentate's decision to set his alpha at .01 realized that this meant that if the ten officers had even one daughter the potentate would be committed to retain the null hypothesis and as a consequence he would be committed to divorce her. She pleaded with her lord and master, "Loving husband, since the choice of a value for alpha is arbitrary, and since for me the consequence of a Type II error would be far more serious than would the consequence of a Type I error, pray consider the possibility of selecting a less rigorous value for alpha."

Moved by her pleas and tears and by the logic of her argument the potentate graciously declared his willingness to take one chance in ten of a Type I error. That is, he agreed to set his alpha at .10. He therefore declared that if the officers and their wives had eight or more sons he would buy and use the Yankee peddler's potion. However, if they had seven or fewer male children and null hypothesis would be retained.

There is no formula for determining alpha. Alpha is an arbitrary probability point which divides those probabilities that will lead to retention of the null hypothesis from those probabilities that will lead to its rejection. One chooses an alpha by weighing the relative seriousness of the consequences of Type I and Type II errors.

Conclusion The officers and their wives had nine sons and one daughter. Since the probability of nine or more sons occurring through chance alone is .011, less than the agreed alpha of .10, the potentate rejected the null hypothesis and declared that the evidence was sufficient to tentatively conclude that the potion did actually increase the chances of producing a son.

However, in the meantime, the queen had become pregnant again and the potentate saw no reason to spend $100,000 until the result of this pregnancy was known. The queen triumphantly gave birth to twin sons, to the delight of the potentate but to the dismay of the Yankee peddler and the wicked cousin.

The potentate forthwith sent the peddler on his way. However, despite his disappointment over the loss of a profitable sale the peddler was not completely discouraged. The experiment with the officers and their wives had, at no cost to himself, provided him with reasonable evidence of the effectiveness of his potion. To this day he is seeking a sonless potentate with an alpha of .011 or greater.

The peddler and the potentate story illustrates the basic rationale of inferential statistics.

1. In scientific inquiry nothing is regarded as an absolute certainty. Scientific conclusions are probability statements rather than "facts."
2. The use of the null hypothesis, which is usually the statement that chance alone can account for any apparent relationship between variables, sets the stage for determining whether observed events provide sufficient evidence for declaring that a relationship probably does exist between variables.
3. Before running an experiment, the investigator determines how unlikely the null hypothesis must be before he will reject it and therefore tentatively declare that a genuine relationship between variables probably exists. He sets this required level of probability (alpha) according to the relative seriousness of Type I and Type II errors.

In our example a Type I error was deemed to be more serious than a Type II error. This is usually but not always the case. For example, the U.S. Department of Defense has frequently argued that to incorrectly conclude that a weapons system does not work and therefore fail to spend a few billion dollars for it (Type II error) is a more serious mistake than to conclude the system does work and buy it when it in fact does not work (Type I error).

Alpha is always the probability of Type I error one is willing to risk. If a Type I error is thought to be very serious a "rigorous" alpha such as .01, .001, or less is selected. If a Type I error is not thought to be very serious a more "relaxed" alpha such as .10, .20 or greater is selected. The majority of research in education and psychology has employed alphas of .05 or .01. These alphas give reasonable protection against Type I errors without excessive risk of Type II errors.

As alpha is decreased the risk of a Type I error is decreased but the risk of a Type II error is increased and vice versa. For example, changing alpha from .05 to .01 decreases the risk of a Type I error but increases the probability of a Type II error, while changing alpha from .01 to .05 increases the probability of a Type I error but decreases the probability of a Type II error.

Journal editors in the field of education and psychology generally expect investigators reporting research to have employed alphas of .05 or less. Given the competition for space in professional journals, there is some reluctance to publish findings which have a high probability of being a result of chance.

268 INFERENTIAL STATISTICS

4. If the probability of the observed events occurring when the null hypothesis is true is equal to or less than the predetermined alpha, the investigator rejects the null hypothesis and declares that there probably is a genuine relationship between variables. If the probability of the observed events occurring when the null hypothesis is true is greater than alpha, the investigator retains the null hypothesis and consequently declares that the evidence is not sufficient for concluding that a genuine relationship exists between the variables.

Exercises

1. Inferential statistics enable researchers to:
   (a) reach infallible conclusions
   (b) reach reasonable conclusions with incomplete information
   (c) add an aura of legitimacy to what is really sheer guesswork
2. What do we call the statement that there is no genuine relationship between variables and any apparent relationship between variables is a result of chance?
3. What two conditions are necessary for a Type I error to occur?
4. Which of the following statements describes the role of the null hypothesis in research?
   (a) It enables us to determine the probability of an event occurring through chance alone when there is no real relationship between variables.
   (b) It enables us to prove there is a real relationship between variables.
   (c) It enables us to prove there is no real relationship between variables.
5. A Type II error occurs when one:
   (a) rejects a false null hypothesis
   (b) rejects a true null hypothesis
   (c) has already made a Type I error
   (d) retains a false null hypothesis
   (e) retains a true null hypothesis
6. The term "alpha" refers to:
   (a) the probability of an event being due to chance alone which is calculated after the data from an experiment are analyzed
   (b) the probability of a Type I error that an investigator is willing to accept
   (c) the actual probability of a Type II error
   (d) the probability of a Type II error that an investigator is willing to accept
7. How does one determine alpha in an experiment?
8. A cigarette manufacturer has employed researchers to compare the lung cancer rate among smokers and nonsmokers. Considering the results of previous research on this question, the manufacturer would

probably urge his researchers to be especially careful to avoid making a:
   (a) Type I error
   (b) Type II error
9. When an investigator rejects the null hypothesis he concludes that:
   (a) the results are necessarily meaningful and useful
   (b) the results are probably not a function of chance alone
   (c) the results were what was predicted by the hypothesis
   (d) the size of the difference between the groups was large
   (e) there is absolute proof of a relationship
10. The level of probability required for rejecting the null hypothesis is determined:
   (a) before running the experiment
   (b) while running the experiment
   (c) after the test of significance has been made
   (d) by randomly selecting from the possible levels of significance available

Solutions

1. (b)
2. Null hypothesis
3. The null hypothesis must be true and the investigator must reject it.
4. (a)
5. (d)
6. (b)
7. By weighing the consequences of Type I and Type II errors.
8. (a)
9. (b)
10. (a)

Binomial Probability

There are various ways we can employ deductive logic to arrive at probabilities based on a priori assumptions. In our example, for instance, if we assume that male and female heirs are equally likely we could construct a tree diagram showing the likelihood of three couples having various combinations of sons and daughters. (See Figure 10.1.)

The tree is based on the assumption that in any birth there is a 50 percent chance of a son and a 50 percent chance of a daughter. Deductive logic tells us that given a 50 percent chance of the first couple having a son and a 50 percent chance of the second couple having a son the probability of both having sons is $(.50)(.50) = .25$, and so on. When we look at the possible

270 INFERENTIAL STATISTICS

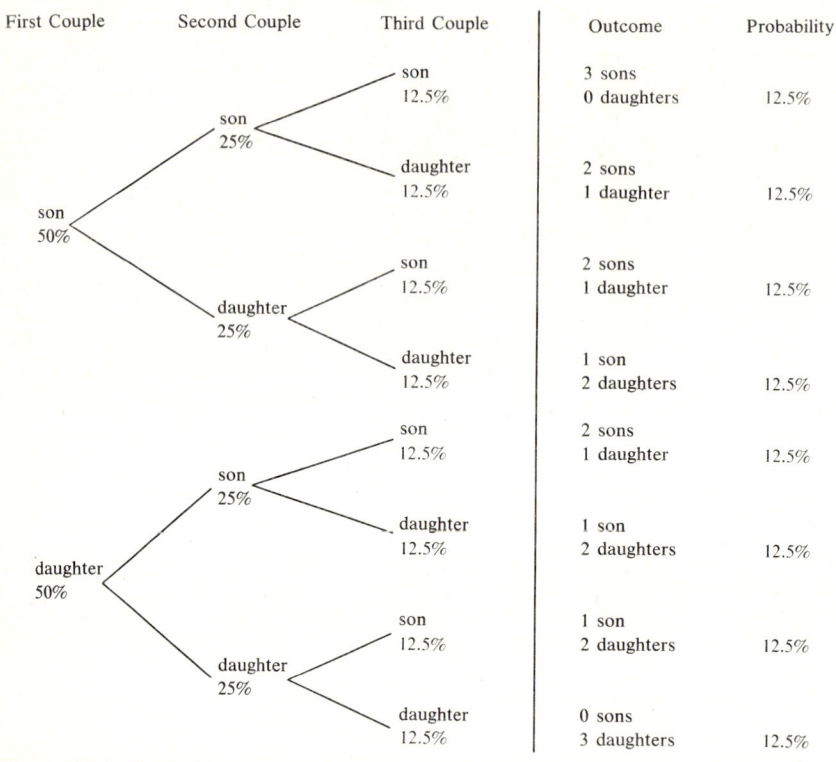

First Couple	Second Couple	Third Couple	Outcome	Probability
son 50%	son 25%	son 12.5%	3 sons 0 daughters	12.5%
		daughter 12.5%	2 sons 1 daughter	12.5%
	daughter 25%	son 12.5%	2 sons 1 daughter	12.5%
		daughter 12.5%	1 son 2 daughters	12.5%
daughter 50%	son 25%	son 12.5%	2 sons 1 daughter	12.5%
		daughter 12.5%	1 son 2 daughters	12.5%
	daughter 25%	son 12.5%	1 son 2 daughters	12.5%
		daughter 12.5%	0 sons 3 daughters	12.5%

Figure 10.1 Probabilities of Sons and Daughters for Three Couples

outcomes for three couples we find there is only one of the eight possible outcomes which has three sons and zero daughters. The probability of three couples having three sons and no daughters is one chance in eight, or 12.5 percent. There are three outcomes which yield two sons and one daughter, so the probability of this observation is three chances in eight, or 37.5 percent. In the same manner we find the probability of one son and two daughters is 37.5 percent, and the probability of no sons and three daughters is 12.5 percent. The probabilities of the four different observations total 100 percent.

```
      3 sons  0 daughters:  12.5 percent
      2 sons  1 daughter:   37.5 percent
      1 son   2 daughters:  37.5 percent
      0 sons  3 daughters:  12.5 percent
               Total:       100 percent
```

Let us construct a tree to show the likelihood of getting from zero to three items correct through sheer guessing in three five-option multiple choice test items. Here we assume the probability of getting any item right

The Logic of Inferential Statistics 271

through luck alone is one chance in five (20 percent) and the probability of getting an item wrong through luck alone is four chances in five (80 percent). (See Figure 10.2.) Adding the outcomes we get:

3 right 0 wrong	.8 percent
2 right 1 wrong	9.6 percent
1 right 2 wrong	38.4 percent
0 right 3 wrong	51.2 percent
Total	100.0 percent

Note that the sum of the probabilities of all possible outcomes is always equal to unity (1.000), or 100 percent.

The tree diagram for these probabilities looks like the tree diagram for probabilities of $P = .5$ (Figure 10.1) but it differs in that the "branches" are not equally probable.

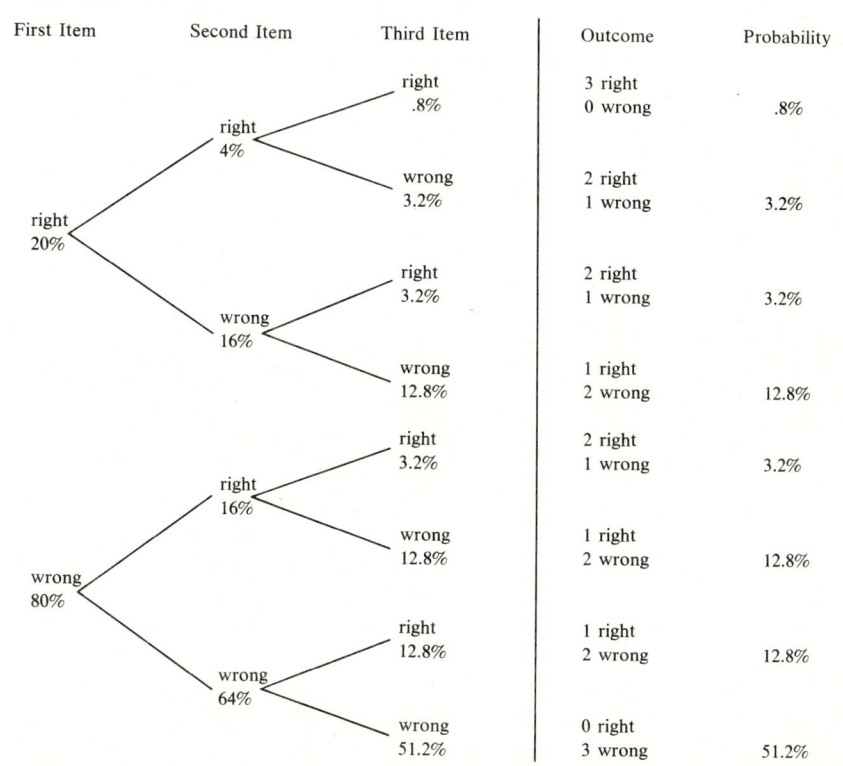

First Item	Second Item	Third Item	Outcome	Probability
right 20%	right 4%	right .8%	3 right 0 wrong	.8%
		wrong 3.2%	2 right 1 wrong	3.2%
	wrong 16%	right 3.2%	2 right 1 wrong	3.2%
		wrong 12.8%	1 right 2 wrong	12.8%
wrong 80%	right 16%	right 3.2%	2 right 1 wrong	3.2%
		wrong 12.8%	1 right 2 wrong	12.8%
	wrong 64%	right 12.8%	1 right 2 wrong	12.8%
		wrong 51.2%	0 right 3 wrong	51.2%

Figure 10.2 Probabilities of Various Scores on a Three-Item Multiple Choice Test

272 INFERENTIAL STATISTICS

The Binomial Expansion

Using a tree diagram to determine probabilities can become terribly unwieldy whenever more than a very few cases are to be considered. In our potentate and potion example, for instance, in order to diagram a tree to show the probabilities in just ten cases we would need to have two to the tenth power, 2^{10}, or 1,024 branches in the right-hand column. Fortunately, there are mathematical procedures available that enable us to calculate probabilities without constructing a diagram.

One of these procedures is the formula for determining the frequency of various combinations that would occur in our tree diagram. In our example this formula will enable us to determine how many of the 1,024 branches in our tree diagram would be ten sons and zero daughters, how many would be nine sons and one daughter, and so on.

The formula for combinations is:

$$C = \frac{n!}{r! \cdot (n-r)!}$$ where: C = the number of times an outcome will occur
n = number of trials. In our example we are concerned with the likelihood of various combinations of sons and daughters in ten births or ten trials.
r = the number of "desired" outcomes in a series of trials. If we ask the probability of four sons, $r = 4$; if we ask the probability of nine sons, $r = 9$. The value of r may be set at any whole number between zero and n.
! = factorial

Formula (10.1)
Combinations

The term factorial (!) means multiply the number by the next smaller whole number, then multiply that product by the next smaller whole number, and so on, until we finally multiply by 1. For example,

$$\begin{aligned} \text{five factorial } (5!) &= 5 \cdot 4 \cdot 3 \cdot 2 \cdot 1 \\ &= 20 \cdot 3 \cdot 2 \cdot 1 \\ &= 60 \cdot 2 \cdot 1 \\ &= 120 \cdot 1 = 120 \end{aligned}$$

The factorial of a number tells us the number of arrangements that can be made with that number. For example, a hostess assigning six guests to six chairs has 6! (six factorial) possible ways to seat them. For the first guest she has six chairs from which to choose, for the second guest there are only five chairs from which to choose since the first guest is already seated. Therefore, she has $6 \times 5 = 30$ different ways to seat the first two guests. For the third guest there are four possible chairs remaining, for the fourth there is a choice among three chairs, for the fifth the choice is between two chairs, and for the last

The Logic of Inferential Statistics 273

guest only one chair remains. So there are $6 \cdot 5 \cdot 4 \cdot 3 \cdot 2 \cdot 1 = 720$ possible arrangements for seating six people in six chairs. One can seat ten guests in ten chairs 10! or 3,628,800 different ways.

Zero factorial (0!) by definition is always equal to one. In our example the frequency of the various combinations would be:

ten sons: $\dfrac{10!}{10! \cdot (10-10)!} = \dfrac{10!}{10!0!} = \dfrac{10!}{10!} = 1$ combination

nine sons: $\dfrac{10!}{9! \cdot (10-9)!} = \dfrac{10!}{9! \cdot 1!} = \dfrac{10 \cdot 9!}{9! \cdot 1!} = \dfrac{10}{1} = 10$ combinations

eight sons: $\dfrac{10!}{8! \cdot (10-8)!} = \dfrac{10!}{8! \cdot 2!} = \dfrac{10 \cdot 9 \cdot 8!}{8! \cdot 2 \cdot 1} = \dfrac{90}{2} = 45$ combinations

seven sons: $\dfrac{10!}{7! \cdot (10-7)!} = \dfrac{10!}{7! \cdot 3!} = \dfrac{10 \cdot 9 \cdot 8 \cdot 7!}{7! \cdot 3 \cdot 2 \cdot 1} = \dfrac{720}{6} = 120$ combinations

six sons: $\dfrac{10!}{6! \cdot (10-6)!} = \dfrac{10!}{6! \cdot 4!} = \dfrac{10 \cdot 9 \cdot 8 \cdot 7 \cdot 6!}{6! \cdot 4 \cdot 3 \cdot 2 \cdot 1} = \dfrac{5{,}040}{24} = 210$ combinations

five sons: $\dfrac{10!}{5! \cdot (10-5)!} = \dfrac{10!}{5! \cdot 5!} = \dfrac{10 \cdot 9 \cdot 8 \cdot 7 \cdot 6 \cdot 5!}{5! \cdot 5 \cdot 4 \cdot 3 \cdot 2 \cdot 1} = \dfrac{30{,}240}{120} = 252$ combinations

four sons: $\dfrac{10!}{4! \cdot (10-4)!} = \dfrac{10!}{4! \cdot 6!} = \dfrac{10 \cdot 9 \cdot 8 \cdot 7 \cdot 6 \cdot 5 \cdot 4!}{4! \cdot 6 \cdot 5 \cdot 4 \cdot 3 \cdot 2 \cdot 1} = \dfrac{15{,}120}{720} = 210$ combinations

three sons: $\dfrac{10!}{3! \cdot (10-3)!} = \dfrac{10!}{3! \cdot 7!} = \dfrac{10 \cdot 9 \cdot 8 \cdot 7 \cdot 6 \cdot 5 \cdot 4 \cdot 3!}{3! \cdot 7 \cdot 6 \cdot 5 \cdot 4 \cdot 3 \cdot 2 \cdot 1} = \dfrac{720}{6} = 120$ combinations

two sons: $\dfrac{10!}{2! \cdot (10-2)!} = \dfrac{10!}{2! \cdot 8!} = \dfrac{10 \cdot 9 \cdot 8 \cdot 7 \cdot 6 \cdot 5 \cdot 4 \cdot 3 \cdot 2!}{2! \cdot 8 \cdot 7 \cdot 6 \cdot 5 \cdot 4 \cdot 3 \cdot 2 \cdot 1} = \dfrac{90}{2} = 45$ combinations

one son: $\dfrac{10!}{1! \cdot (10-1)!} = \dfrac{10!}{1! \cdot 9!} = \dfrac{10 \cdot 9 \cdot 8 \cdot 7 \cdot 6 \cdot 5 \cdot 4 \cdot 3 \cdot 2 \cdot 1!}{1! \cdot 9 \cdot 8 \cdot 7 \cdot 6 \cdot 5 \cdot 4 \cdot 3 \cdot 2 \cdot 1} = \dfrac{10}{1} = 10$ combinations

zero sons: $\dfrac{10!}{0! \cdot 10!} = \dfrac{10!}{10!} = 1$ combination

In our example the tree would have 2^{10} or 1,024 branches or possible outcomes; one of these is ten sons, zero daughters, so there is one chance in 1,024 of this outcome. Ten branches have nine sons and one daughter so the probability of this outcome is 10/1,024, the probability of eight sons is 45/1,024, and so on. In this example the probability of a particular outcome is the number of combinations associated with that outcome divided by the total of all possible outcomes.

If we add the probabilities of each combination from ten to zero sons we have 1,024/1,024, or 1. It is always true that whenever we add the probabilities of all possible outcomes in a series of trials the sum is unity (one).

The above procedure can be used to calculate the probabilities of various outcomes in a number of trials whenever the probability of a "desired" outcome in each single trial is .5, as would be the case in the potion and potentate example. Whenever the probability of a "desired" outcome in each

274 INFERENTIAL STATISTICS

single trial is other than .5 we must add an additional step to the combinations formula in order to calculate the probabilities of various outcomes in a number of trials. Such a situation would be encountered by a crap shooter attempting to throw a three on a single roll of a die, because there are five other possible outcomes, thus the probability of a three is only one in six.

The formula used whenever the probability of a "desired outcome" in each trial is other than .5 is the binomial probability formula:

$$\text{Probability} = \frac{n!}{r! \cdot (n-r)!} \cdot P^r \cdot Q^{n-r}$$

where: n = the number of trials
r = the number of "desired" outcomes in a series of trials
P = the probability of the "desired" outcomes in a single trial
Q = the probability of an outcome other than the "desired" outcome in a single trial
Q is always equal to $1 - P$

Formula (10.2)[2]
Binomial Probability

Notice that this formula incorporates the formula for combinations.

To illustrate the use of this formula let us compute the probabilities shown on Figure 10.2 for a three-item multiple choice test where each item has five options. In this example:

$P = .2$ There is one possibility in five of getting a correct answer by sheer guessing in each single five-option multiple choice item.

$Q = .8$ There are four possibilities in five of choosing the wrong option in each single item through guessing.

$n = 3$ We are concerned with the probabilities of various outcomes with three items.

$r = 0$ to 3   We can compute the probability of from zero to three items correct.

The probability of getting all three correct is:

$$\text{Probability} = \frac{3!}{3! \cdot (3-3)!} \cdot .2^3 \cdot .8^{3-3}$$

$$= \frac{3!}{3! \cdot 0!} \cdot .008 \cdot 1*$$

$$= .008$$

Computing the probabilities of all possible outcomes we have:

---

[2] *This formula works also when P and Q both equal .5. However, in this case it is not necessary to calculate $P^r \cdot Q^{n-r}$ as this value always equals $1/2^n$ when P and Q both equal .5.*

*In mathematics any number to the zero power is equal to one.

The Logic of Inferential Statistics 275

Three correct: $\dfrac{3!}{3! \cdot (3-3)!} \cdot .2^3 \cdot .8^{3-3} = \dfrac{3!}{3! 0!} \cdot .2^3 \cdot .8^0 = 1 \cdot .008 = .008$

Two correct: $\dfrac{3!}{2! \cdot (3-2)!} \cdot .2^2 \cdot .8^{3-2} = \dfrac{3!}{2! \cdot 1!} \cdot .2^2 \cdot .8^1 = 3 \cdot .032 = .098$

One correct: $\dfrac{3!}{1! \cdot (3-1)!} \cdot .2^1 \cdot .8^{3-1} = \dfrac{3!}{1! \cdot 2!} \cdot .2^1 \cdot .8^2 = 3 \cdot .128 = .384$

Zero correct: $\dfrac{3!}{0! \cdot (3-0)!} \cdot .2^0 \cdot .8^{3-0} = \dfrac{3!}{0! \cdot 3!} \cdot .2^0 \cdot .8^3 = 1 \cdot .512 = .512$

Given this formula we can find the probability of any number of "desirable" outcomes within n trials.

We can calculate the probability of guessing exactly three (r) correct answers in a ten (n) item test when the student has .2 (P) chances of getting any item right and .8 (Q) chances of getting any item wrong.

$$\text{Probability of three correct} = \dfrac{10!}{3! \cdot (10-3)!} \cdot .2^3 \cdot .8^7$$

$$= \dfrac{10 \cdot 9 \cdot 8 \cdot 7!}{3 \cdot 2 \cdot 1 \cdot 7!} \cdot .008 \cdot .2097152$$

$$= \dfrac{720}{6} \cdot .0016777216$$

$$= .201326592$$

The probability of getting all ten correct by guessing is:

$$\text{Probability} = \dfrac{10!}{10! \cdot (10-10)!} \cdot .2^{10} \cdot .8^0$$

$$= 1 \cdot (.2)^{10} \cdot 1$$

$$= .0000001$$

It is not impossible for a student to get all ten items correct through guessing but he has only about one chance in ten million of doing so.

Thus, we can compute the exact probability of each r number of successes within the n events. If we compute the probabilities of all r's the sum of these probabilities is unity (1.000).

We see this by inspecting the table below which shows the probability of getting 0 to 10 items correct. These probabilities were calculated with the binomial formula assuming that $P = .2$ and $Q = .8$.

NUMBER CORRECT	BINOMIAL PROBABILITY
0	.1073000
1	.2680000
2	.3015000
3	.2013000

276 INFERENTIAL STATISTICS

NUMBER CORRECT	BINOMIAL PROBABILITY
4	.0880000
5	.0247000
6	.0051000
7	.0007800
8	.0000730
9	.0000041
10	.0000001
	Total = .9967572

(The probabilities do not sum to exactly one because of rounding error in the calculation.)

As an example of an application of the use of binomial probability let us picture a bootlegger who has six jugs of moonshine and 18 jugs of water on his truck when the sheriff stops him. The sheriff randomly selects five jugs to open and sniff. What is the probability that none of the jugs of moonshine will be among the five selected?

In this example:

$$P = \frac{18}{24} = \frac{3}{4}$$

$$Q = 1 - \frac{3}{4} = \frac{1}{4}$$

$$n = 5$$

$$r = 5$$

$$\text{Probability} = \frac{5!}{5! \cdot (5-5)!} \cdot \frac{3^5}{4} \cdot \frac{1^0}{4}$$

$$= 1 \cdot \frac{3^5}{4} \cdot 1$$

$$= \frac{243}{1,024}$$

$$= .2373$$

Exercises

1. How many ways could you arrange six eggs in a carton with six spaces?
2. A contractor advertises to fill three vacancies in his construction crew. Thirty percent of his applicants are men and 70 percent are women. (a) What is the probability that purely by chance two of the three jobs will be offered to men and the other to a woman? (b) What is the probability that by chance all three jobs will be offered to men?

Solutions

1. $6! = 6 \cdot 5 \cdot 4 \cdot 3 \cdot 2 \cdot 1 = 720$

2. (a) $\frac{3!}{2!1!} \cdot .3^2 \cdot .7^1 = 3 \cdot .09 \cdot .7 = .189$

 (b) $\frac{3!}{3!0!} \cdot .3^3 \cdot .7^0 = 1 \cdot .027 \cdot .1 = .027$

Aids Available for Determining Binomial Probabilities

You have probably noticed that the computation of binomial probabilities for other than very small values of N is quite laborious. Fortunately, it is not necessary to do them oneself. Extensive tables of binomial distributions from $P = .01$ to $P = .50$ have been published by the National Bureau of Standards.[3] We have included a binomial probability table (Table A3) for $P = .50$ in the Appendix to illustrate how such tables are used.

Column n represents number of trials. Row r represents outcomes. The numbers in each row represent *cumulative* outcomes rounded to the nearest thousandth. If we go to $n = 10$ we find in cumulative form the values we worked out in the potion and potentate example, that is, the probability of zero sons is .001, the probability of zero or one son is .011, the probability of zero or one or two sons is .055, and so on.

If we toss twelve coins into the air what are the probabilities of various numbers of heads occurring? Since the probability of a head occurring in one coin is .50 we can use Table A3 to determine the various probabilities. Reading across from 12 in column n we find a blank in the 0 column. This tells us that the probability of zero heads is less than .001. Reading across we see that the cumulative probability of zero or one head is .003, the cumulative probability of zero or one or two heads is .019. We can get a specific rather than a cumulative probability by subtracting the value to the left of the desired r from the value of r in the appropriate row. For example, if we want the probability of exactly five heads in our twelve coins we first find the tabled value for five in row 12. The cumulative probability is .387. The value to the left is .194. Subtracting we get .193 the probability of exactly five heads.

Space does not permit the completion of the table for values of r greater than 15. Since the probability curve is symmetrical the needed values not in the table can be found by finding the cumulative probability of $n - (r + 1)$ and subtracting this probability from one. For example, note in row 25 the cumulative probability for 15 is equal to one minus the cumulative frequency of nine $(1 - .115 = .885)$. To find the cumulative frequency of 16 in 25 take the

---

[3] *National Bureau of Standards. Applied Mathematics Series 6, Tables of Binomial Probability Distribution, Washington, D.C., 1949, U.S. Government Printing Office.*

cumulative frequency of $25 - (16 + 1) = 8(.054)$, subtract this value from one to get $1 - .054 = .946$, the cumulative frequency of 17 in 25 is the cumulative frequency of 7(.022) subtracted from one to get .978, and so on.

Exercises

1. If the Yankee peddler had had 20 couples in his experiment instead of ten, what is the probability that although the null hypothesis is true they would have had two or fewer daughters, four or fewer daughters?
2. Given that the potentate is testing his hypothesis at the .05 level how many daughters could occur in 16 births and still the potentate would reject the null hypothesis?

Solutions

1. less than .001, .006
2. four or less

Normal Curve Approximation of the Binomial

The computation of binomial probabilities is a rather time-consuming endeavor unless one has ready access to a high speed calculator or a computer. As n increases this task becomes more and more tedious. Fortunately, there is a quick method for approximating binomial probabilities. And even more fortunately, the accuracy of the approximation increases as n increases.

If we graph the probabilities of zero to ten sons in the potion and potentate example under the assumption that the potion does not work, we have a distribution as shown in Figure 10.3. Note that this figure strongly resembles the normal curve. If we graphed the probabilities associated with zero to 20 sons in 20 births our graph would be even more similar to the normal curve. If we graphed probabilities of zero to 50 sons in 50 births it would be very, very similar to the normal curve. As n increases the similarity between the binomial distribution and the normal curve increases. In fact, the normal curve distribution is identical to the binomial when the number of cases is infinity. Histograms of binomial distributions of finite numbers are discrete and will have "stairsteps" as in Figure 10.3 while the binomial distribution will assume the continuous shape of the normal curve when number equals infinity.

By computing the mean and standard deviation of a binomial distribution we can approximate probabilities through the use of the normal curve table.

The mean of a binomial distribution is equal to the number of trials (n)

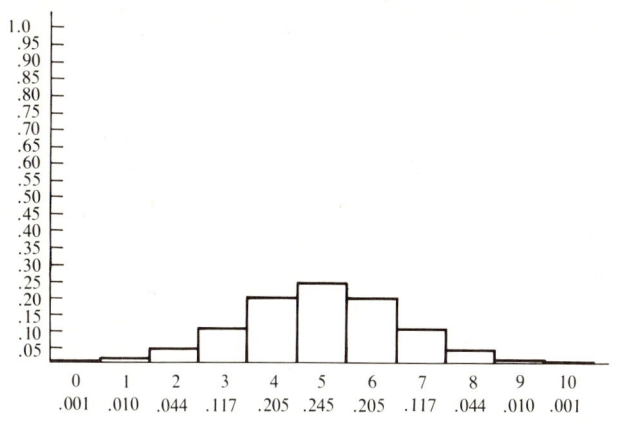

Figure 10.3 Probabilities Graph of Zero to Ten Sons

multiplied by the probability that the "desired outcome" (*P*) could occur:

$$\mu = nP$$

Formula (10.3)
Mean of Binomial

In our potentate example ten couples each have a .5 chance of having a son:

$$\mu = nP$$
$$= (10)(.5)$$
$$= 5$$

The standard deviation of a binomial distribution is the square root of the product of *n*, *P*, and *Q*:

$$\sigma = \sqrt{nPQ}$$

Formula (10.4)
Standard Deviation of Binomial

In the example:

$$\sigma = \sqrt{nPQ}$$
$$= \sqrt{(10)(.5)(.5)}$$
$$= \sqrt{2.5}$$
$$= 1.58$$

We can use the *z*-score procedure to approximate the probabilities of various scores. For example, if we wish to approximate the probability of eight sons in ten births we would first calculate the *z*-scores for 7.5 and 8.5. (Recall from Chapter 2, a score of eight has real limits from 7.5 to 8.5.)

$$z = \frac{X - \mu}{\sigma} \qquad \frac{7.5 - 5.0}{1.58} = 1.58 \qquad \frac{8.5 - 5.0}{1.58} = 2.22$$

Consulting the normal curve table (Table A2) we find the area beyond a z-score of 1.58 is .0571. The area beyond a z-score of 2.22 is .0132. Subtracting we get .0439. This is our approximation of the probability of eight sons in ten births. It is very similar to .044, the exact probability.

To take another outcome, five sons, we compute z-scores for 4.5 and 5.5 and approximate the probability by deriving the area between the associated z-values in the normal curve.

$$z = \frac{\bar{X} - \mu}{\sigma} \qquad \frac{4.5 - 5.0}{1.58} = -.32 \qquad \frac{5.5 - 5.0}{1.58} = +.32$$

Consulting the normal curve table again, we find the area between $z = -.32$ and $z = +.32$ is equal to .251. This is a reasonable approximation of the actual probability, .245.

A more common question concerns the probability of scores above or below a certain point. For example, we might ask the probability of seven or more sons in ten births. To do this we find the z-score for 6.5, in this case .96, and find the area beyond this z-score in the normal curve, .1685. This is our approximate probability of seven or more sons. Compare this with the combined probabilities of 7, 8, 9, and 10 in the binomial distribution

$$.117 + .044 + .010 + .001 = .172$$

The normal curve approximation can also be used in cases where P is greater or less than .5. In a ten-item multiple choice test with five options for each item, $P = .2$ and $Q = .8$. If we graph the binomial probabilities of getting from zero to ten correct through guessing we get a distribution as graphed in Figure 10.4. Although this figure is skewed to the right, it still bears some resemblance to the normal curve as shown in Figure 10.5.

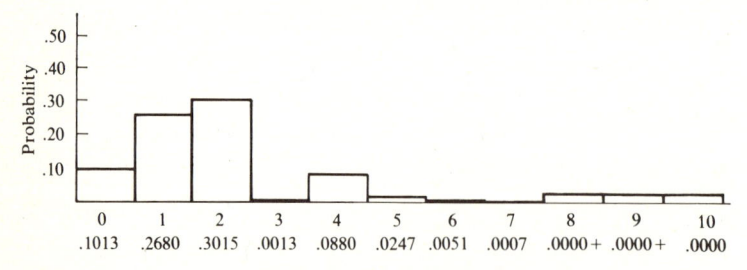

Figure 10.4 Binomial Probabilities for Zero to Ten Correct in a Ten-Item Multiple Choice Test

	0	1	2	3	4	5	6	7	8	9	10
	.1013	.2680	.3015	.0013	.0880	.0247	.0051	.0007	.0000+	.0000+	.0000

The table below allows us to compare the binomial and normal curve probabilities for zero to ten items correct, assuming $P = .2$ and $Q = .8$.

The normal curve approximations become increasingly less useful as the values of P and Q move farther away from .5 and .5 and nearer to 1.0 and 0 or 0 and 1.0. As a rule-of-thumb when P and Q are both near .5 the normal curve

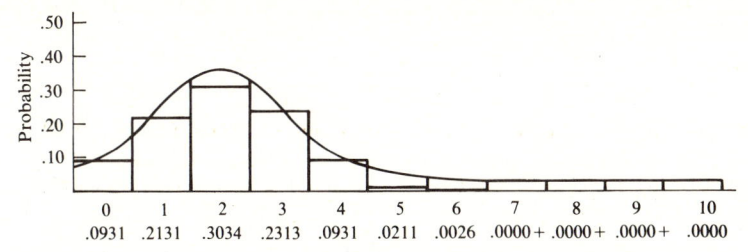

Figure 10.5 Binomial Model with Normal Curve Comparison

approximations are acceptable for most purposes if n equals 10 or more. When P or Q is about .2 these approximations are acceptable when n equals 25 or more. As n increases the binomial distribution becomes increasingly more similar to the normal curve.

NUMBER CORRECT	NORMAL CURVE PROBABILITY APPROXIMATION	EXACT BINOMIAL PROBABILITY
0	.0931	.1073000
1	.2313	.2680000
2	.3034	.3015000
3	.2313	.2013000
4	.0931	.0880000
5	.0211	.0247000
6	.0026	.0051000
7	.0000[4]	.0007800
8	.0000[4]	.0000730
9	.0000[4]	.0000041
10	.0000[4]	.0000001

In this chapter we have introduced procedures used to test hypotheses concerning data divided into two categories. Other hypothesis-testing procedures will be introduced in subsequent chapters.

SUMMARY

The outcomes of scientific inquiry are usually probability statements rather than "facts." The use of the null hypothesis enables one to determine the credibility of an apparent relation between variables. One can calculate the probabilities of various outcomes in a sample when the null hypothesis is true in the population. Then one can decide whether to reject or retain the null hypothesis.

[4] *These probabilities are less than $P = .0001$ and are not found in the standard normal curve table.*

FLOWCHART 10

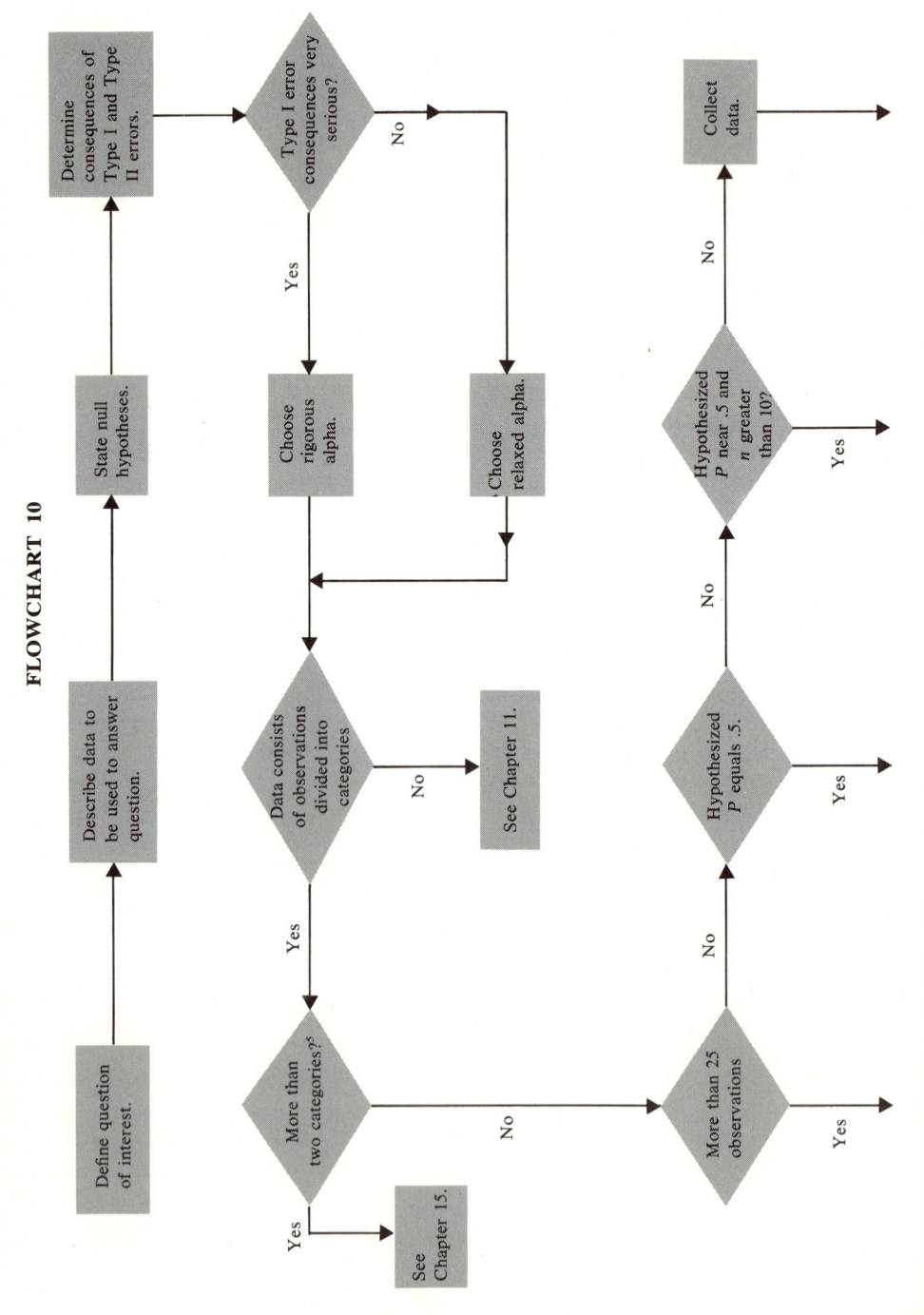

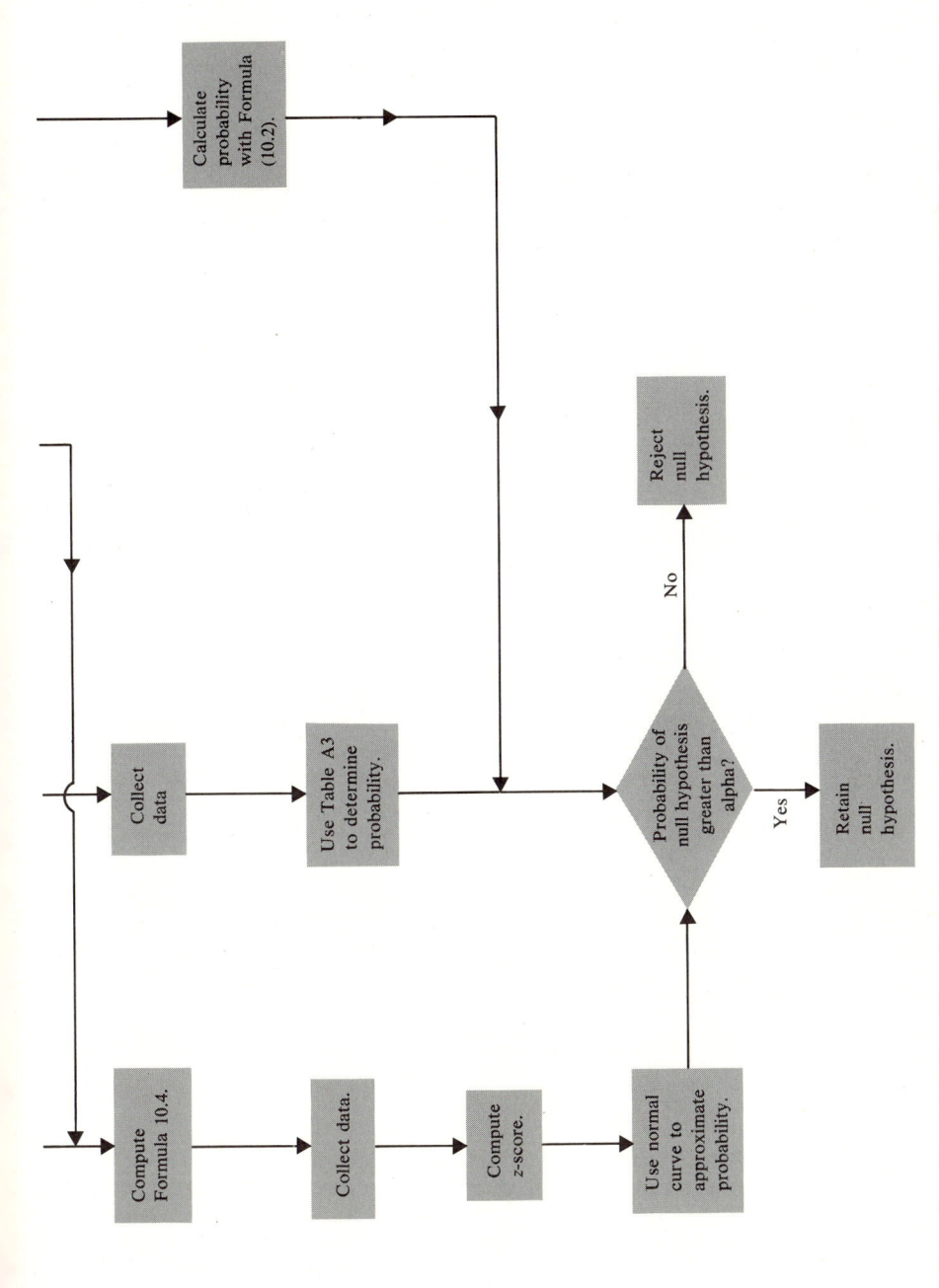

[5]*Techniques described in Chapter 15 can be used with two categories as well as more than two categories. When data consist of observations divided into two categories one can choose between the methods described here and the methods described in Chapter 15.*

When the question involves a relationship between variables, rejecting the null hypothesis leads to the tentative conclusion that the variables are related. Retaining the null hypothesis leads to the conclusion that there is not sufficient evidence for declaring there is a relationship between variables.

A Type I error occurs when one declares the null hypothesis to be false when it is actually true. A Type II error occurs when one does not reject a null hypothesis that is actually false. An investigator sets his alpha, required probability level, according to the relative seriousness of Type I and Type II errors.

If the probability of the observed events occurring by chance is equal to or less than alpha, the investigator tentatively concludes that there is a genuine relation between variables and rejects the null hypothesis. If the probability of an event occurring when the null hypothesis is true is greater than alpha, the null hypothesis is retained and the evidence is not sufficient for concluding that there is a genuine relation between the variables.

The probability of events occurring under a true null hypothesis can be calculated using binomial probability procedures or approximated through the use of the normal curve.

Flowchart 10 is an attempt to illustrate the steps in hypothesis testing. When the question of interest has been defined one determines the data to be used to answer the question. Then the null hypothesis is stated and the consequences of Type I and Type II errors are determined. Through weighing the relative seriousness of consequences of Type I and Type II errors an alpha (level of significance) is selected.

The procedure used to determine probabilities varies according to the nature of the data and the nature of the hypothesis. If the data consist of observations divided into two categories one next asks if there are more than 25 observations. If there are more than 25 observations, Formula (10.4) is used to compute the standard deviation and a z-score is computed. This z-score is referred to the normal curve table to approximate the probability.

With 25 or fewer observations involved, Table A3 can be used if the hypothesized proportions are .50–.50. For other proportions Formula (10.2) is used to calculate probabilities.

Finally, once the probabilities have been established the null hypothesis is tested and either rejected or retained.

EXERCISES

1. An investigator wishes to study the question "Do blondes have more fun?"
   (a) What would be the null hypothesis in this question?
   (b) What would be the result of a Type I error?
   (c) What would be the result of a Type II error?

(d) If one investigator uses an alpha of .05 in investigating this question and another investigator uses an alpha of .001, which would be more likely to make a Type I error?

(e) If one investigator uses an alpha of .05 in investigating this question and another investigator uses an alpha of .001, which would be more likely to make a Type II error?

2. Using the binomial probability procedure (or Table A3) determine the probability of getting all five questions correct on a true-false exam through sheer luck.
3. Using the normal curve approximation determine the probability of getting 60 questions or more correct through luck alone in a 100-item true-false test.
4. Three judges are asked to choose the best of four brands of tea. Using the binomial probability formula, what is the probability that exactly two of the judges will select Brand X through chance alone?
5. If in the casinos at Las Vegas the rules stated you could win $1,000 for every $5.00 you wager if you rolled five 6's in a row at the dice table, what would be the probability of winning the $1,000?
6. The local city council is composed of 11 members, 6 Republicans and 5 Democrats. The governor has asked that a randomly drawn committee of 5 be sent to the state capital to represent the feelings of the city's residents. What is the probability of the mayor's picking randomly a committee composed of 4 Republicans and 1 Democrat? Do you think the mayor has "stacked" the committee?
7. An oriental potentate has been offered 20 maidens by his admirers and he is to choose 4 from the 20 to add to his harem. Thirteen (13) of the 20 are dark-haired and 7 are blondes. The potentate chose 3 blondes and 1 dark-haired girl and has been told by his statistician that he has a bias. What do you think?
8. How many different bridge hands (that is, 13 cards apiece) are possible from a deck of 52 cards? (Leave answer in factorial form.)

ANSWERS

1. (a) Blondes and nonblondes have an equal amount of fun.
   (b) Declaring that blondes have more fun than nonblondes when in fact the two groups have an equal amount of fun.
   (c) Failing to conclude that blondes have more fun when in fact they do.
   (d) The investigator with the .05 alpha.
   (e) The investigator with the .001 alpha.

2. .031

3. $\sigma = \dfrac{X - \mu}{\sqrt{nPQ}} = \dfrac{59.5 - 50}{\sqrt{(100)(.5)(.5)}} = \dfrac{9.5}{\sqrt{25}} = 1.9$

   Probability of z of $+1.9$ or greater $= .0287$

286 INFERENTIAL STATISTICS

4. $n = 3$
   $r = 2$
   $p = \frac{1}{4}$
   $q = \frac{3}{4}$
   $C = \frac{3!}{2! \cdot 1!} \quad 3 \cdot \left(\frac{1}{4}\right)^2 \cdot \left(\frac{3}{4}\right)^1 = \frac{9}{64} = .14$

5. $n = 5 \quad C = \frac{5!}{5!(0!)} = 1$
   $p = .17$
   $q = .83$ Probability of winning $= 1(.17)^5(.83)^0 = .0001$

6. $n = 5 \quad C = \frac{5!}{4!1!} = 5 \quad 5\left(\frac{6}{11}\right)^4\left(\frac{5}{11}\right)^1$
   $p = \frac{6}{11}$ $5(.0885)(.45)$
   $q = \frac{5}{11}$ (Probability of 4 Republicans and 1 Democrat) $= .20$

   There is a substantial likelihood of this arrangement occurring by chance.

7. $n = 4$ (Probability of 3 blondes and 1 dark-haired) $= \frac{4!}{3!1!}\left(\frac{7}{20}\right)^3\left(\frac{13}{20}\right)^1$
   $r = 3$
   $p = \frac{7}{20}$ $= 4(.35)^3(.65)^1$
   $q = \frac{13}{20}$ $= .11$

   With a probability of .11 it would seem unwise to conclude that the evidence is sufficient for concluding that the potentate has a bias.

8. $n = 52$
   $r = 13$ $\frac{52!}{13!(39!)}$ $(= 635,013,559,600)$
   $n - r = 39$

   There are more than 600 billion possible bridge hands.

CHAPTER ELEVEN
RANDOMNESS AND SAMPLING ERROR IN HYPOTHESIS TESTING

OBJECTIVES

After studying this chapter, the reader should be able to:

1. Distinguish and illustrate the difference between a population and a sample.
2. Distinguish between statistics and parameters.
3. Define random sample and describe the steps involved in drawing a random sample.
4. Discuss the relationship between random sampling and statistical inference.
5. Define sampling error.
6. List the characteristics of sampling errors of the mean.
7. Define and compute the standard error of the mean.
8. Use the standard error of the mean and the normal curve to test hypotheses about the value of population means.
9. Distinguish between one-tailed and two-tailed tests and know when each is used.

Introduction

The purpose of most educational research is to investigate the characteristics of populations. A *population* is *all* the cases (persons, objects, events) that constitute an identified group. Some examples of populations are: all sixth graders in U.S. schools, the student body of a particular university, the registered voters in the state of California, 35-year-olds, people who smoke more than one pack of cigarettes per day, and corn grown with Brand X fertilizer. Typically, the researcher is looking for information about certain characteristics of these populations that are of particular interest to him, such as the mean weight, median IQ, standard deviation of test scores, and so on. Such population characteristics are called *parameters*. In most cases it is impossible to study all the members of a population. For this reason the parameters cannot be calculated directly and must be estimated from data derived from a small portion of the population, that is, a *sample*. In practice, therefore, much research involves the study of a subgroup, a sample, from a population. The information derived from the sample is then used to infer what is likely to be true of the entire population.

Sample characteristics are known as *statistics*. Inferential statistics involve generalizing or making inferences from sample statistics to population parameters. For example, a clothing manufacturer has obtained the neck size, waist size, and so on, of a sample of junior high school pupils (statistics) because he needs to get an idea of what these measurements are for all junior high school pupils (parameters) before embarking upon a new production line. Basically, inferential statistics involve a logical process of going from particular and incomplete data to broad conclusions. Inferential statistical techniques enable us to make reasonable decisions on the basis of incomplete information.

Random Sampling

If one is to use sample statistics to make inferences about population parameters, then it is extremely important that an appropriate method of sampling be used. A researcher is entitled to make inferences only about that population from which he has drawn a representative sample. The best way to insure a representative sample is to employ some random method of selecting the sample from the population. If a random method is used, then each member of the population has an equal chance of being selected for the sample.

For a college survey, one would not want to include only students found in the library on a certain evening. Neither would one include only the students visiting the student union building on that evening. Such a sample would not likely be representative of the population of students at that college. Rather the sample would most likely be biased. A biased sample is one selected in a way that favors the inclusion of certain elements. For example, if one wanted a representative sample of the general population in a large city, one would not select the sample from automobile registration lists or the telephone directory. Such a procedure would introduce bias and it would be dangerous to generalize from such a sample to the general population. Since one does not always even know what biasing influences are operating, the best procedure to follow, in order to avoid bias, is *random sampling*. A random sample is defined as a sample drawn in such a way as to insure that all members of the population have an equal and independent chance of being selected. Thus, in order for a sample to be *random*, the individuals making up that sample must be *selected* by a random procedure.

A random sample is defined by the *process* used to draw the observations and not by any characteristics of the observations themselves. The basic rule of random sampling is that chance and chance alone determines which members of the population are included in the sample. Such a sample has no *systematic bias* and should be representative of the entire population.

A convenient method of drawing a random sample is to use a table of random numbers to select the members of the population that are to be included. Such a table contains columns of digits that have been mechanically generated, usually by a computer, to approximate a random order.[1] One assigns a number to each member of the population for identification purposes. One then enters the table of random numbers and reads numbers directly from the table until the desired N is obtained. The numbers will identify which members of the population should be included in the sample. Thus, the selection of each member in the sample is strictly a chance event. In using a table of random numbers, it is customary to determine *by chance* the point at which one enters the table. One way to do this is to touch a page without looking and begin wherever the page is touched. Or another method is to randomly select numbers from the table, which are then used to determine the number of the page, column, and row at which the table is entered.

Random sampling is the sampling procedure that is the basis for inferential statistical reasoning. The random selection guarantees that any differences between the sample and the population are a function of chance and do not represent bias on the part of the researcher. Any differences between random samples and their parent population are thus not systematic ones. For example, the mean intelligence score of a random sample of fourth graders may be higher than the mean intelligence score of the population of fourth graders, but it is equally likely that the mean for the sample will be lower than the mean for the parent population.

Statistical theorists have, through deductive reasoning, shown how much one can expect observations derived from random samples to differ from the

---

[1] *Tables of random numbers may be found in many statistics texts or in R. A. Fisher and F. Yates, Statistical Tables for Biological, Agricultural, and Medical Research (Edinburgh: Oliver and Boyd Ltd., 1963).*

290   INFERENTIAL STATISTICS

observations in the population. The methods of statistical inference may be employed to provide an estimate of these differences.

Figure 11.1 shows the steps in the statistical inference process.

Define Population → Draw Random Sample → Calculate Statistics from Random Sample → Estimate Parameter → Generalize to Population

Figure 11.1 Steps in the Process of Statistical Inference

Random Assignment

The same inferential statistical techniques that can be used when we have random sampling can also be used when we have random assignment. In random assignment an experimenter begins with a pool of subjects and employs a chance procedure to assign subjects to treatment groups. This pool may be a sample from a larger group or it may be the entire population.

For example, a chicken feed producer wants to see if a hormone added to his feed will accelerate weight gain. He has 100 young chickens available for his experiment. He labels each with an identification number, then uses a table of random numbers to assign 50 to the experimental group and 50 to the control group. The two groups are raised under conditions that are identical except that the experimental group has the hormone added to its feed while the control group does not. A record of weight gains is kept.

Here the question is not how well the original pool of 100 chickens represents the population of all chickens. The question being asked here is, "Do we have evidence from which we can infer that feeding hormones to chickens will influence weight gain either positively or negatively?" It could be that, in the experiment, by chance a greater proportion of healthy chickens were assigned to the experimental group and a greater proportion of sickly chickens to the control group. However, it is equally likely that the reverse would be true.

Random assignment, like random sampling, avoids systematic bias. The only difference is that with random sampling we were asking, "How likely is my sample to differ from the population by chance," whereas with random assignment we will ask, "How likely are my groups to differ from one another through chance?"

Exercises

1. Match the definition on the left with the term on the right.
   (a) all cases that constitute an   statistic
       identified group   population

(b) a sample characteristic
(c) a population characteristic
(d) using a chance procedure to select a sample from a population
(e) using a chance procedure to assign subjects to groups
(f) a subgroup of a population

random sampling
random assignment
sample
parameter

2. Which of the following would be a random sample of Central School's freshmen class:
   (a) those freshmen enrolled in the 9 A.M. freshmen English class
   (b) a subgroup drawn through a chance procedure from all students enrolled in freshmen English
   (c) a subgroup drawn through a chance procedure from a list of all students classified as freshmen

Solutions

1. (a) population   (b) statistic   (c) parameter   (d) random sampling
   (e) random assignment   (f) sample
2. (c)

Concept of Sampling Error

When an inference is made from a sample to a population a certain amount of error is involved because even random samples can be expected to vary from one to another. The mean intelligence score of one random sample of fourth graders may be different from the mean intelligence score of another random sample of fourth graders from the same population. Such differences are called sampling errors and they result from the fact that one has observed only a sample and not the entire population.

Sampling error is defined as the difference between a population parameter and a sample statistic. For example, if one knows the mean of the entire population (symbolized μ) and also the mean of a random sample (symbolized $\bar{X}$) from that population, the difference between these two, $\bar{X} - \mu$, represents sampling error (symbolized e). Thus, $e = \bar{X} - \mu$. For example, if we know that the mean intelligence score for a population of 10,000 fourth graders is $\mu = 100$ and a particular random sample of 200 has a mean of $\bar{X} = 99$, then the sampling error is $\bar{X} - \mu = 99 - 100 = -1$. Because we usually depend on sample statistics to estimate population parameters, the notion of how samples are expected to vary from populations is a basic element in inferential statistics. However, instead of trying to determine the discrepancy between a sample statistic and the population parameter (which is not often known), the approach in inferential statistics is to estimate the variability that could be expected in the statistics from a number of different random samples drawn from the same population. Since each of the sample statistics is considered to be an estimate of the same population parameter, then any variation among sample statistics must be attributed to sampling error.

The Lawful Nature of Sampling Errors

Given that random samples drawn from the same population will vary from one another, is using a sample to make inferences about a population really any better than just guessing? Yes, it is, because sampling errors behave in a lawful and predictable manner. The laws concerning sampling error have been derived through deductive logic and have been confirmed through experience.

Although we cannot predict the nature and extent of the error in a single sample, we can predict the nature and extent of sampling errors in general. Let us illustrate this with reference to sampling errors connected with the mean.

Sampling Errors of the Mean

Some sampling error can always be expected when a sample mean $\bar{X}$ is used to estimate a population mean μ. Although, in practice, such an estimate is based on a single sample mean, assume that one drew several random samples from the same population and computed a mean for each sample. We would find that these sample means would differ from one another and would also differ from the population mean (if it were known). This variation among the means is due to the sampling error associated with each random sample mean as an estimate of the population mean. Sampling errors of the mean have been studied carefully and it has been found that they follow regular laws.

The Expected Mean of Sampling Errors Is Zero Given an infinite number of random samples drawn from a single population, the positive errors can be expected to balance the negative errors so that the mean of the sampling errors will be zero. For example, if the mean height of a population of college freshmen is 5 feet 9 inches, and several random samples are drawn from that population, we would expect some samples to have mean heights greater than 5 feet 9 inches and some to have mean heights less than 5 feet 9 inches. In the long run, however, the positive and negative sampling errors will balance. If we had an infinite number of random samples of the same size, calculated the mean of each of these samples, then computed the mean of all these means, the mean of the means would be equal to the population mean.

Since positive errors equal negative errors, a single sample mean is as likely to underestimate a population mean as to overestimate it. Therefore, we can justify saying that a sample mean is an unbiased estimate of the population mean, and is a reasonable estimate of the population mean.

Sampling Error Is an Inverse Function of Sample Size As the size of a sample increases there is less fluctuation from one sample to another in the value of the mean. In other words, as the size of a sample increases the expected sampling error decreases. Small samples are more prone to sampling error than large ones. One would expect the means based on samples of 10 to fluctuate a great deal more than the means based on samples of 100. In our height example it would be much more likely that a random sample of four had three above-average freshmen and only one below-average freshman than that of a random sample of 40 had 30 above average and ten below. As sample size increases the likelihood that the mean of the sample is near the population mean also increases. There is a mathematical relationship between sample size and sampling error. We will show later how this relationship has been incorporated into inferential formulas.

Sampling Error Is a Direct Function of the Standard Deviation of the Population The more spread or variation we have among members of a population, the more spread or variation we expect in sample means. For example, the mean weights of random samples of 25 each selected from a population of professional jockeys would show relatively less sampling error than the mean weights of samples of 25 each selected from a population of school teachers. The weights of professional jockeys fall within a narrow range, the weights of school teachers do not. Therefore, for a given sample size, the expected sampling error for teachers' weights would be greater than the expected sampling error for jockeys' weights.

Sampling Errors are Distributed in a Normal or Near Normal Manner Around the Expected Mean of Zero Sample means near the population mean will occur more frequently than sample means far from the population mean. As we move farther and farther from the population mean we find fewer and fewer sample means occurring. Both theory and experience have shown that the means of random samples are distributed in a normal or near normal manner around the population mean.

Since a sampling error in this case is the difference between a sample mean and the population mean, the distribution of sampling errors is also normal or near normal in shape. The two distributions are by definition identical except that the distribution of sample means has a mean equal to the population mean while the mean of the sampling error is zero.

294 INFERENTIAL STATISTICS

The distribution of sample means will resemble a normal curve even when the population from which the samples are drawn is not normally distributed. For example, in a typical elementary school we find about equal numbers of children of the various ages included, so a polygon of the children's ages would be basically rectangular. If we take random samples of 40 each from a school with equal numbers of children aged 6 through 11 we would find many samples with means near the population mean of 8.5, sample means of about 8 or 9 would be less common, and sample means as low as 7 or as high as 10 would be rare.

Standard Error of the Mean

Since the extent and the distribution of sampling errors can be predicted, we can use sample means with predictable confidence to make inferences concerning population means. However, we need an estimate of the magnitude of the sampling error associated with the sample mean when it is used as an estimate of the population mean. An important tool for this purpose is the *standard error of the mean.*

It has been stated that sampling error manifests itself in the variability of sample means. Thus, if one calculates the standard deviation of a collection of means from random samples from a single population, one would have an estimate of the amount of sampling error. It is possible, however, to obtain this estimate on the basis of only one sample. We have seen that two things affect the size of sampling error, the size of the sample and the standard deviation in the population. When these two things are known, one can predict the standard deviation of sampling errors. This expected standard deviation of sampling errors of the mean is called the standard error of the mean and is represented by the symbol $\sigma_{\bar{x}}$. It has been shown through deductive logic that the standard error of the mean is equal to the standard deviation of the population (σ) divided by the square root of the number in each sample ($\sqrt{n}$). In formula form:

$$\sigma_{\bar{x}} = \frac{\sigma}{\sqrt{n}}$$

where: $\sigma_{\bar{x}}$ = standard error of the mean

σ = standard deviation of the population

n = number in each sample   Formula (11.1)
Standard Error of the Mean

In Chapter Four we saw that standard deviation (σ) is an index of the degree of spread among individuals in a population. In the same way standard error of the mean ($\sigma_{\bar{x}}$) is an index of the spread expected among the means of samples drawn randomly from a population. As we will see, the interpretation of σ and $\sigma_{\bar{x}}$ is very similar.

Exercises

1. What do we call the difference between a sample statistic and the equivalent parameter for the population from which the sample was drawn?
2. (a) If we construct a polygon of the means of an infinite number of samples drawn from a single population, what shape do we expect that polygon to assume?
   (b) What do we expect the value of the mean of this distribution to be?
3. (a) What is the relationship between sample size and sampling error associated with sample means?
   (b) What is the relationship between the standard deviation of a population and sampling error associated with sample means?

Solutions

1. sampling error
2. (a) similar to the normal curve
   (b) the mean of the population
3. (a) as sample size increases the sampling error associated with the mean decreases
   (b) as population standard deviation increases the sampling error associated with the mean increases

Using the Normal Curve To Test Hypotheses about a Population Mean

We have seen in Chapter Ten that the normal curve can be used as a probability model for testing hypotheses concerning the binomial distribution. Since the means of random samples have approximately normal distributions we can also use the normal curve model to test hypotheses concerning population means.

Given that the expected mean of sample means is equal to the population mean, and that the standard deviation of these means is equal to the standard error of the mean, and that the means of random samples are distributed normally, one can compute a z-score for a sample mean and refer that z to the normal curve table to approximate the probability of a sample mean occurring through chance that far or farther from the population mean. The z is derived by subtracting the population mean from the sample mean and then dividing this difference by the standard error of the mean. In formula form this

296 INFERENTIAL STATISTICS

becomes:

$$z = \frac{\bar{X} - \mu}{\sigma_{\bar{x}}}$$

Formula (11.2)
z-Score for Deviation of a Sample
Mean from a Population Mean

To illustrate, let us consider a college admissions officer who wonders if his population of applicants are average, or below average on the College Board examination. The national mean for College Board scores is 500 and the standard deviation is 100. He pulls a random sample of 64 from his population and finds the mean of the sample to be 470. He asks the question, "How probable is it that a random sample of 64 with a mean of 470 would be drawn from a population with a mean of 500?" Using Formula (11.1), the admissions officer calculates the standard error of the mean as 12.5:

$$\sigma_{\bar{x}} = \frac{\sigma}{\sqrt{n}}$$

$$= \frac{100}{\sqrt{64}}$$

$$= 12.5$$

Calculating the z-score for his sample mean with Formula (11.2) he has:

$$z = \frac{\bar{X} - \mu}{\sigma_{\bar{x}}}$$

$$= \frac{470 - 500}{12.5}$$

$$= -2.4$$

Thus, his sample mean deviates from the population mean by 2.4 standard error units. What is the probability of having a sample mean that deviates by this amount (2.4$\sigma_{\bar{x}}$'s) or more from the population mean? It is only necessary to refer to the normal curve in order to express this deviation (z) in terms of probability.[2] Referring a z of -2.4 to the normal curve table, one finds that the probability of a z that low or lower is .0082. This means that a z-score that low or lower would occur by chance only about 8 times in 1,000. Since the probability of getting a sample mean that far from the population mean is remote, he concludes that his sample mean probably did not come from a population with a mean of 500 and therefore the mean of his population, applicants to his college, is probably less than 500. (See Figure 11.2.)

[2] This is possible because the z-score (representing the deviation of a sample mean from a population mean in standard units) can be interpreted in the same way as a z-score which represents the deviation of a raw score from a mean in terms of standard deviation units. Having a normally distributed z-scale, one is able to refer to the normal curve and determine probability based on the relationship between the z and areas under the normal curve.

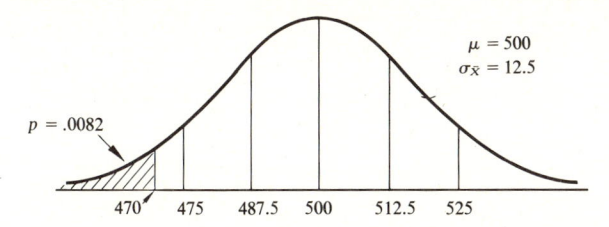

Randomness and Sampling Error in Hypothesis Testing 297

Figure 11.2 Variability of Sample Means Based on a Sample of $n = 64$.

Exercises

1. In the above example, what is the probability of obtaining a sample mean that deviates by 20 points or more either above or below the population mean? In other words, what is the probability of obtaining a sample mean greater than 520 or less than 480?
2. Consider a population with $\mu = 150$ and $\sigma = 30$ from which one randomly selects 100 scores. What is the probability that the mean of the random sample will be equal to or greater than 153?
3. Let us assume that we draw a random sample of 36 from the population of Exercise 2. What is the probability that the mean of this random sample will be equal to or greater than 153?

Solutions

1. We first convert the sample mean into z-score form in terms of the standard error of the mean.
Thus,

$$z = \frac{\bar{X} - \mu}{\sigma_{\bar{x}}} \quad z = \frac{520 - 500}{12.5} \quad z = 1.60; \quad z = \frac{480 - 500}{12.5} \quad z = -1.60$$

Referring to the normal curve table, one finds that the probability of a z-score greater than 1.60 or less than -1.60 is .0548. That is, 5.48 percent of the sample means would fall above 520 and 5.48 percent below 480. This gives about 11 chances in 100 of finding sample means outside the range of 480–520 and about 89 in 100 of finding sample means within this range. (See Figure 11.3.)
2. With a sample size of 100 and a population standard deviation of 30, the standard error of the mean is:

$$\sigma_{\bar{x}} = \frac{\sigma}{\sqrt{n}} = \frac{30}{\sqrt{100}} = 3$$

Next we determine the value of z corresponding to a sample mean of 153:

$$z = \frac{\bar{X} - \mu}{\sigma_{\bar{x}}} \quad z = \frac{153 - 150}{3} = \frac{3}{3} = 1.00$$

298 INFERENTIAL STATISTICS

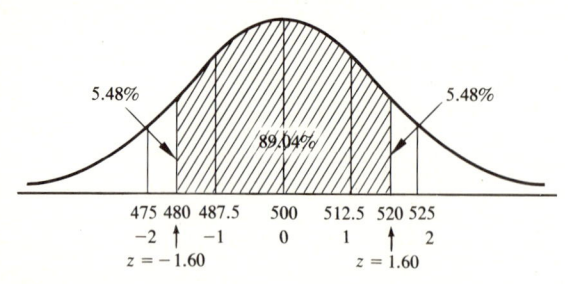

Figure 11.3 A Normal Curve with Mean 500 and $\sigma_{\bar{x}}$ of 12.5.

Thus the sample mean is one standard error above the mean of the population. From the normal curve table we see that 34.13 percent of the sample means will fall between the population μ and a z of 1 and 15.8 percent will fall beyond a z of 1.00. Therefore, there are approximately 16 chances in 100 of obtaining a sample mean equal to or greater than 153 from this population when $n = 100$.

3. $$\sigma_{\bar{x}} = \frac{30}{\sqrt{36}} = \frac{30}{6} = 5.00 \qquad z = \frac{153 - 150}{5} = .6$$

From the normal curve, we see that 27.43 percent of the sample means fall at or above $z = .60$. Therefore, there are approximately 27 chances in 100 of obtaining a sample mean equal to or greater than 153 from this population when $n = 36$.

Note above that the error involved in sampling increases as sample size decreases. When the sample size was 100, the standard error of the mean was 3. Decreasing the sample size to 36 resulted in a standard error of 5. If the sample size were further reduced to 25, the standard error would equal 6 whereas a sample size of 121 would yield a standard error of 2.7. Larger samples result in smaller errors. Also note the effect on the probability statement as the sample size varies. When $n = 100$, the probability was 16 chances in 100 of getting a sample mean greater than 153. When $n = 36$, the probability was 27 chances in 100 of getting a sample mean greater than 153. The larger the sample, the closer the sample mean is expected to approach the population mean. Larger samples provide better estimates of population parameters.

Let us consider an example involving the practical application of some of these concepts.

The Principal and the Superintendent

In their book *Elementary Statistical Methods*, Blommers and Lindquist[3] introduced an example of the use of the standard error of the mean that has become a classic. They tell of an elementary school principal who goes to his superintendent requesting that his school be given more money per pupil than the other elementary schools in the system because the children in his school are not as intelligent as the children in the other schools. He points out that his graduates have done less well in middle school than graduates of other schools.

The principal argues that since the pupils in his school are less intelligent than the pupils in the other schools he needs additional money for compensatory education: additional teachers in order to have a smaller teacher-pupil ratio, special equipment, and so on.

The superintendent agrees that if indeed the children in the principal's school are below average it would be appropriate to give him the additional money he requested. However, there are no data available to indicate how intelligent the children are. It is possible that the intelligence of the children in the principal's school is average. In that case it would be inappropriate to give his school more money than the other schools in the system. It is also possible that the children in the principal's school are above average. If this is the case perhaps something is wrong with the principal or his faculty or both. If the children are above average in intelligence, yet do less well than children from other schools when they go on to middle school, perhaps the best course would be to fire the principal and send in a new man to straighten things out.

Let us put these alternatives into the general form employed in hypothesis testing.

The Research Hypothesis (H_1) "The average intelligence in the principal's school is less than 100." This is the statement of the expected relationship between the variables of intelligence and school attended. In formula form the research hypothesis is:

$$\mu < 100$$

The Null Hypothesis (H_0) "There is no difference between the mean intelligence for pupils in the principal's school and the mean of the population (100), that is, no relationship exists between the variables of intelligence and school attended. In formula form:

$$\mu = 100$$

---

[3] Paul Blommers and E. F. Lindquist, *Elementary Statistical Methods* (Boston: 1960) Houghton Mifflin.

The Alternate Hypothesis (H_2) "The mean intelligence in the principal's school is greater than 100." This is the statement that there is a relationship between variables in the opposite direction from that proposed in the research hypothesis. In formula form:

$$\mu > 100$$

The superintendent has no faith in any of the group intelligence scores commonly administered in schools. He will only accept Binet IQ test scores as a legitimate definition of intelligence. The Binet must be individually administered at a cost of about twenty dollars per child. Therefore, it would be prohibitively expensive to test every pupil in the principal's school. However, the school district does have a couple of psychometricians who have some spare time. The superintendent decides to have them test 25 pupils randomly selected from the principal's school.

Since the superintendent will have scores for a *sample* of the pupils rather than for all of them he must reach a decision on the basis of incomplete information. This is what inferential statistics is all about. By chance the sample may have a mean higher or lower than the population mean, and thereby lead the superintendent to an incorrect decision. Since he does not know the population mean but has to make his decision on the basis of a sample mean, he may reach the correct decision or he may make a Type I or Type II error. He will probably never know which of these three possible things he has done. Let us consider again the chart we first encountered in Chapter 10.

	THE NULL HYPOTHESIS IS IN FACT	
	False	True
THE INVESTIGATOR REJECTS THE NULL HYPOTHESIS	Correct	Type I Error
THE INVESTIGATOR RETAINS THE NULL HYPOTHESIS	Type II Error	Correct

If the null hypothesis is in fact true, that is, the population mean in the principal's school is 100, no action should be taken. It would be inappropriate to give the principal additional money for pupils whose intelligence is average. For the sake of argument we will say that it would also be inappropriate to fire the principal even though one might make a good case that if his pupils are as intelligent as those in other schools they ought to do equally well in middle school. If the null hypothesis is in fact true and the superintendent retains it, he is correct. He will make no change and no change is justified.

If the null hypothesis is in fact true but the superintendent rejects it he has two ways of doing this. The first would be to erroneously reject H_0 in favor of H_1; the second would be to erroneously reject H_0 in favor of H_2. In the first case he would give the principal the money when in fact the pupils did not need additional help. In the second case he would fire the principal without just cause. In either case he would have taken action without justification. Such is the consequence of a Type I error.

If the null hypothesis is in fact false, the superintendent is correct if he rejects it. He will either accept H_1 and justifiably give the principal the requested money or he will accept H_2 and justifiably fire the principal.

If in fact the null hypothesis is false but the superintendent retains it he will make a Type II error. He will take no action when action is called for. Such is the consequence of a Type II error. Either he fails to give the principal needed money or he fails to fire the principal when he should have.

In general, Type I and Type II errors might be compared to sins of commission and sins of omission. Type I leads to unwarranted action, Type II leads to failure to act when action is warranted. Either type of error can lead to unhappy results. When one is making a decision concerning a population on the basis of information from a sample the relative seriousness of the two types of error must be weighed against each other. The superintendent must decide on the possible consequences of the two types of error.

The superintendent decided that for him a Type I error would be more serious than a Type II error. His school board is very stingy and would be dismayed with an unwarranted expenditure of money. They would also wish to avoid an unwarranted firing of a principal. In either case making a Type I error would cause great trouble for the superintendent. The board would not be happy with a Type II error which would result in a failure to meet a genuine need of students but the superintendent is certain that a Type I error would have much more serious consequences for him than a Type II error.

Of course, the superintendent could be absolutely sure of avoiding a Type I error if he takes no action. However, if he retains the null hypothesis no matter what the true state of affairs may be, he will be running a very high risk of a Type II error. What he really wants is a balance between the probabilities of the two types of errors where he has some protection against both types of error but is keeping the risk of a Type I error to an acceptable minimum.

He finally decides that, given his penny-pinching school board, he is only willing to take one chance in 100 for a Type I error. In other words, he sets his alpha at .01. He is in effect saying that he will only reject the null hypothesis if the evidence against it is quite convincing.

The choice of alpha is a judgmental rather than a mathematical problem. In other words, there is no formula for determining alpha. In each situation one weighs the relative seriousness of Type I and Type II errors and then

302 INFERENTIAL STATISTICS

uses one's best judgment to select an alpha. When the consequences of a Type I error would be very serious, alpha may be set at .0001 (one chance in 10,000) or at .001 (one chance in 1,000).

Alphas of .01 (one chance in 100) or .05 (one chance in 20) are the most commonly used in educational research. They provide considerable protection against Type I errors without excessive risk of Type II errors. In cases where Type I errors seem only slightly more serious than Type II errors, more relaxed alphas such as .10 (one chance in ten), or .20 (one chance in five) may be employed.

By choosing the .01 alpha the superintendent is requiring that there be really strong evidence before he will reject the null hypothesis. He will not take action unless he has only a small probability of a Type I error.

Having decided on his alpha, the superintendent can determine which z-scores will lead to a retention of the null hypothesis and which z-scores will lead to its rejection.

The superintendent can make a Type I error in two different ways, either by incorrectly concluding on the basis of his sample that the population mean IQ is less than 100 or by incorrectly concluding on the basis of his sample that the population mean IQ is more than 100. To guard against both possibilities he divides his alpha into two parts. He will take a .005 chance of the former error and a .005 chance of the latter for a total of .01. Having decided on the probability of a Type I error that he will tolerate he must now translate that figure into the z-score values that will lead him to retain or reject his null hypothesis. He does this by consulting the normal curve table where he finds that a z-score of $+2.58$ or greater has a .005 probability and a z-score of -2.58 or less has a .005 probability. Figure 11.4 shows the

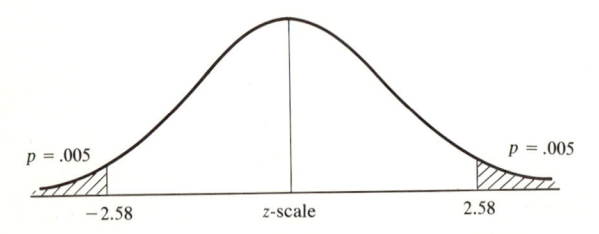

Figure 11.4 Regions of Normal Curve for Rejection of the Null Hypothesis at the .01 Level

z-values that would leave .5 percent of the sample means in each tail of the curve; that is, the probability of occurrence of sample means in these areas is .01.

If the obtained z is equal to or greater than ($\geq$) $+2.58$ or equal to or less than ($\leq$) -2.58, he will reject the null hypothesis at the .01 level. A z-score between $+2.58$ and -2.58 will lead to the retention of the null hypothesis.

One-tailed and Two-tailed Tests

The superintendent is making what is called a *two-tailed test*. He has decided that he will reject the null hypothesis under two different circumstances. If the evidence from his sample indicates that the population mean in the principal's school is either above or below 100, he will take the appropriate action. In other words, the superintendent will reject the null hypothesis if his z-score falls either in the lower tail or the upper tail of the normal curve.

In the potion and potentate story a *one-tailed test* was used. The potentate will reject the null hypothesis only if the number of sons observed exceeds the number determined by his alpha. The chances of getting ten daughters in ten births are as small as the chances of getting ten sons in ten births. However, the potentate is completely uninterested in a potion that evidence suggests will increase the likelihood of daughters. He will only take action (buy the potion) if the evidence is sufficient to suggest that the potion will increase the proportion of *sons*. He will only reject the null hypothesis if the observation falls in the upper tail of the distribution. Therefore, he has a one-tailed test.

When a two-tailed test is used, the alpha is divided between the two extreme areas of the probability distribution. For example, the superintendent with an .01 alpha is taking a .005 chance of a Type I error arising from an observed mean greater than 100 and a .005 chance of a Type I error arising from an observed mean less than 100.

When a two-tailed test is used the alpha can be divided any number of ways. For example, an alpha of .01 could be divided to put .001 at one tail and .009 at the other, or .002 and .008, and so on. However, the usual procedure is to divide the alpha equally and we will assume this to be the case in all subsequent discussions.

For any given alpha a smaller z-score is needed to reject the null hypothesis when a one-tailed test is used than when a two-tailed test is used. For example, a one-tailed test at the .05 level requires a z of 1.645 or greater in the direction of interest in order to reject the null hypothesis. (With a one-tailed test any difference in the opposite direction no matter how great leads to the retention of the null hypothesis.) With a two-tailed test a z of 1.960 in either direction is required to reject the null hypothesis at the .05 level.

One decides whether to do a one-tailed or a two-tailed test according to whether both alternatives to the null hypothesis are of interest or not. For example, consider the null hypothesis, "Method A and Method B are equally effective for teaching spelling." A student running an experiment as part of a dissertation on this subject would prefer a two-tailed test in order to be able to reject the null hypothesis if the evidence is sufficient to conclude either that Method A is better than Method B or that Method B is better than Method A. A school system which is presently using Method A and is running an experiment comparing it to Method B might prefer a one-tailed test. The school system would change from Method A to Method B if there is sufficient evidence that B is better than A. If the experiment suggests that A is better than B they would stay with A, which is the same decision they would make if the retained null hypothesis suggests that both methods are equally effective.

Exercises

Which of the following would call for a two-tailed test and which would call for a one-tailed test?
1. An investigator wants to determine if the grade-point average of drinking students differs from the grade-point average of teetotalers.
2. A rabbit breeder is running an experiment to determine if a hormone influences the fertility of rabbits. He will purchase the hormone only if it increases the fertility.
3. A physical education instructor wants to compare the effectiveness of two muscle building programs.

Solutions

1. two-tailed
2. one-tailed
3. two-tailed

Interpreting a Rejected or Retained Null Hypothesis

With a two-tailed test, whenever the null hypothesis is rejected, one can also reject the alternative hypothesis that is in the *opposite direction* from the obtained sample statistic. In our example, if the mean of the sample from the principal's school is three standard errors below the hypothesized mean $[z = (\bar{X} - \mu)/\sigma_{\bar{x}} = -3.0]$ the superintendent will not only reject the null hypothesis which states that the population mean is 100, he will also reject the alternate hypothesis that the population mean is greater than 100.

If the null hypothesis is retained, both the other hypotheses are retained as well. In our example, if the observed z is $-.5$, the evidence is not sufficient for the superintendent to reject the hypothesis that the population mean is 100. He retains the null hypothesis ($\mu = 100$) and he retains the two other hypotheses, $\mu < 100$ and $\mu > 100$, as well. A retained null hypothesis means that there is not enough evidence for concluding the null hypothesis is false. It does *not* mean that the null hypothesis is true. In our example, if the superintendent retains the null hypothesis he neither gives the principal the requested money nor fires the principal, not because he has positive evidence that the null hypothesis is true but because he does not have sufficient evidence for taking either action. One should *not* interpret a retained null hypothesis as positive evidence of the truth of the null hypothesis. A retained null hypothesis means that there is insufficient evidence for drawing a conclusion.

When the null hypothesis is rejected the sample statistic is said to be *statistically significant*. This phrase means that if the null hypothesis is true the probability of the observed statistic occurring through chance alone is equal to or less than the predetermined acceptable probability of a Type I error (alpha). The familiar meaning of the word *significant* is "important" or "meaningful." In statistics this word has the very specific meaning: "less likely to be a function of chance than some predetermined probability." Results of investigations can be statistically significant without being inherently meaningful or important.

Exercises

1. A retained null hypothesis means:
    (a) the null hypothesis is true
    (b) both the other hypotheses can be rejected
    (c) one of the hypotheses can be rejected but not the other
    (d) there is not sufficient evidence for drawing a conclusion
    (e) the null hypothesis is false
2. An investigator is making a two-tailed test of the hypothesis $\mu = 75$ with the alternatives $\mu < 75$ and $\mu > 75$ and with an alpha of .05. The population standard deviation is known to be 24 and with a random sample of 64 subjects he obtains a mean of 83. What decision does he make?

Solutions

1. (d)
2. $z = \dfrac{83 - 75}{24/\sqrt{64}} = \dfrac{8}{3} = 2.67$

The z of 2.67 is greater than the z of 1.96 required to reject the null hypothesis. He also rejects the hypothesis $\mu < 75$ and concludes that the evidence is sufficient to accept the hypothesis that the population mean is greater than 75.

Areas of Rejection and Retention for Raw Score Means

The superintendent can identify which values of the sample mean will lead to the rejection or the retention of the null hypothesis. The first step is to compute the standard error of the mean. Given the number in his sample, 25, and the standard deviation of Binet scores, 16, using Formula (11.1) he would calculate his standard error of the mean to be:

$$\sigma_{\bar{x}} = \frac{\sigma}{\sqrt{n}}$$

$$= \frac{16}{\sqrt{25}}$$

$$= 3.2$$

He multiplies the .005 z-score (2.58) by his standard error of the mean (3.2) and gets 8.256. Adding this value to the hypothesized mean of 100 he gets 108.256. Subtracting it from 100 he gets 91.744. (See Figure 11.5.) If the mean

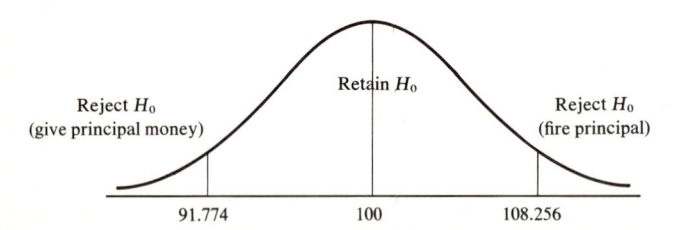

Figure 11.5 Superintendent's Areas for Retention and Rejection of H_0 for $\alpha = .01$, $n = 25$

for the 25 pupils from the principal's school is equal to or less than 91.744, the superintendent will assume that the population is actually below 100. The null hypothesis is rejected and the superintendent will give the principal the requested money. The evidence is sufficient to warrant such action under the predetermined conditions (probability < .005).

If the mean of the 25 pupils is equal to or greater than 108.256, there is sufficient evidence (probability < .005) to decide that the IQ mean of the children in the principal's school is probably greater than 100. Again, the null hypothesis will be rejected and the superintendent will fire the principal.

If the sample mean is between 91.774 and 108.256, the evidence is *not* sufficient to conclude that the population differs from 100 and therefore no action is warranted. The null hypothesis is retained.

The Effect of Numbers on the Likelihood of Type II Errors

Let us see what would have happened if the Binet test had been administered to a random sample of 100 children from the principal's school. We can do this by substituting 100 for 25 throughout our calculations:

$$\sigma_{\bar{x}} = \frac{16}{\sqrt{100}} = 1.6$$

Multiplying 1.6 by 2.58 we have 4.128 which is added to and subtracted from 100. Our distribution is shown in Figure 11.6. In this situation the null

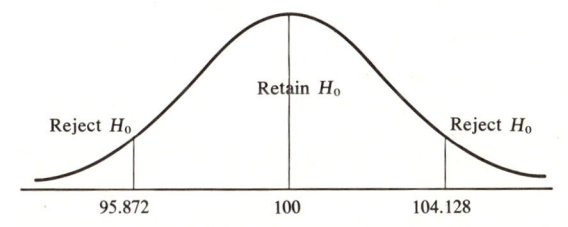

Figure 11.6 Superintendent's Area of Retention and Rejection: $\alpha = .01$, $n = 100$

hypothesis will be rejected if the mean for the sample is less than 95.872 or greater than 104.128. The superintendent is less likely to retain the null hypothesis than he was where there were only 25 in the sample. With 25 subjects he needed 91.744 or less or 108.256 or more before rejecting H_0. However, he still has the same protection against a Type I error regardless of numbers since the alpha of .01 remains the same. As the number in the sample increases, the probability of a Type II error decreases but the probability of a Type I error is not changed.

Exercise

1. What scores would lead to rejection or retention of the null hypothesis if the superintendent only had a sample of 16 pupils from the principal's school?

Solution

1. Either 89.68 or less or 110.32 or greater would lead to rejection. Scores between those two values would lead to retention.

Steps in Testing a Hypothesis about a Population Mean

Let us summarize the steps for testing a hypothesis about a population mean that was used in the principal and superintendent example.

1. Specify the hypothesized value of the population mean (the null hypothesis).
2. Determine whether the test of the null hypothesis will be one-tailed or two-tailed.
3. Describe the consequences of Type I and Type II errors, weigh the relative seriousness of these consequences, and set alpha.
4. Identify the z-score in the normal curve table that corresponds to one's alpha for the one-tailed or two-tailed test.
5. Draw a random sample from the population.
6. Calculate the standard error of the mean by dividing the population standard deviation by the square root of the number in the sample.
7. Compute a z-score by subtracting the hypothesized mean from the sample mean and dividing the difference by the standard error of the mean.
8. If the test is two-tailed retain the null hypothesis if the observed z-score is nearer zero than the z-scores required to reject the null hypothesis. Reject the null hypothesis if the observed z-score is as far from or farther from zero than the required z-score. If the test is one-tailed the null hypothesis is retained if the observed z-score is nearer zero than the required z-score *or* if the sign of the z-score is opposite the sign hypothesized in the research hypothesis. To reject the null hypothesis with a one-tailed test the observed z-score must be as far or farther from zero than the required z-score and in the same direction.

SUMMARY

Inferential statistics are used to make reasonable decisions on the basis of incomplete data. Samples are drawn to make inferences about populations when samples are randomly drawn from a population (random sampling) or when subjects from an available population are randomly assigned to treatments. In sampling, errors will occur which conform to the laws of probability. Although these laws do not enable us to predict the error in a specific sample, they do enable us to predict the nature and extent of sampling errors in general.

We know that the sampling errors related to the mean conform to the following rules: (1) The expected mean of sampling errors is zero; (2) sampling error is an inverse function of sample size; (3) sampling error is a direct function of the standard deviation of the population; and (4) sampling

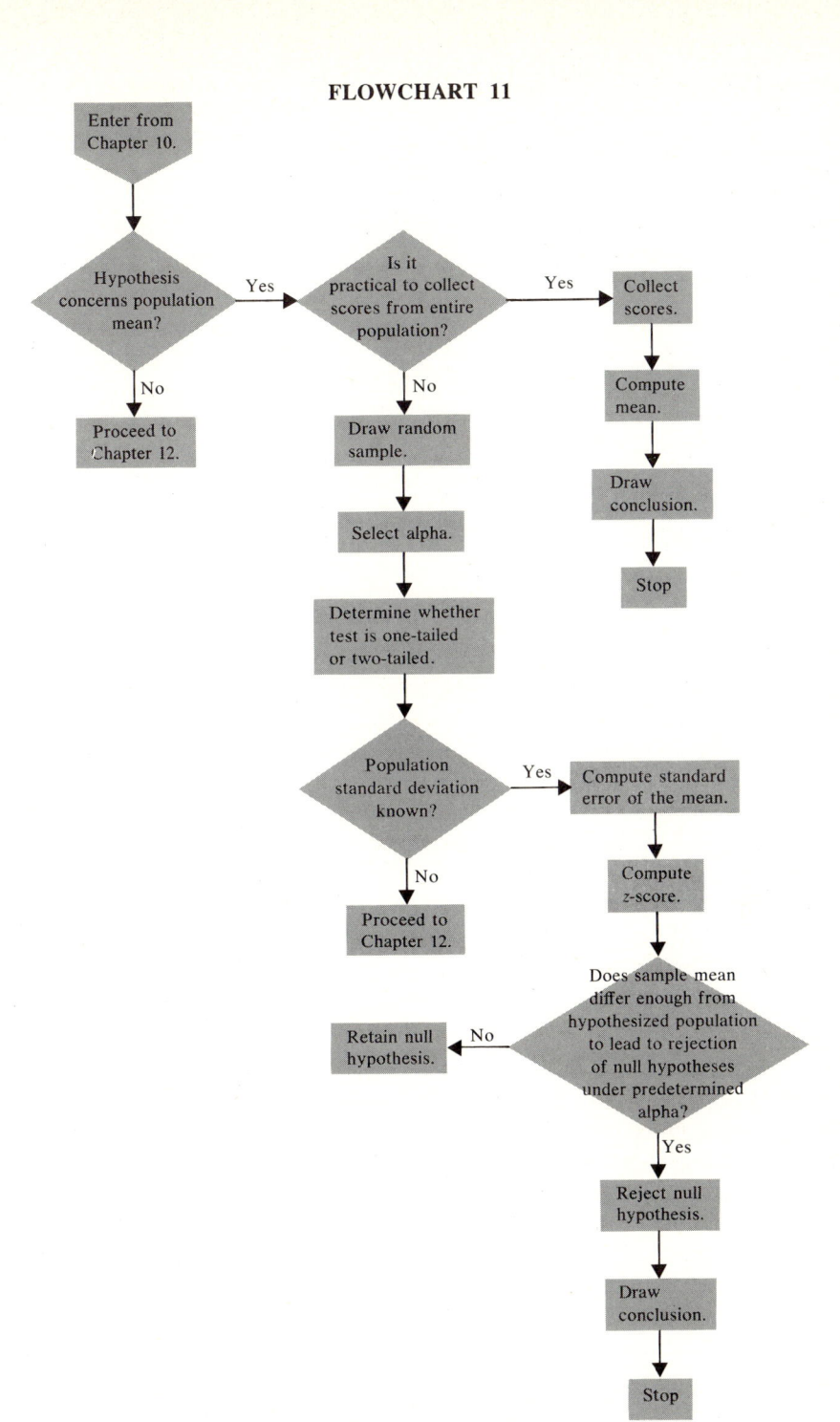

FLOWCHART 11

- Enter from Chapter 10.
- Hypothesis concerns population mean?
  - No → Proceed to Chapter 12.
  - Yes → Is it practical to collect scores from entire population?
    - Yes → Collect scores. → Compute mean. → Draw conclusion. → Stop.
    - No → Draw random sample. → Select alpha. → Determine whether test is one-tailed or two-tailed. → Population standard deviation known?
      - No → Proceed to Chapter 12.
      - Yes → Compute standard error of the mean. → Compute z-score. → Does sample mean differ enough from hypothesized population to lead to rejection of null hypotheses under predetermined alpha?
        - No → Retain null hypothesis.
        - Yes → Reject null hypothesis. → Draw conclusion. → Stop.

errors are distributed in a normal or near normal manner around their expected mean of zero.

The expected standard deviation of all means of random samples drawn from a population is known as the standard error of the mean. It is calculated by dividing the standard deviation of the population by the square root of the number in each sample ($\sigma_{\bar{x}} = \sigma/\sqrt{N}$).

One can use the standard error of the mean and the normal curve to approximate which values of a sample mean will lead to the retention of the null hypothesis and which values will lead to its rejection at a given alpha. If it has been decided beforehand that either positive or negative differences between the sample mean and the hypothesized population mean will lead to rejection of the null hypothesis, the test is two-tailed. If only positive or only negative differences will lead to rejection of the null hypothesis, the test is one-tailed.

As can be seen from the preceding flowchart, if the hypothesis concerns a population mean, one first determines whether it is practical to collect scores for the entire population, in which case one would simply compute the population mean and draw his conclusion. If it is not practical to collect scores for the entire population, a random sample is drawn. If the population standard deviation is known, the next steps are to select alpha, decide whether the test will be one-tailed or two-tailed, and compute the standard error of the mean. Then one determines whether the sample mean departs sufficiently from the hypothesized population mean to lead to the rejection of the null hypothesis under the predetermined alpha.

EXERCISES

1. What is a random sample? Tell how one would proceed to draw a random sample of 200 students from a large high school.
2. Assume that you draw a random sample of 64 cases from a population where the mean test score is 25 and the standard deviation is 10.
    (a) What is the probability of obtaining a sample mean between 24 and 26 for the random sample?
    (b) What would be the probability of a sample mean between 24 and 26 if the sample size were reduced to 25?
3. For a population with $\mu = 75$ and $\sigma = 15$, what is the probability that, in a sample of 100, the sample mean $\bar{X}$ will be:
    (a) equal to or less than 79?
    (b) equal to or more than 77?
4. An instructor has given a particular examination for a number of years and has determined that the population mean on the examination is $\mu = 82$ and $\sigma = 8$. His present class of 25 obtains an $\bar{X} = 86$. Should he retain the hypothesis that this class represents the same population the previous classes represented?

5. Assume a sample of 49 children who have a mean score on a test of 77. How probable is it that this is a random sample from a population whose mean is 80 and standard deviation is 7?
6. A random sample of 144 college freshmen had a mean score of 520 on a college aptitude examination.
   (a) Test the hypothesis that the population mean on the examination is 500 ($\sigma = 100$). Test at the 5 percent level.
   (b) Test the hypothesis that the population mean is 510.
7. A sample of 36 scores is randomly selected from a population where $\mu = 50$ and $\sigma = 18$. The probability is .95 that the mean would fall between what two points equidistant from 50?

ANSWERS

1. A random sample is a sample drawn by a chance procedure from a population. To obtain a random sample of 200 from a large high school one could assign I.D. numbers to each student and use a table of random numbers to determine which 200 students to include in the sample.

2. (a) $\sigma_{\bar{x}} = 1.25$
$$z = \frac{24-25}{1.25} = \frac{-1}{1.25} = -.8 \qquad z = \frac{26-25}{1.25} = .8$$
$P = .2881 + .2881 = .5762$ that the sample mean will fall between 24 and 26.

   (b) $\sigma_{\bar{x}} = 2$
$$z = \frac{24-25}{2} = -.5 \qquad z = \frac{26-25}{2} = .5$$
$P = .1915 + .1915 = .3830$ that the sample mean will fall between 24 and 26.

3. (a) $\sigma_{\bar{x}} = 1.5$
$$z = \frac{79-75}{1.5} = 2.67$$
$P = .9962$ that $\bar{X} \leq 79$

   (b) $z = \frac{77-75}{1.5} = 1.33 \qquad P = .0918$ that $\bar{X} \geq 77$

4. $\sigma_{\bar{x}} = 1.6$
$$z = \frac{86-82}{1.6} = 2.5$$
Using the .05 level of significance, he would conclude that this class represents a different population than the previous classes. Fewer than 1 percent of samples from the original population would be expected to have means of 86 or more.

5. $\sigma_{\bar{x}} = 1 \qquad z = \frac{77-80}{1} = -3$
Since $z = -3$, it is not likely that the mean of a random sample would differ this much from the population mean. Fewer than 1 percent of random samples would have a mean as low as 77.

6. (a) $\sigma_{\bar{x}} = 8.33 \qquad z = \frac{520-500}{8.33} = 2.4$

Therefore, reject null at .05 level. It is unlikely that the population mean is 500.

(b) $z = \dfrac{520 - 510}{8.33} = 1.2$

Therefore, retain null. Since $\bar{X}$ and σ are only $1.2\sigma_{\bar{x}}$'s apart, it is quite likely that the population mean is 510.

7. $44.12 - 55.88$

$\bar{X} = \dfrac{18}{\sqrt{36}} = 3 \quad \bar{X} = 50 \pm 1.96(3)$

CHAPTER TWELVE
THE t-TESTS

OBJECTIVES

After completing this chapter the reader should be able to:

1. Describe the general nature of t-curves and tell how they differ from the normal curve.
2. Describe the general meaning of the concept "degrees of freedom" and its application to t-tests.
3. Identify situations in which a t-test is an appropriate method for hypothesis testing.
4. Find needed values in the t-table.
5. Calculate a t-test for hypotheses concerning a population mean.
6. Calculate a t-test for hypotheses concerning differences between means both for independent and for nonindependent samples.
7. Identify situations which require an independent t-test and situations that call for a nonindependent t-test.
8. Compare the standard error of the mean of the independent and nonindependent t-tests.
9. Explain the logic behind the t-tests.
10. Test null hypotheses which state that the correlation in a population is zero.

t-Curves

In Chapter Eleven we discussed the use of the normal curve to determine the probability of differences between sample means and population means.

314 INFERENTIAL STATISTICS

All of our examples involved cases where the population standard deviation (σ) was known. When σ is known, we saw that it was possible to describe the form of a distribution of means from random samples (a sampling distribution). We learned that the sampling distribution of sample means was a normal distribution with standard error of the mean $\sigma_{\bar{x}} = \sigma/\sqrt{n}$. By using the relationship between z-scores and the normal distribution, we tested hypotheses using $z = (\bar{X} - \mu)/\sigma_{\bar{x}}$.

However, in most practical research situations, the population standard deviation is *not* known. In fact, if one knew the population mean and standard deviation one would not have to bother with sampling at all. Thus, when the population standard deviation is not known, it has to be estimated from sample statistics. Subsequently, the estimated population standard deviation is used to estimate the standard error of the mean. As we saw in Chapter 4, the best estimate of the population standard deviation available from sample statistics is the square root of the sum of squares (Σx^2) divided by $n - 1$, that is, $s = \sqrt{\Sigma x^2/(n-1)}$ [Formula (4.11)]. In order to estimate the standard error of the mean, one takes this estimate of the population standard deviation s and divides it by the square root of n. The estimated standard error of the mean (a statistic) is symbolized $s_{\bar{x}}$ to distinguish it from $\sigma_{\bar{x}}$, the true standard error of the mean (a parameter):

$$s_{\bar{x}} = \frac{s}{\sqrt{n}} = \frac{\sqrt{\frac{\Sigma x^2}{n-1}}}{\sqrt{n}}$$

where: $s_{\bar{x}}$ = estimated standard error of the mean
s = estimated standard deviation of the population $\sqrt{\frac{\Sigma x^2}{n-1}}$
n = number in the sample

Formula (12.1)
Estimated Standard Error of the Mean

When the standard error of the mean is based on a standard deviation, s, that is, only an estimate of the population standard deviation rather than on the known parameter, considerable sampling error may be introduced. William Gosset, who wrote under the pseudonym "Student," showed that when the standard error of the mean is calculated using sample statistics ($s_{\bar{x}} = s/\sqrt{n}$) the sampling distribution of the mean is not identical to the normal curve. Consequently, when this estimated standard error of the mean is used as the denominator in the ratio $(\bar{X} - \mu)/s_{\bar{x}}$, the ratio is not normally distributed like the z ratio $[(\bar{X} - \mu)/\sigma_{\bar{x}}]$ that we used in the previous chapter. Furthermore, the statistic calculated from the ratio $(\bar{X} - \mu)/s_{\bar{x}}$ is no longer a z, but rather it is labeled a *t*-statistic.

Thus, when the population standard deviation σ is not known, one estimates the standard error of the mean $s_{\bar{x}}$ and calculates a *t* instead of a *z*.

The t-ratio may be used in the testing of hypotheses just as the z was used. Gosset described a distribution, or actually a family of distributions, which permits the testing of hypotheses with samples when the population is unknown. These distributions are known as the t-*distributions*, or t-*curves*, or Student's t. When testing hypotheses one calculates the t-ratio: $t = (\bar{X} - \mu)/s_{\bar{x}}$ and then refers to the t-curves in order to determine the probability of the t-statistic in the same way that the normal curve is used to determine probability values of z.

As indicated above, the t-distribution is a family of t-curves rather than just a single curve as was the case with the normal distribution. Theoretically there is a different t-distribution for every possible size of sample. We express this by saying that the t-distribution varies as a function of the number of *degrees of freedom* for the sample.

Some of the t-curves are shown in Figure 12.1 along with the normal

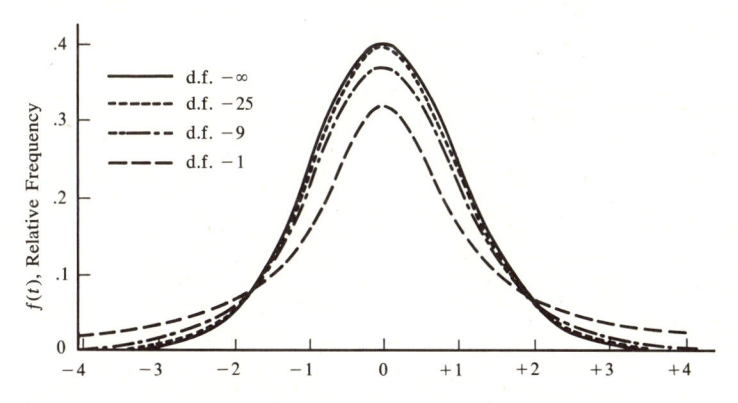

Figure 12.1 t-Curves for Various Degrees of Freedom

curve, the solid line labeled d.f. = ∞. The t-curves are labeled according to their degrees of freedom, abbreviated d.f. Before further discussion of the characteristics of t-curves, let us turn our attention to the concept of degrees of freedom.

Degrees of Freedom

The term "degrees of freedom" (d.f.) refers to the number of values that are free to vary after certain restrictions have been placed on the data. For example, if one had three numbers with the restriction that the sum of those numbers had to equal 25, then two of those numbers could take on any number of values, that is, they are free to vary, but the third would be fixed. That is, if any two of the numbers are known, the third one is determined. If two of the values are 8 and 11, the third must be 6 in order that the sum equals 25. The number of degrees of freedom is 2. In a sample such as this on which a single restriction has been placed, the number of degrees of freedom is $N-1$. Consider a case where one has ten numbers with the restrictions that the first three numbers must add to 25 and all ten numbers must add to 50. With ten numbers and two restrictions, there are $N-2=8$ degrees of freedom.

When one is testing hypotheses about the distribution of sample means around a population mean, the degrees of freedom are the number in a sample minus one (d.f. = $N-1$). The t-curve in Figure 12.1 labeled d.f. = 25 is for a sample of 26, d.f. = 9 is for a sample of ten, and so on. (The labeling of t-curves by degrees of freedom rather than by N may seem pointless at this stage but you will find later that the use of the degrees of freedom concept enables one to apply the t-curves to a wider variety of uses.) Degrees of freedom for the t-curves are a direct function of sample size. As sample size increases the degrees of freedom increase correspondingly.

Characteristics of t-Distributions

The t-distributions are similar in appearance to the normal curve; that is, they are symmetrical and bell-shaped with a mean equal to 0 and a standard deviation slightly greater than 1.[1] The lower case t is the symbol used to represent the standard deviation unit when we use t-curves in the same way that a lower case z is used to represent a standard deviation unit in the normal curve. Just as a z of +1.25 represents 1.25 standard deviations above the mean in the normal curve, a t of +1.25 represents 1.25 standard deviations above the mean in a t-curve. However, the t-curve is lower and more pointed at the peak (leptokwrtic) than the normal curve. The greater the d.f., the more the t-distributions resemble the normal curve. For infinite degrees of freedom, the t-distribution is identical to the normal distribution. Compare the curve for 25 d.f. with the curve for 9 d.f. and for 1 d.f. in Figure 12.1. For 25 d.f., the t-curve is only slightly different from the normal curve. For d.f. = 9, the difference becomes more obvious; the curve is more narrow and the tails are larger than the normal curve. Then for d.f. = 1, the difference between the t-curve and the normal curve is quite apparent. In fact, it is as different from the normal curve as a t-curve could ever be. Thus, as the degrees of freedom become less, the curve spreads out more and has more of its area in the tail of the distribution than does the normal curve. Therefore, the proportion of the area beyond a specific t is greater than the proportion of area beyond the corresponding value of z. The thick tails reflect the greater variability present when the sample size and d.f. are small.

[1] *The standard deviation of the t-curve is actually $\sqrt{d.f./(d.f.-2)}$ which means that as degrees of freedom increase the standard deviation approaches one.*

The implication of the greater area in the tails of the curve is that extreme values are not as unusual in the t-distribution as they are in the normal curve.

Using the t-Distribution for Hypothesis Testing

The t-curves are used to determine probabilities in the same way the normal curve is used. For each value of degrees of freedom a table similar to the normal curve has been derived. In practice complete tables are not required for hypothesis testing. We only need the t-values associated with the commonly used levels of significance for each d.f. value.

In using the normal curve we were particularly interested in certain landmarks such as $z \pm 1.96$ which includes 95 percent of the curve, and $z \pm 2.58$ which includes 99 percent of the curve since these are used to test two-tailed hypotheses at the .05 and .01 levels of significance. Since the principal use of the t-curves is in hypothesis testing, only the landmarks needed for such purposes are included in the t-tables in most statistics texts. Thus, in contrast to the normal curve table, the t-table shows only the critical values that are needed for *significance* at the commonly used levels (.05 and .01). Significance means that the obtained result is not likely to be due to chance. A significant result is one that leads to the rejection of the null hypothesis.

Table A4 in the Appendix gives the t-values for different degrees of freedom and for the commonly used levels of significance. One locates the appropriate number of degrees of freedom in the first column and then proceeds across that row to the column containing the desired probability level. The intersection of row and column will give the value of t needed for significance. For example, with d.f. = 10, the value of t needed (for significance) at the .05 level with a two-tailed test is 2.23. At the .01 level it is 3.17. This means that when d.f. = 10, 95 percent of the total area under the curve falls within ± 2.23 t-units from the mean, and 99 percent of the total area will fall within ± 3.17 t-units from the mean. In the normal curve the distances from the mean that would include 95 and 99 percent of the total area are 1.96 and 2.58 standard deviation units, respectively. The t-curves have a greater area in the two tails of the distribution than does the normal curve. This means that for a given alpha a higher value of t is required than would be required if one were using the normal curve. It follows that in some cases a null hypothesis could be rejected when using the normal curve but would be retained using the t-distribution. For example, a z of 1.96 would lead one to reject the null hypothesis at the .05 level with a two-tailed test and to conclude that the finding was significant and not due to chance.

However, a t of 1.96 with 30 degrees of freedom would lead to the retention of the null hypothesis at the .05 level with a two-tailed test. The smaller the degrees of freedom the greater the value of t required to reject a null hypothesis. The larger the degrees of freedom the smaller the value of t required until on reaching d.f. = ∞ at the bottom of the table, the values are identical to those of the normal curve.

When degrees of freedom reach 30 the t-curve is not very different from the normal curve and increases in degrees of freedom above 30 result in only small decreases in the value of t required to reject a null hypothesis. For that reason, in Table A4 the d.f. column skips from 30 to 40, then to 60, then to 120. When the d.f. value in a study falls between two of those values it is customary to use the tabled t-values for the *lower* of the two numbers. In this way one is being conservative and is minimizing the risk of rejecting a null hypothesis that is true. Thus, if one wanted the t-value for a two-tailed test at the .05 level for 35 d.f., one would go to d.f. = 30 and use t = 2.04. Similarly, the critical value of t to use for 75 d.f. at the two-tailed .01 level would be 2.66 which is the value listed for 60 d.f.

Exercises

1. What difference between the t-curves and the normal curve is most important when the t-curves are used for hypothesis testing?
2. With six numbers and one restriction on those numbers how many degrees of freedom are there?
3. Which t-curves least resemble the normal curve?
4. With 38 degrees of freedom which row in Table A4 would be used?
5. Why is the t-table used instead of the normal curve table for hypothesis testing?
6. Which t-curve is identical to the normal curve?
7. What proportion of the area of the t-distribution falls:
    (a) above t = 2.947 when d.f. = 15
    (b) below t = −1.812 when d.f. = 10
    (c) between t = ±2.086 when d.f. = 20
    (d) above t = 2.228 when d.f. = 10
8. Find the value of t for d.f. = 25 such that the proportion of the area
    (a) to the left of t is .025
    (b) to the right of t is .005
    (c) to the left of t is .0005
    (d) between the mean and t is .45

Solutions

1. There is a greater proportion of area in the extreme ends of the t-curves.
2. 6 − 1 = 5

3. those for small degrees of freedom
4. d.f. = 30
5. The t-curves are used when the standard error of the mean is calculated from s rather than σ.
6. The curve for d.f. = ∞
7. (a) .005 (b) .05 (c) .95 (d) .025
8. (a) −2.06 (b) 2.787 (c) −3.725 (d) either 1.708 or −1.708

Testing Hypotheses Concerning a Population Mean

The *t*-test provides a way to test hypotheses concerning the location of a population mean. One hypothesizes a specified value for the population mean and then determines if that hypothesized value is consistent with the data obtained from the sample. That is, one determines from the *t*-table if the obtained *t*-statistic is probable or improbable. To test hypotheses using the *t*-distribution, one follows the same general procedure that was used with the normal distribution. The first step is to calculate the *t*-ratio for the data. This ratio tells one how far the sample mean is from the hypothesized population mean in terms of *t*-units.

$$t = \frac{\bar{X} - \mu}{s_{\bar{X}}}$$

where: t = the *t*-ratio
$\bar{X}$ = the sample mean
μ = the hypothesized population mean
$s_{\bar{X}}$ = the estimated standard error of the mean

Formula (12.2)
t-Ratio

The *t*-ratio resulting from this calculation is then compared with the tabled value of *t* for the degrees of freedom and specified alpha. If the absolute value of the calculated *t* equals or is larger than the tabled value, the difference between the sample mean and population mean is significant and the null hypothesis of no difference between the two means is rejected. If the calculated value is smaller than the tabled value, the difference is not significant and the null hypothesis is retained. The table values of *t* are the cut-off points needed to reject the null hypothesis at the predetermined probability level.

Let us consider a practical application of this process. Assume that we want to make a two-tailed test of the hypothesis that the mean reading achievement score of a population of fourth graders is 90. We draw a random sample of 25 fourth graders and find that the mean of this sample is 92 and the sum of the squared deviation scores (Σx^2) is 1536. Using Formula

320 INFERENTIAL STATISTICS

(4.11) to estimate the standard deviation of the population (s) we have:

$$s = \sqrt{\frac{\Sigma x^2}{n-1}} = \sqrt{\frac{1{,}536}{25-1}} = \sqrt{64} = 8$$

Now we can calculate the estimated standard error of the mean:

$$s_{\bar{x}} = \frac{s}{\sqrt{n}} = \frac{8}{\sqrt{25}} = 1.6$$

The null hypothesis (H_0) states: The mean of the population from which the sample was drawn is equal to 90; H_0: $\mu = 90$. Alternative hypotheses are H_1: $\mu > 90$ and H_2: $\mu < 90$. Next we calculate the value of the t-ratio using Formula (12.2):

$$t = \frac{\bar{X} - \mu}{s_{\bar{x}}} = \frac{92 - 90}{1.6} = 1.25$$

The mean of the random sample is 1.25 t-units from the hypothesized mean of the population. For 24 ($n-1$) degrees of freedom, the t-table shows that a t of 2.064 is required for significance at the .05 level. Our calculated t-ratio of 1.25 is less than the tabled value, 2.064. This means that there are more than five chances in 100 of obtaining this much difference between the sample and the hypothesized population mean through chance alone. In other words, it is quite probable that a random sample mean could deviate by that much from a population mean through chance alone. We must conclude that the evidence is insufficient for rejecting the null hypothesis. Rather, the null would be retained and we would also retain both the alternative hypotheses. The population mean may be $<$ or $>$ 90 or it may $= 90$. In Figure 12.2 we see that the sample mean was not different enough from the hypothesized population mean to fall in the area of rejection but rather it fell within the area of retention.

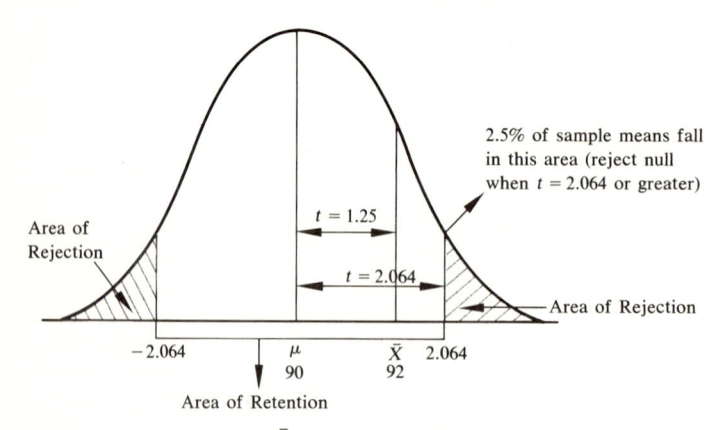

Figure 12.2 t-Curve for $\bar{X}$ when H_0: $\mu = 90$ is True and d.f. $= 24$

Let us consider another example. A high school football coach takes umbrage when some of his colleagues maintain that the boys who try out for football tend to be less competent in verbal skills than the average. The coach maintains the opposite is true and that those who try out for football are above average in verbal skills. Fortunately, a verbal skills test has been administered to all the students in the school and the mean score for the school has been calculated and is known to be 92. The standard deviation for the scores of all the students has not been calculated. The coach takes a random sample of 17 subjects from a list of all the boys who have tried out for football. For his sample he calculates a mean $\bar{X}$ of 104 and a standard deviation [using Formula (4.11), $\sqrt{\Sigma x^2/(n-1)}$] of 15. He decides to do a two-tailed test with an alpha of .05. The null hypothesis (H_0) states: The mean of the population from which the sample was drawn (all boys who tried out for football) equals 92: $H_0 = 92$. The alternative hypotheses are $H_{a_1}: \mu > 92$ and $H_{a_2}: \mu < 92$.

$$s_{\bar{X}} = \frac{s}{\sqrt{n}} = \frac{15}{4.12} = 3.64 \qquad t = \frac{\bar{X} - \mu}{s_{\bar{X}}} = \frac{104 - 92}{3.64} = 3.3$$

The obtained $t = 3.3$ indicates that the sample mean and the population mean are 3.3 t-units apart. The t-table indicates that with $N - 1$ or 16 degrees of freedom, the t needed for significance for a two-tailed test is 2.120. Since $3.3 > 2.12$, one rejects the null hypothesis that $\mu = 92$ and also rejects H_2 that $\mu < 92$. That is, it is unlikely that a random sample mean would deviate by 3.3 t-units from the population mean; thus, it is unlikely that the sample with a mean of 104 was drawn from a population with mean = 92. One accepts H_1 and concludes that the population mean μ is greater than 92 ($\mu > 92$).

The football coach could conclude that he has sufficient evidence to maintain that the population of boys who try out for football score higher on verbal skills than the general population in the school.

Exercises

1. Test the hypothesis that $\mu = 75$ with a random sample of 16 cases that yields an $\bar{X}$ of 81 and $s = 8$. Do a two-tailed test at the .05 level of significance.
2. An instructor gives his class of 25 an examination which, based on years of use, has been shown to have mean = 80. His class obtains an $\bar{X} = 84$ and $s = 10$ on the test. Is the difference between this class' mean and 80 statistically significant at the .05 level with a two-tailed test?
3. An investigator wishes to test the hypothesis that the mean of a certain population is 80. A random sample of ten is drawn whose scores are: 15, 45, 60, 60, 75, 75, 90, 105, 105, 120. Do a two-tailed test of the null hypothesis that $\mu = 80$. Use the .05 level of significance.

Solutions

1. $s_{\bar{x}} = \dfrac{s}{\sqrt{n}}$ $s_{\bar{x}} = \dfrac{8}{4} = 2$

   $t = \dfrac{81 - 75}{2} = 3$ ($t_{.05}$ for 15 d.f. = 2.13)

   Therefore, reject null hypothesis and accept the alternate that $\mu > 75$.

2. $t = \dfrac{84 - 80}{2} = 2$ ($t_{.05}$ for 24 d.f. = 2.06)

   Retain null hypothesis. Class' mean of 84 is not significantly different from μ of 80.

3. $\Sigma X = 750$ $\bar{X} = 75$
   $\Sigma x^2 = 9,000$, $s_{\bar{x}} = 10$

   $t = \dfrac{75 - 80}{10} = -.5$ ($t_{.05}$ for 9 d.f. = 2.26)

   Retain null hypothesis.

The *t*-Test for Significance of the Difference between Means

Thus far we have considered the use of the *t*-test for the testing of hypotheses concerning the value of population means. All of the problems that were discussed involved only a single sample. However, there are few practical applications of these single sample procedures. Only rarely is research concerned with a single sample; much more frequently researchers work with two or more groups and attempt to ascertain the effects of different experimental treatments. The main purpose of the discussion of the one sample procedures was to illustrate the concepts and the reasoning involved in the use of the *t*-curves in hypothesis testing.

Most behavioral research involves the comparison of two or more means or other statistics derived either from samples from two populations or samples from the same population under different experimental treatments. For example, suppose we randomly assign students to two groups in order to compare the effectiveness of two methods of teaching reading. The groups are placed in different instructional situations for a semester. They are then tested in reading and their mean scores determined. The two groups will most probably show some difference in their mean achievement scores. Since we know that some difference in the means of random samples from a population may be expected due simply to sampling error, we need a way to determine if the difference that appears between our two groups is merely a result of chance factors, or sampling error, or if it is such that it can be attributed to the differential teaching situation. The essential problem is one of determining whether the difference in sample means is large enough to justify the conclusion that one group is truly different from the other as a result of the treatment. The remainder of this chapter will be concerned with the statistical procedures that can be used to determine whether the difference in the means of samples is a *significant* difference, that is, one that is *not* likely to be due to chance.

Distribution of the Differences between Means

In order to test hypotheses concerning the differences between means we need to know how much difference between two means should be credited to chance alone. We know that if we pull two samples at random from the same population their means will rarely be identical. Through mathematical deduction it has been shown that differences between an infinite number of pairs of sample means, drawn from a sample population, form a predictable distribution of differences. Assume the simultaneous drawing of an infinite number of *pairs* of random samples from a population. The mean score for each sample is determined and the *difference* between the means ($\bar{X}_1 - \bar{X}_2$) in each pair of samples is found. Assume that a frequency distribution of the differences in means is set up. This distribution is called *the distribution of differences between sample means*. As is the case with the means of single samples pulled from a population, the distribution of the differences between pairs of means follows the laws of probability and can therefore be predicted.

1. *The expected mean of the differences between pairs of means drawn from a common population is zero.*

   Since both samples are drawn at random from a single population the number of cases where the first mean would be larger than the second ($\bar{X}_1 > \bar{X}_2$) would be expected to equal the number of cases where the second is larger than the first ($\bar{X}_1 < \bar{X}_2$); therefore given an infinite number of samples we expect the mean of the $\bar{X}_1 - \bar{X}_2$ differences to be zero.

2. *The variability of the differences between pairs of means drawn from a common population is an inverse function of sample size.*

   In Chapter Eleven we saw how sampling error decreases as sample size increases. As a result of this both sample means will vary less and less from the population mean as sample size increases. Therefore the expected difference between sample means decreases as sample size increases.

3. *The variability of the differences between pairs of means is a direct function of the standard deviation of the population.*

324 INFERENTIAL STATISTICS

We have seen in Chapter Eleven that the variation of sample means is a direct function of the standard deviation of a population; that is, the greater the variability in the population the greater the variability among sample means. The variability of differences between pairs of means drawn from a common population is also a direct function of the standard deviation of the population.

4. *The distribution of differences between pairs of means drawn from a common population assumes the shape of the normal curve.*

Small differences between sample means will occur more frequently than large differences between sample means. The larger the difference the less likely it is to occur by chance. It has been shown through mathematical deduction that a distribution of the differences between sample means assumes the shape of the normal curve.

In summary, the distribution of the *differences* between pairs of means drawn from a common population is normal with a mean of zero. This distribution also has a standard deviation called the *standard error of the difference*.

Standard Error of the Differences between Two Means

The standard deviation of the distribution of the differences between pairs of means from a common population is called the *standard error of the difference between two means*. It takes into account the variability of scores in both samples $[\sqrt{\sum x_1^2/(n_1-1) + \sum x_2^2/(n_2-1)}]$ and the number in the samples (n_1 and n_2). These elements are combined into one formula:

$$s_{\bar{x}_1 - \bar{x}_2} = \sqrt{\frac{\sum x_1^2 + \sum x_2^2}{n_1 + n_2 - 2} \left(\frac{1}{n_1} + \frac{1}{n_2}\right)}$$

where: $s_{\bar{x}_1 - \bar{x}_2}$ = the estimated standard error of the difference between two means

$\sum x_1^2$ = the sum of the squared deviations in sample 1

$\sum x_2^2$ = the sum of the squared deviations in sample 2

n_1 = the number of cases in sample 1

n_2 = the number of cases in sample 2

Formula (12.3)
Estimated Standard Error of the Difference between Two Means

The *t*-Test for the Difference between Two Means

Once the standard error of the difference for the population is obtained, that is, the estimated variability of the difference between the means of two random samples that can be expected due to chance factors, one can use this to determine whether the *observed* difference between two means is likely to be a function of chance or not. That is, one creates a ratio by dividing the observed difference by the variation of differences that can be expected due to chance factors. This ratio is known as the *t-ratio* or the *t-test* for the significance of the difference between means:

$$t = \frac{\bar{X}_1 - \bar{X}_2}{s_{\bar{X}_1 - \bar{X}_2}}$$

where: $\bar{X}_1 - \bar{X}_2$ = the observed difference between two means
$s_{\bar{X}_1 - \bar{X}_2}$ = the standard error of the difference between means (expected difference due to chance)

Formula (12.4)
t-Test for the Difference between Two Means (definition formula)

It should be pointed out that if the population standard deviation σ was known, one could calculate the standard error of the differences between mean $\sigma_{\bar{X}_1 - \bar{X}_2}$ and then a z-ratio $[z = (\bar{X}_1 - \bar{X}_2)/\sigma_{\bar{X}_1 - \bar{X}_2}]$. In this case, the normal curve would be used to determine probability rather than the *t*-curves. However, it is a rare situation when σ is known. So again we must estimate the standard error of the difference and use *t*.

To convert Formula (12.4) to a computational formula one substitutes the computation for $s_{\bar{X}_1 - \bar{X}_2}$ [Formula (12.4)] into Formula (12.5) as follows:

$$t = \frac{\bar{X}_1 - \bar{X}_2}{\sqrt{\frac{\Sigma x_1^2 + \Sigma x_2^2}{n_1 + n_2 - 2}\left(\frac{1}{n_1} + \frac{1}{n_2}\right)}}$$

Formula (12.5)
t-Test for the Difference between Two Means (deviation scores formula)

In the most widely used formula, the *t*-test is calculated directly from raw scores:

$$t = \frac{\bar{X}_1 - \bar{X}_2}{\sqrt{\frac{[\Sigma X_1^2 - ((\Sigma X_1)^2/n_1)] + [\Sigma X_2^2 - ((\Sigma X_2)^2/n_2)]}{n_1 + n_2 - 2}\left(\frac{1}{n_1} + \frac{1}{n_2}\right)}}$$

where: $\bar{X}_1 - \bar{X}_2$ = the difference between the two means
ΣX_1^2 = the sum of the squares of each *X*-score in group 1
ΣX_2^2 = the sum of the squares of each *X*-score in group 2
$(\Sigma X_1)^2$ = the sum of the raw scores in group 1 squared
$(\Sigma X_2)^2$ = the sum of the raw scores in group 2 squared
n_1 = the number of cases in group 1
n_2 = the number of cases in group 2

Formula (12.6)
t-Test for the Difference between Two Means (raw score computation formula)

The t-ratio indicates how far apart the two sample means are in terms of standard error of the difference between means ($s_{\bar{X}_1-\bar{X}_2}$). When we have calculated the t-ratio we can go on to determine the probability of getting various size differences between means as a result of sampling error. We refer the t-ratio calculated through Formula (12.5) to the t-table in order to establish whether the difference is likely to be a function of chance or not.

A typical application of the t-test can be illustrated by the experiment mentioned earlier in which one wished to determine the effectiveness of a new method of teaching reading. One would draw two random samples from a population and one (the experimental group) would receive the new method of instruction and the other (the control group) would receive the conventional method of reading instruction. The random selection of samples from the same population should result in no *significant* differences between the two group means at the beginning of the experiment. This is because the mean of the differences between means in pairs of random samples drawn from the same population tends to be zero. The two groups are then exposed to different methods of reading instruction. After a semester some criterion measure of reading achievement is administered to both groups and the two means computed. Thus, one will have $\bar{X}_1$ for the experimental group and $\bar{X}_2$ for the control group. If there is a difference between $\bar{X}_1$ and $\bar{X}_2$, one must decide if that difference is "real," that is, if it is a statistically *significant* difference or if it is a difference that could be accounted for by chance alone. As you have learned, one can expect some differences between random samples due to chance even in the absence of any experimental treatment.

The Null Hypothesis

The *null hypothesis* states that the two samples to be tested for mean difference are nothing more than one of many pairs of random samples from the same population where the mean of the differences in means is zero; that is, the difference between the means is a function of chance alone. If the difference between the two means could easily be a result of chance this null hypothesis is retained. If the difference is unlikely to be a result of chance the null hypothesis is rejected and it is reasonable to assume that the difference between means is due to the difference in treatment.

Although the researcher started with two random samples from the same population, he has, through the application of two different treatments, made them into random samples from *different treatment populations*. So the sample means $\bar{X}_1$ and $\bar{X}_2$ are really estimates of the means of the populations, that is, the hypothetical populations that would exist if all such subjects were given these particular methods of instruction. Thus, $\bar{X}_1$ and $\bar{X}_2$

are estimates of μ_1 and μ_2. The null hypothesis states that there is no difference between the population means μ_1 and μ_2; that is, H_0: $\mu_1 = \mu_2$ or $\mu_1 - \mu_2 = 0$. In other words, the null hypothesis states that the experimental treatment will have no effect and that the two groups will not be significantly different. If this is true, then $\mu_1 = \mu_2$.

The Logic of the *t*-Test

The magnitude of the difference between the means of the two samples must be evaluated so that the researcher can decide whether the observed difference probably is or probably is not a result of chance. The *t*-ratio $(\bar{X}_1 - \bar{X}_2)/s_{\bar{X}_1 - \bar{X}_2}$ is used for this purpose.

The numerator of the *t*-ratio $(\bar{X}_1 - \bar{X}_2)$ is the actual difference that has been observed between two groups. The denominator $(s_{\bar{X}_1 - \bar{X}_2})$ is an estimate of how much the distribution of the mean difference varies by chance alone. This denominator is a function of: (1) the number in the samples, $n_1 + n_2$. The larger the number, the less random variation to be expected in sample means; (2) the variation within the groups, $\Sigma x_1^2/(n_1 - 1) + \Sigma x_2^2/(n_2 - 1)$. The greater the variation *within* groups, the greater the random differences to be expected between groups. Because the denominator is a measure of how much apparent difference can be expected through chance alone, it is called the *error term* of the *t*-test.

We examine the ratio of the observed difference between the two means (numerator) to the estimated standard error of the difference between the means (denominator), that is, we determine the ratio between the observed difference and the variation that can be accounted for by chance alone. If the null hypothesis of no difference between the means is true, the expected value of *t* will be one. As the value of *t* becomes larger, it becomes more likely that the two sample means do differ significantly. One refers to the *t*-table in order to determine the probability associated with different values of *t* as obtained from the *t*-ratio. The obtained *t* is compared with the tabled *t* at the appropriate d.f.. For the *t*-test, the d.f. = $n_1 + n_2 - 2$. If the obtained value of *t* equals or exceeds the tabled value at the desired level of significance, the null hypothesis is rejected and one concludes that the observed difference between the means is significant at the predetermined alpha. One can conclude that there is reasonable evidence that the experimental treatment resulted in the difference between the group means. If the obtained *t*-value is smaller than the tabled value, the null hypothesis is retained because there is not sufficient evidence that the difference between the means is other than a function of chance. One cannot, in such a case, conclude that the experimental treatments had any differential effect on the groups.

To summarize: The researcher must determine the probability of obtaining a difference in means equal to or greater than the obtained difference $\bar{X}_1 - \bar{X}_2$ when drawing samples at random from populations. Assuming the null hypothesis to be true, one may use the known theoretical distribution of t to find the probabilities associated with different values of t.

If the probability of such a difference as a result of chance is small, then the null hypothesis may be rejected. In other words, the observed difference in the two means is greater than the difference expected through chance and may be attributed to the effects of the experimental treatment. The t is said to be significant.

If the probability of such a difference in means as a result of chance is great, then one retains the null hypothesis and concludes that the difference in means could be attributed to sampling error instead of the experimental treatment. The t is said to be not significant.

Knowing the degrees of freedom, one may determine from the t-table the t-values associated with certain probabilities. For example, assume a situation with 60 degrees of freedom. Entering the table with 60 d.f., one finds a t of 2.00 in the column headed .05. This indicates that with a true null

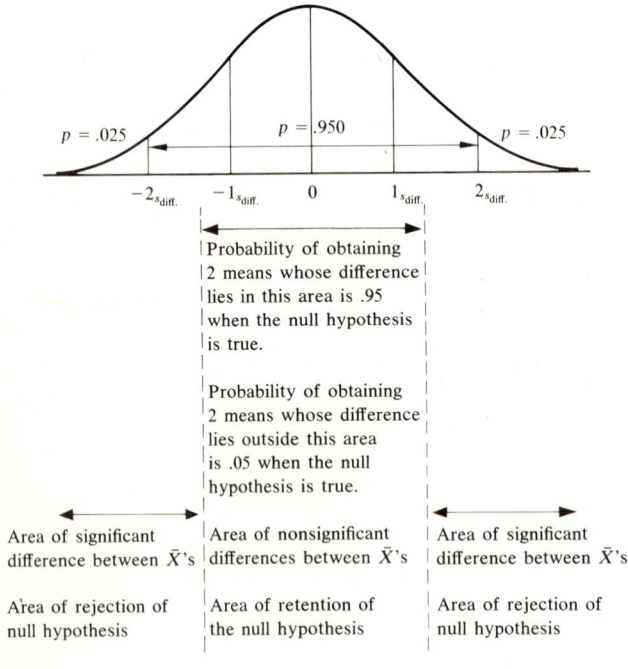

Figure 12.3 t-Distribution of Differences between Means for d.f. = 60

hypothesis and 60 d.f., a t of $+2.00$ or more, or -2.00 or less, will occur by chance 5 times in 100, or 5 percent of the time. In other words, with d.f. = 60 a t of ± 2.00 is required for significance at the .05 level and hence necessary to reject the null hypothesis when a two-tailed test is being used.

Figure 12.3 shows the t-distribution of differences between means for 60 degrees of freedom.

Assumptions Necessary for *t*-Test

Some basic assumptions must be made before a t-test can be used to analyze a difference between means. The populations from which the samples are drawn are assumed to be normal and to have equal standard deviations. However, provided that the samples are not too small and that the numbers in each sample are about equal, violation of these assumptions is not usually serious.[2]

Exercise

Two classes of 25 students each obtained the following results on the final examination in statistics:

$$\text{Class 1:} \quad \bar{X}_1 = 85, \quad \Sigma x_1^2 = 400.15$$
$$\text{Class 2:} \quad \bar{X}_2 = 79, \quad \Sigma x_2^2 = 989.85$$

Test the hypothesis that the two classes do not differ in average performance on the exam, that is, $\mu_1 = \mu_2$, employing $a = .05$.

Solution

$$s_{\bar{x}_1 - \bar{x}_2} = \sqrt{\frac{\Sigma x_1^2 + \Sigma x_2^2}{n_1 + n_2 - 2}\left(\frac{1}{n_1} + \frac{1}{n_2}\right)} = \sqrt{\left(\frac{1{,}390}{48}\right)\left(\frac{1}{25} + \frac{1}{25}\right)}$$

$$= \sqrt{(28.95)(.08)} = \sqrt{2.3160} = 1.52$$

$$t = \frac{\bar{X}_1 - \bar{X}_2}{s_{\bar{x}_1 - \bar{x}_2}} = \frac{85 - 79}{1.52} = 3.95 \text{ (significant)}$$

Reject the null hypothesis and conclude that the two classes do differ significantly in their performance.

---

[2] For additional discussion concerning the assumptions of the t-test see: W. L. Hays, *Statistics* (New York: Holt, Rinehart and Winston, 1963), pp. 321–322.

Significance of the Difference between Two Means: Independent Samples

In our discussion thus far, we have assumed that the subjects in the two samples were *independently* and randomly drawn from the same population or from two populations. That is, we were considering independent samples. However, some samples are not independent but are related to each other. Thus we must make a distinction between independent and dependent or related samples. This distinction is important because there are two different procedures for testing the significance of the difference between means—one procedure used for independent samples and another used for correlated or dependent samples.

Independent samples are those in which the selection of cases in one sample is not influenced by the selection of cases in the second sample. As the name implies, they are completely independent of each other; there is no logical pairing of subjects in the two groups or any reason to connect any given measure in one sample with measures in the other sample. There are two types of independent samples typically used in research: (1) The independent samples may be random samples from *two different populations* where one wishes to determine whether the two populations differ significantly on some criterion variable. For example, one may wish to determine whether male or female subjects have the higher mean score on a verbal reasoning test. Thus, two distinct populations will be compared using random samples drawn from the two populations. These would be independent samples because the selection of the male subjects in the one sample is not influenced by the female subjects chosen in the other sample. Furthermore, knowing any particular male's score would not tell us anything about any particular female's score. One would then compare the mean of the male sample with the mean of the female sample and attempt to infer whether the means of the populations they represent are also different or whether the two populations (male and female) have the same mean with respect to verbal reasoning. (2) The independent samples may also be two random samples drawn from the *same* original population but exposed to two different treatments or experimental conditions. If all factors are similar except the difference in treatment, then any significant difference in the sample means on a postexperimental measure may be attributed to the difference in treatment.

The procedure for the analysis of the data is the same regardless of which of the above methods is used for selecting the samples. The important aspect is that they are *independent samples*. The measures from independent samples are symbolized X_1 and X_2 and one wishes to compare their means $\bar{X}_1$ and $\bar{X}_2$. The formulas for $s_{\bar{X}_1-\bar{X}_2}$ and the t-ratio that we have already presented in this chapter are the ones appropriate for use with independent samples. Let us consider another example involving the t-test for the significance of the difference between means of independent samples.

Example A researcher is interested in determining the effects of two methods of reinforcement on the learning of children. Two random samples from a population of third graders are chosen. One sample receives Method A reinforcement and the second sample receives Method B reinforcement. After a period of time, the two samples are compared on a criterion measure of learning. The results are as follows:

TABLE 12.1 Scores of Two Groups on a Measure of Learning

GROUP 1 METHOD A REINFORCEMENT		GROUP 2 METHOD B REINFORCEMENT	
X_1	X_1^2	X_2	X_2^2
16	256	10	100
12	144	7	49
11	121	7	49
8	64	6	36
9	81	5	25
4	16		
$\Sigma X = 60$	$\Sigma X^2 = 682$	$\Sigma X = 35$	$\Sigma X^2 = 259$
$\bar{X}_1 = 10$, $n_1 = 6$		$\bar{X}_2 = 7$, $n_2 = 5$	

The null hypothesis (H_0) states: There is no difference between the population means of the Method A group and the Method B group on the learning test. That is, $\mu_1 = \mu_2$ or $\mu_1 - \mu_2 = 0$. The alternative hypothesis (H_1) states: There is a difference between the population means of the two groups on the learning test. That is, $\mu_1 \neq \mu_2$. Since the alternative hypothesis does not specify the direction of the difference, a two-tailed test of significance will be used with an alpha of .05.

The first step is to compute an unbiased estimate of the standard error of the difference between means. Since we have independent samples, the $s_{\bar{X}_1 - \bar{X}_2}$ can be found by using Formula (12.3):

$$s_{\bar{X}_1 - \bar{X}_2} = \sqrt{\frac{\Sigma x_1^2 + \Sigma x_2^2}{n_1 + n_2 - 2}\left(\frac{1}{n_1} + \frac{1}{n_2}\right)} = \sqrt{\frac{82 + 14}{6 + 5 - 2}\left(\frac{1}{6} + \frac{1}{5}\right)}$$

$$= \sqrt{\left(\frac{96}{9}\right)\frac{11}{30}} = \sqrt{3.9111} = 1.98$$

The sum of squares for group 1 Σx_1^2 is:

$$\Sigma x_1^2 = \Sigma X_1^2 - \frac{(\Sigma X)^2}{n_1}$$

$$\Sigma x_1^2 = 682 - \frac{(60)^2}{6} = 82$$

The sum of squares for group 2 Σx_2^2 is:

$$\Sigma x_2^2 = \Sigma X_2^2 - \frac{(\Sigma X)^2}{n_2}$$

$$\Sigma x_2^2 = 259 - \frac{(35)^2}{5} = 14$$

The standard error of the difference of 1.98 indicates the difference that would be expected through chance factors if the null hypothesis is true. That is, a difference of 1.98 would be expected between the mean scores of the two groups (random sample from a population) if they had not been exposed to differential treatments. Now let us compare this chance difference with the difference that was observed between the groups after the experiment. This is done with the *t*-ratio:

$$t = \frac{\bar{X}_1 - \bar{X}_2}{s_{\bar{x}_1 - \bar{x}_2}} = \frac{10 - 7}{1.98} = \frac{3}{1.98} = 1.52$$

We have arrived at a *t*-value by steps; that is, we computed the $s_{\bar{x}_1 - \bar{x}_2}$ separately and then substituted this value in the *t*-ratio formula. The usual procedure, as we mentioned before, is to place the computational formula for the $s_{\bar{x}_1 - \bar{x}_2}$ directly into the denominator of the *t*-ratio and proceed with the one big formula, Formula (12.6):

$$t = \frac{\bar{X}_1 - \bar{X}_2}{\sqrt{\frac{[\Sigma X_1^2 - ((\Sigma X_1)^2/n_1)] + [\Sigma X_2^2 - ((\Sigma X_2)^2/n_2)]}{n_1 + n_2 - 2}\left(\frac{1}{n_1} + \frac{1}{n_2}\right)}}$$

$$= \frac{10 - 7}{\sqrt{\frac{[682 - ((60)^2/6)] + [259 - ((35)^2/5)]}{6 + 5 - 2}\left(\frac{1}{6} + \frac{1}{5}\right)}}$$

$$= \frac{3}{\sqrt{\left(\frac{82 + 14}{9}\right)\left(\frac{5}{30} + \frac{6}{30}\right)}}$$

$$= \frac{3}{\sqrt{3.911}} = \frac{3}{1.98} = 1.52$$

The number of degrees of freedom in this problem is $6 + 5 - 2 = 9$. The table indicates that for 9 degrees of freedom a $t \geq 2.26$ is required for significance at the .05 level for a two-tailed test. The difference between the means is not significant at the .05 level. The null hypothesis is retained. The evidence is not good enough to justify concluding that the reinforcement methods had a differential effect on the learning of the two groups.

Exercises

1. Elementary school students are randomly selected and assigned to two groups for a reading experiment. Group A receives individualized instruction and Group B uses the basal reader. The researcher hypothesized that children taught by individualized instruction would achieve at a higher level. A one-tailed test was used. The results on the X reading test given at the end of the study are shown below:

GROUP A	GROUP B
4	0
3	1
6	2
5	3
7	4

(a) State the null hypothesis for this problem.
(b) State the alternative hypothesis for this problem.
(c) The degrees of freedom for this problem are _____
(d) The t-value for these data is _____
(e) Is the t significant at the .05 level?
(f) Is the t significant at the .01 level?
(g) What conclusion would be reached?

2. An experiment investigated the effects of a drug on the learning of nonsense syllables. The following data represent the number of trials required to learn the list for Group 1 (the drug group) and Group 2 (the control [placebo] group). Test H_0 at .05 level.

GROUP 1 (DRUG)	GROUP 2 (PLACEBO)
9	11
9	10
8	10
5	9
5	8
4	8
4	7
3	7
2	6
1	4
$\Sigma X_1 = 50$	$\Sigma X_2 = 80$
$\bar{X}_1 = 5$	$\bar{X}_2 = 8$
$\Sigma x_1^2 = 72$	$\Sigma x_2^2 = 40$

(a) State the null hypothesis and the alternative hypotheses that will be tested in this experiment.
(b) Calculate t.
(c) Draw the appropriate conclusion concerning the effect of the drug on the learning of nonsense syllables.

Solutions

1. (a) Null Hypothesis: There is no difference in the population mean of Group A (those receiving individualized instruction) and Group B (those using the basal reader) on the reading test. That is, $H_0: \mu_A = \mu_B$.
   (b) Alternative Hypothesis: The mean of Group A on the reading test will be higher than the mean of Group B. That is, $H_1: \mu_A > \mu_B$. (The alternative hypothesis specifies the direction; therefore, a one-tailed test of significance will be employed.)
   (c) d.f. = 5 + 5 − 2 = 8
   (d) $t = 3.0$
   (e) yes
   (f) yes. If a one-tailed test, it is significant at .01 level.
   (g) The null hypothesis that $\mu_A = \mu_B$ is rejected and the alternative hypothesis is accepted.

2. (a) Null H_0: There is no difference in the mean number of trials needed to learn list for Group 1 and Group 2; $\mu_1 = \mu_2$.
   Alternative: There is a difference in the mean number of trials needed to learn list for the two groups. $H_1: \mu_1 \neq \mu_2$ (or one could state the alternative hypotheses according to the expected difference. For example, $H_1: \mu_1 > \mu_2$; $H_2: \mu_1 < \mu_2$
   Assuming one chooses the nondirectional alternative hypothesis $\mu_1 \neq \mu_2$, a two-tailed test is employed.
   (b) $t = \dfrac{\bar{X}_1 - \bar{X}_2}{S_{\bar{X}_1 - \bar{X}_2}}$; $t = \dfrac{5 - 8}{1.12}$ $t = -2.67$
   (c) A t of -2.67 with 18 d.f. is significant at .05 level. The null hypothesis is rejected and so is the hypothesis $\mu_1 > \mu_2$. The drug group needed significantly fewer trials. The data suggest that the drug facilitates the learning of nonsense syllables.

Significance of the Difference between Two Means: Correlated or Nonindependent Samples

In the previous section, we considered research designs in which the two samples observed are drawn independently from their respective populations. Some research studies are concerned with two samples that are *not* independent but rather are related to each other. Such samples are referred to as *correlated* or *nonindependent* samples. The following two research designs involve the use of correlated samples:

1. *The Repeated Measurements Design:* In this type of research study, two measurements are made on each of the subjects in a sample. That is, a measurement is made before the experimental treatment and another is made after the experiment, or on the same subjects under different conditions. The researcher then uses a test of significance to determine whether any change has taken place, which presumably could be attributed to the effect of the experimental treatment or the conditions under which the experiment was made. For example, assume that a researcher is interested in the effect of a certain drug on the learning of nonsense syllables. A measure of learning would be obtained from each subject under ordinary learning conditions. The drug would be administered and a second measure of learning obtained from each subject. The before-and-after scores are analyzed for significant differences in the learning of nonsense syllables. This type of design is frequently employed to measure change or gain.

Since the same subjects are being measured in the two situations, one would expect some correlation between the two sets of scores. The two measures are usually designated X and Y to indicate that they are derived from correlated samples. You recall that we used X_1 and X_2 to indicate the scores from independent samples.

2. *The Matched Pairs Design:* In this type of research study, the individuals in the two samples (experimental and control groups) are matched on one or more variables that are known to be correlated with the dependent variable or the criterion variable of the study. For example, if IQ is known to be related to performance on the criterion, then subjects may be matched on the basis of their IQ scores before the beginning of the study. This is done by selecting pairs of subjects on the basis of similarity of their IQ scores and then one member of the pair is randomly assigned to the experimental group and the other is assigned to the control group. As a result, one has two groups equivalent in intelligence, which means that this factor has been controlled. Any postexperimental differences in the two groups could not be attributed to initial differences in intelligence. Of course, the variable on which subjects are matched must be highly correlated with the dependent variable of the study or there is no advantage to the matching process. In a study of school achievement there would be no point in matching subjects on the basis of hair color, for example, since it is not correlated with any of the variables that are being investigated.

Sometimes a research situation will make use of naturally matched pairs already in existence. For example, much research has focused on comparisons of the physical, mental, and personality characteristics of identical twins or of pairs of fraternal twins. In an experimental study, one twin may be randomly assigned to one treatment and the other twin to the other treatment. Fathers and sons or mothers and daughters may be compared on intelligence, height, attitudes, and so on. Heredity is not the only possible basis for natural pairs. For example, husbands and wives or roommates may be compared on attitudes or other behavioral characteristics. Whatever the basis for the pairing, these pairs represent correlated samples since one would expect some relationship between the measures obtained for the two groups.

In the matched pairs research design, the pairing of subjects serves to reduce the sampling error if the variable on which the subjects are paired is related to the dependent variable. However, the matching process does add restrictions to the research and degrees of freedom are lost in the matching process. Matching increases the precision of a study when there is a reasonable relationship between the matching variable and the dependent variable. If there is little or no relationship between these variables more is lost than is gained in matching. For example, in a study where reading comprehension is the dependent variable, it would be a good strategy to match on IQ scores but a bad idea to match on manual dexterity scores.

The *t*-Test for Two Correlated Samples

The logic of the *t*-test for correlated samples is similar to that for independent samples. One compares the actual difference that has been observed between the two samples with the estimate of the difference that could be expected by chance alone, that is, the standard error of the difference. The standard error of the difference between the means for correlated samples is symbolized $s_{\bar{D}}$ to distinguish it from the symbol $s_{\bar{x}_1-\bar{x}_2}$ used for independent samples. As before, the first step is to calculate the estimate of the standard error of the difference which will become the denominator of the *t*-ratio.

Calculation of the Standard Error for a Correlated *t*-Test

The simplest way to calculate the standard error of the difference is known as the direct-difference method. The first step in this method is to find the difference (*D*) between the two scores for each pair, giving $X - Y = D$ for each pair. If this difference is zero, the scores were the same for an individual under both conditions or for both members of the pair. The direction of any difference is indicated by its sign: If *D* is positive, then score-*X* is higher than score-*Y*; if *D* is negative, then score-*Y* is greater than score-*X*. Next, we find the sum of all the *D*-scores and divide by the number of pairs in order to arrive at the mean difference, $\bar{D}$. In formula form this is expressed as $\bar{D} = \Sigma D/N$. The steps in finding the mean difference are illustrated in Table 12.2. (This table illustrates a repeated measures design. The process for matched pairs is identical.) We first find the difference

The t-Tests 337

TABLE 12.2 Repeated Measures Scores of Five Individuals Tested Under Two Different Treatments

INDIVIDUAL	TREATMENT 1 X	TREATMENT 2 Y	DIFFERENCE D	
1	24	25	−1	$\bar{X} = \dfrac{270}{5} = 54$
2	20	23	−3	
3	65	63	2	
4	78	78	0	$\bar{Y} = \dfrac{265}{5} = 53$
5	83	76	7	
	$\Sigma X = 270$	$\Sigma Y = 265$	$\Sigma D = 5$	$\bar{D} = \dfrac{5}{5} = 1$
	$\bar{X} = 54$	$\bar{Y} = 53$	$\bar{D} = 1$	
	$\bar{X} - \bar{Y} = \bar{D}$		$\Sigma X - \Sigma Y = \Sigma D$	
	$54 - 53 = 1$		$270 - 265 = 5$	

between each pair ($X - Y = D$) and then add the differences and divide by the number of pairs to get the mean difference, $\bar{D}$. The mean difference, $\bar{D}$, in Table 12.2 is 1, as you would expect, because $X - Y = D$, then $\bar{X} - \bar{Y} = \bar{D}$. This is an important relationship that will always be found in designs involving paired observations.

The mean difference, $\bar{D}$, is the numerator for the *t*-test for correlated samples, as shown in:

$t = \dfrac{\bar{D}}{s_{\bar{D}}}$ where: $\bar{D}$ = the mean of the difference in scores between pairs of matched observations:

$$\bar{D} = \dfrac{\Sigma D}{N}$$

$s_{\bar{D}}$ = the standard error of the difference between two means when observations are paired   Formula (12.7)
t-Test for Correlated Samples (definition formula)

If you compare this with the independent *t*-test formula, (12.4), you will notice that we have substituted $\bar{D}$ for $\bar{X}_1 - \bar{X}_2$ and $s_{\bar{D}}$ for $s_{\bar{X}_1 - \bar{X}_2}$.

The denominator of Formula (12.7), $s_{\bar{D}}$, the estimated standard error for the mean differences, is computed using:

$s_{\bar{D}} = \sqrt{\dfrac{\Sigma D^2 - \dfrac{(\Sigma D)^2}{N}}{N(N-1)}}$ where: $s_{\bar{D}}$ = standard error of the difference between two means when observations are paired
ΣD^2 = sum of the squared difference scores
ΣD = sum of the difference scores
N = number of pairs   Formula (12.8)
Standard Error of Difference between Paired Scores

After we have the $s_{\bar{D}}$, we are ready to calculate the *t*-ratio: $t = \bar{D}/s_{\bar{D}}$. If we substitute the computational formula for $s_{\bar{D}}$, then we have the computational formula for the *t*-test for the significance of the difference between the means of correlated samples:

$$t = \frac{\bar{D}}{\sqrt{\frac{\Sigma D^2 - ((\Sigma D)^2/N)}{N(N-1)}}}$$

where: $\bar{D}$ = the mean difference actually observed

$\sqrt{\frac{\Sigma D^2 - ((\Sigma D)^2/N)}{N(N-1)}}$ = the standard error of the difference between two means when observations are paired

Formula (12.9)
t-Test for Correlated Samples (computation formula)

When the *t*-value has been calculated we enter the *t*-table with $N-1$ degrees of freedom, where N = the number of pairs of observations. If the calculated *t* equals or exceeds the tabled value, the null hypothesis that the mean difference $\mu_D = 0$ is rejected at the indicated level of significance. The null hypothesis in the *t*-test for correlated means may be written $\mu_D = 0$ or $\mu_x = \mu_y$. The alternative hypotheses may be written:

H_1: $\mu_D > 0$ or $\mu_x > \mu_y$
H_2: $\mu_D < 0$ or $\mu_x < \mu_y$

If one does not wish to specify the direction of the difference, H_1 may be written $\mu_x \neq \mu_y$.

Let us consider an example. A sample of 20 subjects was given an

TABLE 12.3 Scores of Ten Matched Pairs of Subjects in an Experiment

PAIR	METHOD 1 X	METHOD 2 Y	D	D^2
1	10	7	3	9
2	5	3	2	4
3	6	7	−1	1
4	7	5	2	4
5	10	8	2	4
6	6	4	2	4
7	7	5	2	4
8	8	2	6	36
9	6	3	3	9
10	5	6	−1	1
	$\Sigma = 70$ $\bar{X} = 7$	50 5	20 2	76

$$t = \frac{2}{\sqrt{\frac{76 - (20^2/10)}{(10)(9)}}} = \frac{2}{\sqrt{\frac{36}{90}}} = \frac{2}{.632} = 3.16$$

intelligence test and then arranged in matched pairs on the basis of their scores. One member of each pair was assigned to instructional Method 1 and the other to instructional Method 2. Later a dependent variable measure was obtained for each subject. The results are shown in Table 12.3. The null hypothesis states that there is no difference in the scores of students taught by Method 1 and those taught by Method 2. That is, $\mu_x = \mu_y$ or $\Delta = 0$. The alternative hypothesis H_1 is that $\mu_x \neq \mu_y$.

With 9 d.f., the *t*-table indicates that a *t* of 2.262 is required for significance at the .05 level for a two-tailed test. Since the obtained *t*-value is 3.16, one would reject the null hypothesis and conclude that the difference between means is significant.

Exercises

1. What *t*-test is used with a repeated measures design?
2. A physical instructor wants to compare the effectiveness of two methods of training high school freshmen to run the 100-yard dash. He decides to use a matched pairs design. Which of the following would be the most suitable variable for matching subjects? Explain.
    (a) previous performance on 100-yard dash
    (b) socioeconomic status
    (c) reading test scores
    (d) hair length
    (e) vocabulary scores
3. Would you recommend that the physical instructor match subjects on the basis of vocabulary scores if these scores are the only ones that are available to him?
4. An investigator is interested in assessing the effect of a film on the attitudes of students toward a certain issue. A sample of five students is administered an attitude scale before and after viewing the film. The before-and-after scores are as follows:

SUBJECT	BEFORE X	AFTER Y	D	D^2
1	16	20	− 4	16
2	7	27	− 20	400
3	19	32	− 13	169
4	11	33	− 22	484
5	8	14	− 6	36
$\Sigma =$	61	126	− 65	1105
$\bar{X} =$	12.2	25.2		

The value of $\bar{D} = \dfrac{\Sigma D}{N} = \dfrac{-65}{5} = -13$ ($= 12.2 - 25.2$)

Do a two-tailed test at the .05 level of the hypothesis that the mean difference in the population is zero.

Solutions

1. nonindependent t-test
2. One would expect previous performance on the 100-yard dash to be highly correlated with the dependent variable and therefore matching on this variable would reduce the error in the experiment.
3. Previous studies have indicated that there is little correlation between vocabulary and 100-yard dash scores among high school freshmen. Therefore, matching on vocabulary scores would be less desirable than an independent t-test design.
4. $t = \dfrac{\bar{D}}{\sqrt{\dfrac{\Sigma D^2 - [(\Sigma D)^2/N]}{N(N-1)}}} = \dfrac{-13}{\sqrt{\dfrac{1{,}105 - [(-65)^2/5]}{(5)(4)}}} = \dfrac{-13}{\sqrt{\dfrac{260}{20}}} = \dfrac{-13}{3.61} = -3.61$

The subjects' attitudes have changed significantly in a negative direction.

Comparison of the Power of Tests Based on Independent and Correlated Samples

We have indicated earlier in this chapter that the formula used in calculating the standard error of the difference for correlated samples provides a lower standard error than does the formula used with independent samples. Thus, the standard error is less with correlated samples. The lower standard error results in a more sensitive test of the significance of the difference between means and increases the likelihood of rejection of the null hypothesis when it is false. For this reason the test based on correlated samples is described as being a more powerful test. In statistics, the *power* of a test is defined as the probability of rejecting the null hypothesis when it is, in fact, false.

Let us compare the standard error of sample data when computed by the two methods using the same data:

1. Calculation of the standard error of the difference between means assuming independent or unmatched groups using Formula (12.4). (See Table 12.4.)
2. Calculation of the standard error of the difference between means assuming correlated or matched groups using Formula (12.5). (See Table 12.5.)

TABLE 12.4 Estimate of Standard Error of the Difference between Means Assuming Independent or Unmatched Groups

GROUP 1 EXPERIMENTAL		GROUP 2 CONTROL	
X_1	X_1^2	X_2	X_2^2
6	36	20	400
5	25	15	225
4	16	12	144
3	9	10	100
2	4	3	9
$\Sigma = 20$	90	$\Sigma = 60$	878

$n_1 = 5$ $\quad n_2 = 5$
$\bar{X}_1 = 4$ $\quad \bar{X}_2 = 12$
$\Sigma x_1^2 = 10$ $\quad \Sigma x_2^2 = 158$

$$s_{\bar{X}_1 - \bar{X}_2} = \sqrt{\left(\frac{\Sigma x_1^2 + \Sigma x_2^2}{n_1 + n_2 - 2}\right)\left(\frac{1}{n_1} + \frac{1}{n_2}\right)}$$

$$= \sqrt{\left(\frac{10 + 158}{8}\right)\left(\frac{1}{5} + \frac{1}{5}\right)}$$

$$= \sqrt{\left(\frac{168}{8}\right)\left(\frac{2}{5}\right)}$$

$$= \sqrt{8.4}$$

$$= 2.898$$

TABLE 12.5 Estimate of Standard Error of the Difference between Means Assuming Correlated or Matched Groups

MATCHED PAIR	EXPERIMENTAL X	CONTROL Y	D	D^2
1	6	20	−14	196
2	5	15	−10	100
3	4	12	−8	64
4	3	10	−7	49
5	2	3	−1	1
	20	60	−40	410

$$s_{\bar{D}} = \sqrt{\frac{\Sigma D^2 - [(\Sigma D)^2/N]}{N(N-1)}} = \sqrt{\frac{410 - [(-40)^2/5]}{5(4)}} = \sqrt{\frac{90}{20}} = \sqrt{4.5} = 2.121$$

The calculations in Tables 12.4 and 12.5 show clearly that the estimated standard error for data using independent samples is greater than the estimated standard error for correlated samples. In this case, $s_{\bar{x}_1-\bar{x}_2} = 2.898$ while $s_{\bar{D}} = 2.121$. Thus, the effect of matching subjects in this example was to reduce the error. When subjects in an experiment are matched on a variable which is highly related to the dependent variable, the matching will usually reduce the error in the experiment. A reduction of error increases the precision of the experiment resulting in a greater probability of rejecting the null hypothesis when it is false. The reduction of error provided by the pairing of observations is to some degree counterbalanced by a loss of degrees of freedom. You recall that with independent samples the number of degrees of freedom is equal to $n_1 + n_2 - 2$; with correlated samples the number of d.f. is equal to $N - 1$, where N = the number of pairs. Thus, in an experiment using two independent samples of ten subjects each, the number of d.f. = 18; in an experiment involving the same total number of subjects paired into ten pairs, there would only be 9 d.f.. This difference can be very important because as we pointed out earlier, larger t-ratios are required for significance when the degrees of freedom are small. Thus, a matched pairs design is only preferred when the variable on which the subjects are matched is highly related to the dependent variable. In other cases an independent t-test design should be chosen.

The *t*-Tests for Pearson *r* Correlation Coefficients

Another important use for the t-test is in testing hypotheses concerning a population correlation (ρ). The most common null hypothesis in such cases is that the population correlation is zero and that the correlation observed in the sample (r) is a function of chance. For example, we might draw a sample of 30 college freshmen, administer vocabulary and spatial orientation tests to them and find a Pearson r correlation coefficient of .20 between the two measures. The next step is to decide whether this observed coefficient could easily be a result of chance in a population where the true correlation (ρ) is zero. Let us say the investigator has decided to do a two-tailed test at the .05 level.

The t computations have been done for various observed values of the Pearson r and for various alphas and degrees of freedom and compiled in table form. Table A5 in the Appendix shows the required observed values of a sample correlation coefficient r required to reject the null hypothesis.

The degrees of freedom appropriate for tests concerning the Pearson coefficient are the number of paired scores minus two ($N - 2$). In our example the degrees of freedom are $30 - 2 = 28$. We enter Table A5 at the row labeled 28 d.f.. Then we proceed to the column for a two-tailed alpha of .05. At the intersection of this column with the row for 28 d.f., we find the value .361. This tells us that with a two-tailed alpha of .05 an observed Pearson r above +.361 or less than −.361 is required to reject the null hypothesis. The investigator must conclude that there is not sufficient evidence to conclude that there is a correlation between the two variables in the population from which his sample was drawn.

Let us consider another example. A sociologist hypothesizes that among the workers in a particular factory there is a negative correlation between family income and number of children in the family. He draws a random sample of 50 subjects from his population and finds an r of −.31. Since the direction of the correlation is specified, a one-tailed test of significance is used. He is testing his one-tailed hypothesis at the .05 level so he finds that column in Table A5 and then looks for the row for 50 − 2 = 48 d.f.. There is no row for 48 d.f. so he goes to the next lowest value in the table (d.f. = 45). He finds that an r coefficient of .243 or greater is required to reject the null hypothesis. His r of −.31 exceeds that value so he rejects the null hypothesis and concludes that there is sufficient evidence for surmizing that there is a negative correlation between the two variables in the population.

The t-test can also be used to test hypotheses about population correlations other than zero. It can also be used to test the hypothesis that the correlations observed in two samples could have arisen from the same population. Since this is an introductory text we have chosen not to include such tests here. A useful description of these tests may be found in Blommers and Lindquist[3] and in various other texts.

[3] Paul Blommers and E. F. Lindquist, *Elementary Statistical Methods* (Boston: Houghton Mifflin Co., 1960).

Exercises

1. The following correlations were observed in a sample of 25 subjects. Which are statistically significant at the .01 alpha under a two-tailed test?
    (a) $r = .50$
    (b) $r = .44$
    (c) $r = -.62$

2. An investigator hypothesized that age and grade point are negatively correlated among nursing students. He did a one-tailed test with an alpha of .05 with a sample of 75 subjects. Which of the following Pearson r's would lead to rejection of the null hypothesis?
    (a) $r = -.17$
    (b) $r = -.21$
    (c) $r = .25$

Solutions

1. With 23 d.f. (c) is significant, (a) and (b) are not.
2. Only (b) would lead to rejecting the null hypothesis. Correlation (a) is too near zero to justify rejecting the null hypothesis. Since the test was one-tailed only negative r's could lead to rejecting the null hypothesis and any positive r including (c) would lead to the retention of the null hypothesis.

SUMMARY

The t-curves are used when an estimate of the population standard deviation is used to test hypotheses about population means. Different t-curves have been calculated for various values of the degrees of freedom. Degrees of freedom are defined as the number of values that are free to vary after certain restrictions have been placed on the data. When testing the significance of the difference between a sample mean and a hypothesized population mean, the degrees of freedom are $N-1$, the number in the sample minus one. When testing the significance of the difference between two sample means, the degrees of freedom are $n_1 + n_2 - 2$, the number in the first sample plus the number in the second sample minus two. As the degrees of freedom increase, the corresponding t-curve becomes more and more similar to the normal curve.

For testing the difference between a sample mean and a hypothesized population mean, the t-ratio is calculated using the formula:

$$t = \frac{\bar{X} - \mu}{s/\sqrt{N}}$$

This is used when the population standard deviation is not known. The t-ratio can be used to test the significance of the difference between two sample means for both independent and nonindependent (paired) samples. In the former case the formula used is:

$$t = \frac{\bar{X}_1 - \bar{X}_2}{\sqrt{\left(\frac{\Sigma x_1^2 + \Sigma x_2^2}{n_1 + n_2 - 2}\right)\left(\frac{1}{n_1} + \frac{1}{n_2}\right)}}$$

In the latter case it is:

$$t = \frac{\bar{D}}{\sqrt{\frac{\Sigma D^2 - ((\Sigma D)^2/N)}{N(N-1)}}}$$

To test the significance of the observed value of t, find the row in Table A4 for the appropriate degrees of freedom. Then find the column for alpha for

FLOWCHART 12

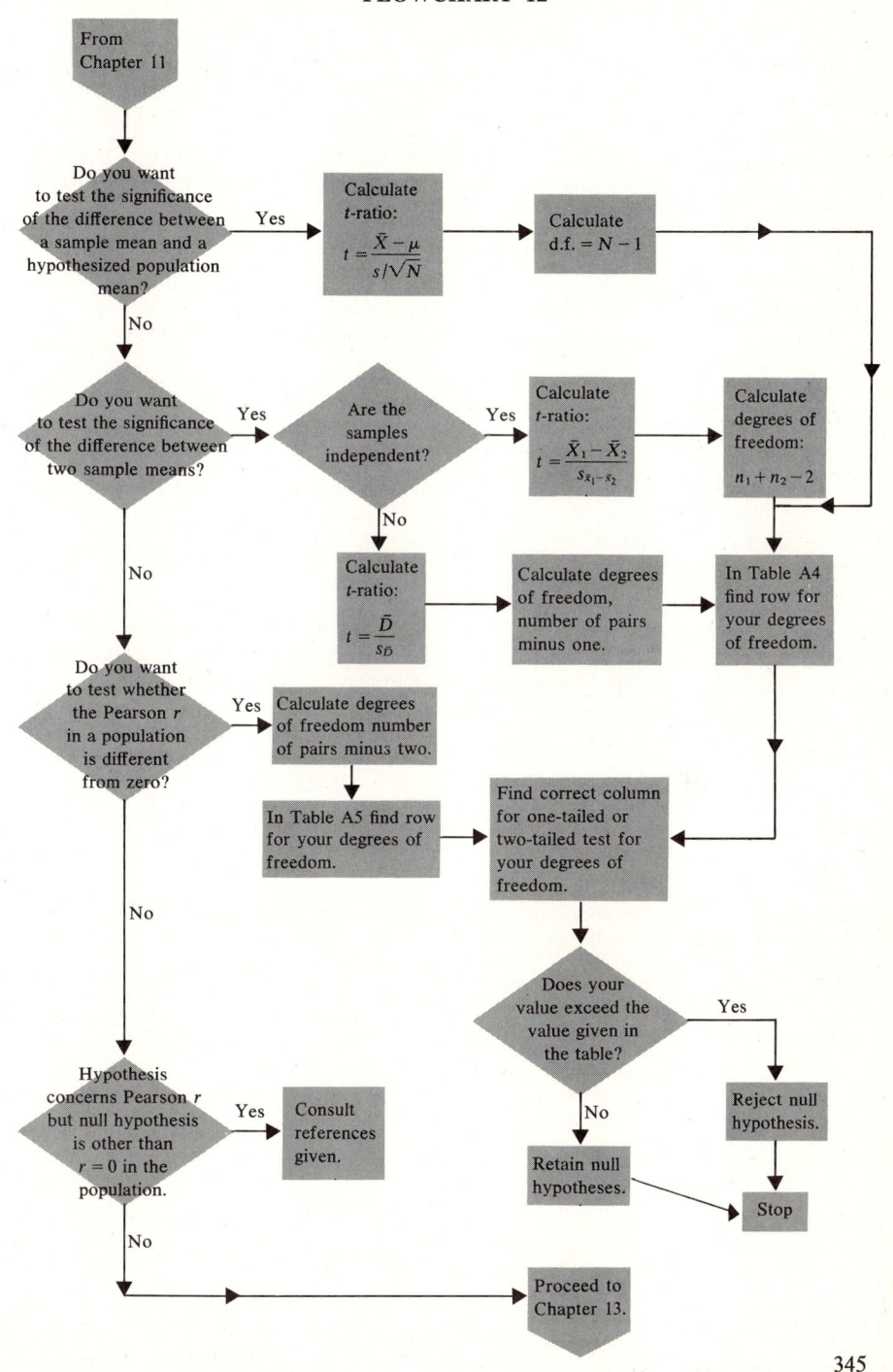

From Chapter 11

↓

Do you want to test the significance of the difference between a sample mean and a hypothesized population mean? — Yes → Calculate *t*-ratio: $t = \dfrac{\bar{X} - \mu}{s/\sqrt{N}}$ → Calculate d.f. = $N - 1$ →

↓ No

Do you want to test the significance of the difference between two sample means? — Yes → **Are the samples independent?**
- Yes → Calculate *t*-ratio: $t = \dfrac{\bar{X}_1 - \bar{X}_2}{s_{\bar{x}_1 - \bar{x}_2}}$ → Calculate degrees of freedom: $n_1 + n_2 - 2$ → In Table A4 find row for your degrees of freedom.
- No → Calculate *t*-ratio: $t = \dfrac{\bar{D}}{s_{\bar{D}}}$ → Calculate degrees of freedom, number of pairs minus one. →

↓ No

Do you want to test whether the Pearson *r* in a population is different from zero? — Yes → Calculate degrees of freedom number of pairs minus two. → In Table A5 find row for your degrees of freedom. → Find correct column for one-tailed or two-tailed test for your degrees of freedom.

↓

Does your value exceed the value given in the table?
- Yes → Reject null hypothesis. → Stop
- No → Retain null hypotheses.

↓ No

Hypothesis concerns Pearson *r* but null hypothesis is other than $r = 0$ in the population. — Yes → Consult references given.

↓ No

Proceed to Chapter 13.

345

346 INFERENTIAL STATISTICS

either a one-tailed test or two-tailed test according to whether the hypothesis was directional or nondirectional. If the value of the observed t equals or exceeds the one in the table at the intersection of the appropriate row and column, the results are statistically significant and the null hypothesis is rejected. If the value of the observed t is less than the value in the table the results are not statistically significant and the null hypothesis is retained.

The t-test is also used to test hypotheses concerning the Pearson r. The most common question concerns whether a Pearson r in a sample is sufficiently different from zero to reject the hypothesis that the correlation between two variables in the population is zero. Table A5 in the Appendix gives the values of r required to reject this hypothesis at different values of alpha for given degrees of freedom.

EXERCISES

1. Two samples of subjects are drawn at random from a population and subjected to two experimental treatments. The following data were gathered. Test the significance of the difference between the means at the .05 level using a nondirectional test.

    Sample 1: 10, 9, 8, 7, 5, 3
    Sample 2: 16, 14, 11, 10, 6, 3

2. The following data were gathered for two independent groups. Test the two-tailed hypothesis that the two groups are similar in ability on the measure used. Use an alpha of .05.

	GROUP 1	GROUP 2
N	13	19
$\bar{X}$	7	9.4
Σx^2	96	112

    $S^2 = \dfrac{\Sigma x^2 - (\Sigma x)^2 / n_1}{n-1}$

3. Indicate whether a one-tailed or a two-tailed test of significance would be used when testing the following research hypotheses:

    (a) There will be a difference in the intelligence test scores of children from differing socioeconomic classes.
    (b) Male students will have a higher final examination grade in statistics than will female students.
    (c) Students exposed to an individualized program of instruction will show greater gains on an arithmetic achievement test than will a comparable group of students taught by the traditional method.
    (d) Drug X will have a significant effect on the psychomotor coordination of subjects receiving it.
    (e) Subjects who view a certain film will have more favorable attitudes toward educational research than will subjects who do not view the film.

4. Twenty subjects were given an intelligence test and paired on the basis of their scores. One member of each pair was then assigned at random to Treatment 1

and the other to Treatment 2. After the experiment, a measure on a dependent variable was obtained for each subject. Results are shown below. Do a two-tailed test of the null hypothesis that $\mu_D = 0$ or that $\mu_1 = \mu_2$ ($\mu_x = \mu_y$) at the .05 level.

PAIR	TREATMENT 1 X	TREATMENT 2 Y
1	13	9
2	6	7
3	8	10
4	7	5
5	9	8
6	9	6
7	10	7
8	5	4
9	8	7
10	5	2

5. A researcher is interested in determining the effects of a special education program on raising the intelligence test scores of disadvantaged children. A random sample of twelve children is chosen to participate in the special program. Their IQ scores are measured before and after the program using equivalent forms of the same instrument. The results are given below. Test the null hypothesis that $\mu_D = 0$ at the .01 level.

SUBJECT	TEST SCORES BEFORE	AFTER
1	96	101
2	89	94
3	81	88
4	85	85
5	96	102
6	95	100
7	87	90
8	79	89
9	80	85
10	90	98
11	92	100
12	99	105

6. For a sample of 25 paired measurements $\Sigma D = 50$; $\Sigma D^2 = 400$. Calculate t.
7. Six subjects are randomly divided into two groups, experimental and control. The scores obtained were as follows:

E: 12, 10, 8   C: 7, 3, 2

Test the significance of the difference between the means. Would you conclude at the .05 level that the population means are equal or unequal?
8. With a sample of 15 subjects and a two-tailed alpha of .01 what values of the Pearson r would lead to rejection of the hypothesis that the Pearson r in the population is zero?

ANSWERS

1. $t = -1.33$ not significant; retain null hypothesis $\left(t = \dfrac{7-10}{2.25} = -1.33\right)$
2. $t = -2.53$; reject null hypothesis $\left(t = \dfrac{-2.4}{.95} = -2.53\right)$
3. (a) two-tailed
   (b) one-tailed
   (c) one-tailed
   (d) two-tailed
   (e) one-tailed
4. $t = 2.46$; reject null hypothesis (two-tailed test) $\left(t = \dfrac{1.5}{.61} = 2.46\right)$
5. $t = -7.64$; reject null hypothesis (one-tailed test) $\left(t = \dfrac{-5.67}{.742} = -7.64\right)$
6. $t = 2.83$ $\left(t = \dfrac{2}{.707} = 2.83\right)$ significant
7. $t = 3.14$ $\left(t = \dfrac{6}{1.91} = 3.14\right)$ significant. Reject null and conclude that population means are unequal.
8. any value from .641 to 1.000 or any value from $-.641$ to -1.000

CHAPTER THIRTEEN
ONE—WAY ANALYSIS OF VARIANCE

OBJECTIVES

After completing this chapter the reader should be able to:

1. Describe the rationale of analysis of variance (ANOVA).
2. Identify situations in which the one-way ANOVA would be the appropriate method for testing a hypothesis.
3. Identify the relationship between sum of squares total, sum of squares between, and sum of squares within groups.
4. Apply the concept of degrees of freedom to the ANOVA procedure.
5. Calculate mean square between, mean square within, and F-ratio in a one-way ANOVA problem.
6. Use the F-table to test hypotheses.
7. Identify the assumptions underlying ANOVA and describe their practical limitations.

Introduction

The t-test introduced in Chapter Twelve is used to test the significance of the difference between the means of two random samples. Frequently, however, research studies involve more than two samples; for example, a study may compare several different methods of instruction in order to determine which is most effective. In such a situation, there are more than two means to analyze for significant differences and the appropriate technique to

350 INFERENTIAL STATISTICS

use is called *analysis of variance* (ANOVA).[1] Analysis of variance is a statistical procedure designed to analyze the differences between the means of *two or more* samples. As such, it may be used like the *t*-test with only two means; when it is used in a two-sample situation, it is mathematically equivalent to the *t*-test. However, it is most commonly employed when there are three or more samples. Thus, analysis of variance is a more versatile technique than the *t*-test and some statisticians prefer it even when there are just two samples. It is one of the most widely used statistical procedures.

In a typical ANOVA research design, n subjects are assigned randomly to k experimental groups. A different form of the independent variable or treatment is applied to each of the k groups. A measure on the dependent variable of the study is obtained for each subject in each group and then the means of the k groups are calculated. The values of these treatment group means are compared. If the experimental treatments have had no differential effects on the dependent variable, then the values of the k sample means are such that $\mu_1 = \mu_2 = \mu_3 = \mu_k$. The experimental treatment has made no difference. This is the null hypothesis to be tested. That is, in ANOVA the null hypothesis states that the samples are simply independent samples drawn from populations having the same mean.

Of course, even in the absence of any effect of the experimental treatment, we know that there will be some variation in the values of the k sample means just as a result of random fluctuation. But certainly if the treatment has had any effect, then the differences between the sample means should be increased. If the difference between the sample means is greater than can be attributed to sampling fluctuation, then the null hypothesis is rejected and it is assumed that the treatments have differed in their effect on the dependent variable. If the difference between the sample means is no greater than the variability to be expected between random samples, then the null hypothesis is retained and it is assumed that the treatments do not differ in their effect.

The general rationale of ANOVA is that the *total variance* of all the data (scores) in an experiment can be separated and attributed to two sources: *variance between* groups and *variance within* groups. (According to the rules of grammar the former should be called variance among groups when more than two groups are involved. However, it has long been the custom to refer to this as "variance between" and we shall use this term in order to be consistent with other texts.) *Variance within* groups reflects the spread of

---

[1] There are three types of ANOVA models: the random, the fixed, and the mixed models. The fixed model is the most widely used in educational and psychological research. All our descriptions in this chapter and the next apply to the fixed model. For a description of the other models, see W. L. Hays, Statistics (New York: Holt, Rinehart and Winston, 1963).

the scores within each of the k groups. Variance within groups represents differences among subjects that have nothing to do with the independent variable in an experiment. For this reason it is frequently termed *error variance*. It is variance due to individual differences in subjects, the fallibility of the instruments used to measure the dependent variable, and so on. *Variance between* groups reflects the size of the difference between the k group means. Variance between groups may be due to the effect of the independent variable or it may be just a function of chance. As we said above, if the experimental treatment has been effective, the between-groups difference will be expected to be greater than can be accounted for by chance. The purpose of ANOVA is to establish whether the variation between groups is likely to be a function of chance or not. Let us use an example to introduce the concepts of total variance, variance within groups, and variance between groups.

A statistics professor has three different programmed instruction units on the use of the t-test. He wants to see if the programs are equally effective. Thus, the null hypothesis to be tested in this study is $\mu_1 = \mu_2 = \mu_3$. The professor randomly divides his 15 students into three groups. The students in Group I are given Program A to work through, those in Group II are given Program B, and those in Group III are given Program C. The programmed instruction unit is the independent variable. When they have finished the units all 15 subjects are given the same criterion test to measure how well they have mastered the concepts. Performance on the test is the dependent variable of the study. The scores on the criterion test are shown in Table 13.1, column (1).

When we look at the scores of the 15 subjects in column (1) we see that the means of the three groups differ from one another. Does this mean that the three programmed instruction units differ in their effectiveness? The differences in the group means may be due to differences in program effectiveness but they may also be due to random error. Even though random assignment was used, it is possible that by chance the most competent students were all assigned to Group I.

In order to get a measure of the extent of the variation in the student scores that is unrelated to the independent variable, we consider the variation of the scores *within* each group. This variation will be a result of general competence in statistics, motivation, and other variables that are *not* related to the programmed instruction unit used. This variation within groups can be used to provide a "yardstick" for determining whether the variation between groups is likely to be a function of chance or not. We first calculate the total variation of the scores and then divide, or partition, this variation into component parts, variation between and within, then use these values for hypothesis testing.

TABLE 13.1 Data for Groups Using Three Different Programmed Instruction Units

	(1)	(2)	(3)	(4)	(5)	(6)	(7)	(8)	(9)
		\multicolumn{2}{c}{FINDING THE TOTAL SUM OF SQUARES}	\multicolumn{3}{c}{FINDING THE SUM OF SQUARES WITHIN GROUPS}	\multicolumn{3}{c}{FINDING THE SUM OF SQUARES BETWEEN GROUPS}
	SCORES	DEVIATION OF SCORES FROM GRAND MEAN OF 9	SQUARED DEVIATION FROM GRAND MEAN	GROUP MEAN	DEVIATION OF SCORES FROM GROUP MEAN	SQUARED DEVIATION FROM GROUP MEAN	DEVIATION OF GROUP MEAN FROM GRAND MEAN OF 9	SQUARED DEVIATION	SQUARED DEVIATION WEIGHTED ACCORDING n, SIZE OF GROUP
Group I	15	6	36	13	2	4	4	16	80
	14	5	25		1	1			(5×16)
	13	4	16		0	0			
	12	3	9		-1	1			
	11	2	4		-2	4			
	$\Sigma = 65$				$\Sigma = 0$	$\overline{10}$			
	$\bar{\bar{X}} = 13$								
Group II	10	1	1	8	2	4	-1	1	5
	9	0	0		1	1			(5×1)
	8	-1	1		0	0			
	7	-2	4		-1	1			
	6	-3	9		-2	4			
	$\Sigma = 40$				$\Sigma = 0$	$\overline{10}$			
	$\bar{\bar{X}} = 8$								

Group III	8	−1	1	6	2	4	−3	45
	7	−2	4		1	1		(5 × 9)
	6	−3	9		0	0		
	5	−4	16		−1	1		
	4	−5	25		−2	4		
	$\Sigma = \overline{30}$	$\Sigma = 0$			$\Sigma = \overline{0}$	$\overline{10}$		
	$\bar{X} = 6$						$\Sigma = 0$	

Sum of All
  Scores   135 $\Sigma = 0$ $SS_t = 160$ $SS_w = 30$ $SS_b = 130$
Grand
  Mean   9

Partitioning the Sum of Squares

The basic ingredient of the ANOVA procedure is the *sum of squares*. This is the measure of *variability* that is analyzed in this procedure. The sum of squares refers to the total squared differences or the deviations between a set of individual scores and a mean. In previous chapters we have used the notation Σx^2 or $\Sigma(X - \bar{X})^2$ to represent sum of squares. By convention the symbol SS is used for this value in the analysis of variance procedure.

1. The first sum of squares that is computed in analysis of variance is called the *total sum of squares*, which refers to the sum of squares of the deviations of each of the observations from the grand mean. The grand mean is the mean of all the scores taken together as a group. The grand mean is symbolized $\bar{X}_t$. Total sum of squares is represented as SS_t:

$$SS_t = \Sigma(X - \bar{X}_t)^2$$

where: SS_t = total sum of squares
X = an individual score
$\bar{X}_t$ = the mean of all the scores (grand mean)

Formula (13.1)
Sum of Squares Total (definition formula)

In our example the mean of all 15 scores (the grand mean) is 9. The difference between each score and this mean is shown in Table 13.1, column (2) and the squares of these differences is shown in column (3). The total sum of squares, SS_t, is 160.

2. A *component* of the total sum of squares that can be calculated separately is the *sum of squares within groups*, SS_w. The SS_w is unrelated to any difference in the treatment and therefore is a basic component of the error term. It is found by computing the deviation of each individual score in each group from the mean of its own group and then squaring and summing the squared deviations:

$$SS_w = \Sigma(X_1 - \bar{X}_1)^2 + \Sigma(X_2 - \bar{X}_2)^2 + \Sigma(X_3 - \bar{X}_3)^2 \ldots \text{through all the } k \text{ groups,}$$

where: SS_w = sum of squares within
$\bar{X}_1$ = the mean of the first group
$\bar{X}_2$ = the mean of the second group, and so on
X_1 = an individual score in the first group
X_2 = an individual score in the second group, and so on
k = number of groups

Formula (13.2)
Sum of Squares Within (definition formula)

In our example in Table 13.1 the mean of each of the three groups is shown in column (4), the difference between each score within each group and its respective group mean is shown in column (5), and the squares of these differences are shown in column (6). The sum of these squared differences (SS_w) is 30.

3. The other component of the total sum of squares is the variability from group to group which is represented by the *sum of squares between groups*, SS_b. The SS_b is the variation that may be due to the experimental treatment. It is sometimes called "treatment SS." It is computed by finding the sum of the squares of the deviation of each separate group mean from the grand mean. Each term $(\bar{X} - \bar{X}_t)^2$ must be weighted by n, the number of cases in each group in order to weight the deviations according to the number of scores which contributed to producing the mean of each subgroup. That is, each subgroup mean is subtracted from the grand mean, this deviation is squared, and each subgroup squared deviation is multiplied by the n for the group:

$$SS_b = n_1(\bar{X}_1 - \bar{X}_t)^2 + n_2(\bar{X}_2 - \bar{X}_t)^2 + n_3(\bar{X}_3 - \bar{X}_t)^2 \ldots \text{ through all the groups,}$$

where: SS_b = sum of squares between
$\bar{X}_1$ = the mean of the first group
$\bar{X}_2$ = the mean of the second group, and so on
n_1 = the number in the first group
n_2 = the number in the second group, and so on
$\bar{X}_t$ = the mean of all the scores   Formula (13.3)
Sum of Squares Between (definition formula)

In our example column (7) shows the difference between each group mean and the grand mean, column (8) shows these differences squared, and column (9) shows these squared deviations weighted according to the number in each group. The total SS_b is 130. This partitioning of the total sum of squares is expressed as: $SS_t = SS_w + SS_b$. That is, the total sum of squares (the sum of squares of all the scores from the grand mean) is equal to the sum of squares within groups (the sum of squares of deviations from their own group means) plus the sum of squares between groups (the sum of squares of deviations of the group means from the grand mean with each squared deviation weighted by n). In our example, $SS_w = 30$ and $SS_b = 130$. These total to 160, the value of SS_t.

Population Variance Estimates (Mean Squares)

In the analysis of variance procedure the between and within sums of squares are used to derive *estimates of the population variance* which are then used to test the null hypothesis. One divides the sum of squares between groups and sum of squares within groups by their appropriate degrees of freedom in order to obtain the two estimates of variance that are needed. The two variance estimates are called *mean squares*. The conventional symbol for mean square is MS.

The estimate of the variance derived from within groups data is called the mean square within groups and is symbolized MS_w:

$$MS_w = \frac{SS_w}{\text{d.f.}_w} = \frac{SS_w}{N-k}$$

where: SS_w = the sum of squares within groups
N = the total number of scores
k = the number of groups

Formula (13.4)
Mean Square Within

The number of degrees of freedom associated with the within-groups sum of squares is $N - k$ because the degrees of freedom for the separate groups are added. That is, the number of d.f. for each group is $n - 1$ and for k groups there would be $(n_1 - 1) + (n_2 - 1) + (n_3 - 1) + \cdots + (n_k - 1)$ or $\Sigma n - k$ or $N - k$ degrees of freedom. In our example we have 15 subjects in three groups so the degrees of freedom within are $15 - 3 = 12$. The mean square within is $SS_w/\text{d.f.}_w = 30/12 = 2.5$.

The estimate of the variance between groups is called the mean square between groups and is symbolized MS_b:

$$MS_b = \frac{SS_b}{\text{d.f.}_b} = \frac{SS_b}{k-1}$$

where: SS_b = the sum of squares between groups
k = the number of groups

Formula (13.5)
Mean Square Between

The number of d.f. associated with the between-groups sum of squares is $k - 1$ because there are k means and 1 degree of freedom is lost by subtracting the grand mean from each group mean. In our example the degrees of freedom between groups are $3 - 1 = 2$. The mean square between is $SS_b/\text{d.f.}_b = 130/2 = 65$.

It is significant that the degrees of freedom (like the sums of squares) are also additive:

$$N - 1 = (N - k) + (k - 1)$$

Total = Within + Between

The *F*-Ratio

A ratio is used to compare the estimate of the population variance derived from the sample between-groups data, MS_b, to the estimate of the population variance estimate derived from the sample within-groups data, MS_w. This ratio is called the *F*-ratio. Thus, *F* is defined as the ratio of:

$$F = \frac{MS_b}{MS_w}$$

where: MS_b = mean square between groups
MS_w = mean square within groups

Formula (13.6)
F-Ratio

The numerator of the *F*-ratio is thus influenced by the observed differences between the groups, while the denominator represents the error term since it is derived from variation within groups. As the difference between the groups increases, the F-ratio increases.

The null hypothesis to be tested in analysis of variance is that the sample means being compared with the F-ratio are no different than what is to be expected from random samples from the same population. *If the null hypothesis is true*, then the variance estimate based on differences between groups and the variance estimate based on differences within groups will tend to be about the same since both are estimates of the *same* common population variance and any difference in the two mean squares would be the result of random variation. Thus, one would expect the ratio of MS_b/MS_w to be about 1.[2]

On the other hand, if there is a genuine difference between the groups, then the mean square between groups (the variance estimate derived from the variance of group means around the grand mean) is markedly greater than the mean square within groups (the variance estimate derived from the variation of scores within each group), and F will be considerably greater than 1. Only values of $F > 1$ would be considered as evidence against the null hypothesis. That is, as the difference between the mean squares increases, the F-ratio increases and the probability that the null hypothesis is correct decreases. In our example the F-ratio equals $65/2.5 = 26$. The next step is to consult the F-table to determine whether our observed F-ratio is statistically significant.

Using the *F*-Table

It has been shown through deductive logic that under a true null hypothesis the F-ratios will fall into a predictable pattern according to the degrees of freedom involved. These patterns are called F-distributions. With F-ratios both values for the degrees of freedom are required to identify the appropriate distribution. These values are those associated with the variance estimates used in the F-ratio, the numerator MS_b and the denominator MS_w. The important landmarks in the F-distributions are given in Table A6 in the Appendix. These are the .01 and .05 probability points. If the obtained value of F equals or exceeds the tabled value, one rejects the null hypothesis and accepts the alternative hypothesis that there are significant differences between the groups, presumably due to the effect of the experimental treatment. If the obtained F is smaller than the tabled value, the null hypothesis is retained and one concludes that the differences in the group means could easily be a function of chance and are therefore not statistically significant.

[2] Actually the ratio of MS_b/MS_w is expected to be $d.f._w/d.f._{w-2}$ under a true null hypothesis. However, unless the number of subjects is quite small, the expected ratio is very near 1.

To find the appropriate value in the F-table we locate the column labeled with the degrees of freedom of the numerator (MS_b) and the row labeled with the degrees of freedom of the denominator (MS_w). At the intersection of this row and this column are the values needed for testing the hypothesis. In our example the numerator degrees of freedom are 2 and the denominator degrees of freedom are 12. At the intersection of column 2 and row 12 we find the value 3.88 in Roman type and 6.93 in bold face type. The Roman type number is the value required to reject the null hypothesis at the .05 level; the boldface type number is the value required to reject the null hypothesis at the .01 level. In our example the observed value of F was 26. This exceeds both the value required to reject the null hypothesis at the .05 level (3.88) and also the value required at the .01 level (6.93). The professor can reject the hypothesis that the three programs are equally effective.

Exercises

1. As the difference between groups increases, the F-ratio
    (a) decreases
    (b) increases
    (c) stays the same
    (d) cannot tell since it depends on the specific situation
2. Which of the following is influenced by the observed differences between the groups?
    (a) MS_w
    (b) MS_b
    (c) ΣN
    (d) $N - k$
3. Variance between groups is presumably due to:
    (a) individual differences
    (b) error
    (c) random factors beyond the control of the researcher
    (d) experimental manipulation
4. The variance estimate derived from the variance of group means around the grand mean is called the:
    (a) MS_w
    (b) MS_b
    (c) F-ratio
    (d) SE_m
5. If experimental manipulation has an effect on the dependent variable this will:
    (a) decrease the variance between group means
    (b) decrease the total variance
    (c) increase the variance between group means
    (d) increase the variance derived from the variation of scores within each group
6. A sum of squares divided by its degrees of freedom is called:

(a) population mean
(b) F-ratio
(c) sample mean
(d) mean square
7. The null hypothesis is rejected if:
  (a) the F-ratio is less than 1
  (b) the group means differ only as a function of chance
  (c) $\dfrac{MS_b}{MS_w} = 0$
  (d) none of the above
8. Which of the following is the null hypothesis to be tested by the analysis of variance?
  (a) $\bar{X}_1 = \bar{X}_2 = \bar{X}_3$
  (b) $\mu_1 = \mu_2 = \mu_3$
  (c) $SS_{t1} = SS_{t2} = SS_{t3}$
  (d) $SS_{b1} = SS_{b2} = SS_{b3}$

Solutions

1. (b) 2. (b) 3. (d) 4. (b) 5. (c) 6. (d) 7. (d) 8. (b)

Computational Formulas for Sums of Squares

It is not necessary to find the deviation scores as we did in Table 13.1 in order to arrive at the sums of squares. A simpler method is available which works directly with the raw scores. Table 13.2 presents the same data as was presented in Table 13.1 for the three groups of subjects with the addition of columns to show the squares of the raw scores.

TABLE 13.2 Raw Scores for Groups Using Three Different Programmed Instruction Units

	GROUP I		GROUP II		GROUP III		
	X_1	X_1^2	X_2	X_2^2	X_3	X_3^2	
	15	225	10	100	8	64	
	14	196	9	81	7	49	
	13	169	8	64	6	36	
	12	144	7	49	5	25	
	11	121	6	36	4	16	TOTAL
$\Sigma X =$	65		40		30		135
$\Sigma X^2 =$		855		330		190	1,375

360 INFERENTIAL STATISTICS

1. This time we will use raw scores and the computational formula to compute the total sum of squares for the 15 cases:

$$SS_t = \Sigma (X - \bar{X}_t)^2 = \Sigma X^2 - \frac{(\Sigma X)^2}{N}$$

where: SS_t = sum of squares total
X = each individual score
$\bar{X}_t$ = grand mean of all the N cases
N = the total number of observations $(n_1 + n_2 + n_3 = N)$
ΣX^2 = sum of the squares of each raw score (each score is squared and the squares are summed)
$(\Sigma X)^2$ = sum of the raw scores and the quantity squared (the scores are summed and the total is squared)

Formula (13.7)
Sum of Squares Total (computational formula)

For the data in Table 13.2, ΣX^2 is equal to $855 + 330 + 190$ or $1,375$. The sum of the scores is equal to $65 + 40 + 30$ or 135, which is then squared to equal $18,225$. Thus, $\Sigma (X - \bar{X}_t)^2 = 1,375 - (135)^2/15 = 1,375 - 18,225/15 = 1,375 - 1,215 = 160$. The total sum of squares for the data is 160, the same value we found using the definition formula.

2. The next step is to determine the sum of squares between groups using the raw scores:

$$SS_b = \Sigma n(\bar{X} - \bar{X}_t)^2 = \frac{(\Sigma X_1)^2}{n_1} + \frac{(\Sigma X_2)^2}{n_2} + \frac{(\Sigma X_3)^2}{n_3} + \cdots + \frac{(\Sigma X_k)^2}{n} - \frac{(\Sigma X)^2}{N}$$

where: SS_b = sum of squares between
ΣX_1 = sum of the first group
ΣX_2 = sum of the second group, and so on
n_1 = number in the first group
n_2 = number in the second group, and so on
ΣX = sum of all the scores
N = total number of scores = $n_1 + n_2 + \cdots + n_k$

Formula (13.8)
Sum of Squares Between (computational formula)

For the data in Table 13.2, the sum of squares between groups is:

$$\Sigma (\bar{X} - \bar{X}_t)^2 = \frac{(65)^2}{5} + \frac{(40)^2}{5} + \frac{(30)^2}{5} - \frac{(135)^2}{15} = (845 + 320 + 180) - 1,215 = 130$$

3. The within-groups sum of squares is found by determining the sum of squares within each group separately and then pooling these sums:

$$SS_w = \Sigma (X_1 - \bar{X}_1)^2 + \Sigma (X_2 - \bar{X}_2)^2 + \cdots + \Sigma (X_k - \bar{X}_k)^2$$

$$= \Sigma X_1^2 - \frac{(\Sigma X_1)^2}{n_1} + \Sigma X_2^2 - \frac{(\Sigma X_2)^2}{n_2} + \cdots + \Sigma X_k^2 - \frac{(\Sigma X_k)^2}{n_k}$$

where: SS_w = sum of squares within
ΣX_1^2 = sum of the squares of the scores in the first group
ΣX_2^2 = sum of the squares of the scores in the second group, and so on
$(\Sigma X_1)^2$ = the sum of scores in the first group squared
$(\Sigma X_2)^2$ = the sum of the scores in the second group squared, and so on
n_1 = the number in the first group
n_2 = the number in the second group, and so on

Formula (13.9)
Sum of Squares Within (computational formula, direct method)

In our example the calculation is as follows:

$$\Sigma (X_1 - \bar{X}_1)^2 = 855 - \frac{(65)^2}{5} = 10$$

$$\Sigma (X_2 - \bar{X}_2)^2 = 330 - \frac{(40)^2}{5} = 10$$

$$\Sigma (X_3 - \bar{X}_3)^2 = 190 - \frac{(30)^2}{5} = 10$$

Pooling the three sums of squares, we have 30 as the within-groups sum of squares (ΣX_w^2).

Although we have illustrated the computation of the within-groups sum of squares, it is not necessary to compute this quantity by Formula (13.9) unless one wants to use it as a check. The sum of squares within groups may be found by subtracting the sum of squares between groups from the total sum of squares. That is, Within SS = Total SS − Between SS, or $30 = 160 - 130$. The usual procedure is to compute the total sum of squares and the between sum of squares and then subtract $SS_t - SS_b$ in order to determine the within sum of squares:

$$SS_w = SS_t - SS_b \quad \text{where: } SS_w = \text{sum of squares within}$$

$$SS_t = \text{sum of squares total}$$

$$SS_b = \text{sum of squares between}$$

Formula (13.10)
Sum of Squares Within (computational formula, subtractive method)

Let us work through another example with the computation formulas. Assume that a large group of subjects was assigned at random to three experimental treatments. The treatment is the independent variable. The results on a dependent variable measure administered after the experiment are given in Table 13.3. Let us use the analysis of variance to determine if the differences between the treatment means are statistically significant. The null hypothesis to be tested is: $\mu_1 = \mu_2 = \mu_3$; the alternative hypothesis is that not all means are equal ($\mu_1 \neq \mu_2 \neq \mu_3$).

362 INFERENTIAL STATISTICS

TABLE 13.3 Hypothetical Data for Three Treatment Groups

TREATMENT 1		TREATMENT 2		TREATMENT 3		
X_1	X_1^2	X_2	X_2^2	X_3	X_3^2	
10	100	12	144	10	100	
8	64	10	100	8	64	
7	49	8	64	7	49	
6	36	7	49	7	49	
6	36	6	36	6	36	
5	25	5	25	6	36	
5	25	5	25	5	25	
4	16	4	16	4	16	
4	16	4	16	4	16	
2	4	3	9	2	4	TOTAL
$\Sigma X = 57$		64		59		180
ΣX^2	371		484		395	1,250

The total sum of squares is:

$$(SS_t) = \Sigma X^2 - \frac{(\Sigma X)^2}{N}$$
$$= 1{,}250 - \frac{(180)^2}{30}$$
$$= 1{,}250 - 1{,}080$$
$$= 170$$

The sum of squares between groups is:

$$(SS_b) = \frac{(\Sigma X_1)^2}{n_1} + \frac{(\Sigma X_2)^2}{n_2} + \frac{(\Sigma X_3)^2}{n_3} - \frac{(\Sigma X)^2}{N}$$
$$= \frac{3{,}249}{10} + \frac{4{,}096}{10} + \frac{3{,}481}{10} - 1{,}080$$
$$= 1{,}082.6 - 1{,}080$$
$$= 2.6$$

The sum of squares within groups is:

$$(SS_w) = \text{Total } SS - \text{Between } SS$$
$$= 170 - 2.6$$
$$= 167.4$$

The mean square between groups is:

$$MS_b = \frac{SS_b}{k-1}$$
$$= \frac{2.6}{2}$$
$$= 1.3$$

The mean square within groups is:

$$MS_w = \frac{SS_w}{N-k}$$
$$= \frac{167.4}{27}$$
$$= 6.2$$

The F-ratio is:

$$\frac{MS_b}{MS_w} = \frac{1.3}{6.2} = .209$$

A useful convention is to show the results of ANOVA in a summary table as in Table 13.4.

TABLE 13.4 Summary Table for the Analysis of Variance of the Data from Table 13.3

SOURCE OF VARIATION	SUMS OF SQUARES	d.f.	MEAN SQUARE	F
Between Groups	2.6	2	1.3	.209
Within Groups	167.4	27	6.2	
Total	170.0	29		

The sums of squares in the second column are divided by the degrees of freedom in the third column to yield the mean squares in the fourth column. The final step in the ANOVA is to divide mean square between by the mean square within to yield the F-ratio.

In our example, the F-ratio is less than 1.0 so we do not need to consult the F-table to know that the F-ratio is not significant. Whenever the F-ratio is less than 1 the null hypothesis is retained and one concludes that there are no significant differences between the means of the treatment groups.

Exercises

1. Complete this summary table for the analysis of variance.

SOURCE OF VARIATION	SUMS OF SQUARES	d.f.	MEAN SQUARE	F
Between Groups		2		
Within Groups	350			
Total	500	99		

2. What is the between-groups SS?
3. What is the within-groups d.f.?
4. What is MS_b?
5. What is MS_w?
6. What is the F-ratio?
7. Is the F significant at the .01 level?

Solutions

2. 150 3. 97 4. 75 5. 3.608 6. 20.787 7. yes

Comparison of Group Means Following the *F*-Test

Analysis of variance tests for the presence of differences in means in a total set of data; it is not designed nor is it able to answer questions about differences between particular means. A significant *F*-ratio indicates only that there are statistically significant differences among the groups contributing to the total set of data but it does not indicate where the significance lies.

In our first example the *F*-ratio was found to be significant. However, there is no indication as to which of the possible differences are statistically significant. The difference between some pairs may be significant, while differences between other pairs may not be significant. For instance, Group I may be significantly different from Group II and also from Group III, but Groups II and III may not differ significantly from each other.

If a significant *F* is obtained in an analysis of variance involving more than two groups and one wishes to locate the statistically significant differences within the data, then additional statistical tests must be applied. When the obtained *F* is not significant, that is, if the analysis of variance indicates no statistically significant differences in the data, as in our second example, then no further analysis is justified.

If an F-ratio is significant it is often important to the researcher to know which group means are significantly different from which other group means. There are a number of statistical tests available for this purpose but they are beyond the scope of this text. Among others, Glass and Stanley[3] have useful descriptions of these tests and their applications.

Assumptions Underlying Analysis of Variance

One-way analysis of variance is used to test the significance of the differences in means when only one independent variable has been manipulated. In the examples in this chapter the independent variable was the method of instruction.

There are certain assumptions involved in one-way analysis of variance. One assumption is that the variables under investigation be normally distributed in the population from which the samples are drawn. It is also assumed that the variances in the populations from which the samples are drawn are equal. As in the case of the t-test, this assumption is referred to as homogeneity of variance. Finally, it is assumed that all individuals included in the study are randomly and independently drawn from the population.

In practice the data of experiments often do not entirely satisfy these assumptions. When these assumptions are not met the actual probabilities will differ from those given in the F-table. It has been found that moderate departures from normality and homogeneity may exist without seriously affecting the validity of the F-test. This is especially true if the number of cases in the various samples are equal. However, if the researcher knows that the data depart markedly from the necessary assumptions, then perhaps a nonparametric test should be considered as an alternative procedure.

SUMMARY

Analysis of variance is used to test the statistical significance of differences between means when more than two groups are used in a study. The technique is based on analyzing the variance of the scores in the different groups. The rationale of the procedure lies in the fact that the total variance in a number of samples can be divided into two component parts: the

---

[3]*Gene V. Glass and Julian C. Stanley, Statistical Methods in Education and Psychology* (Englewood Cliffs, N.J.: Prentice-Hall, 1970).

FLOWCHART 13

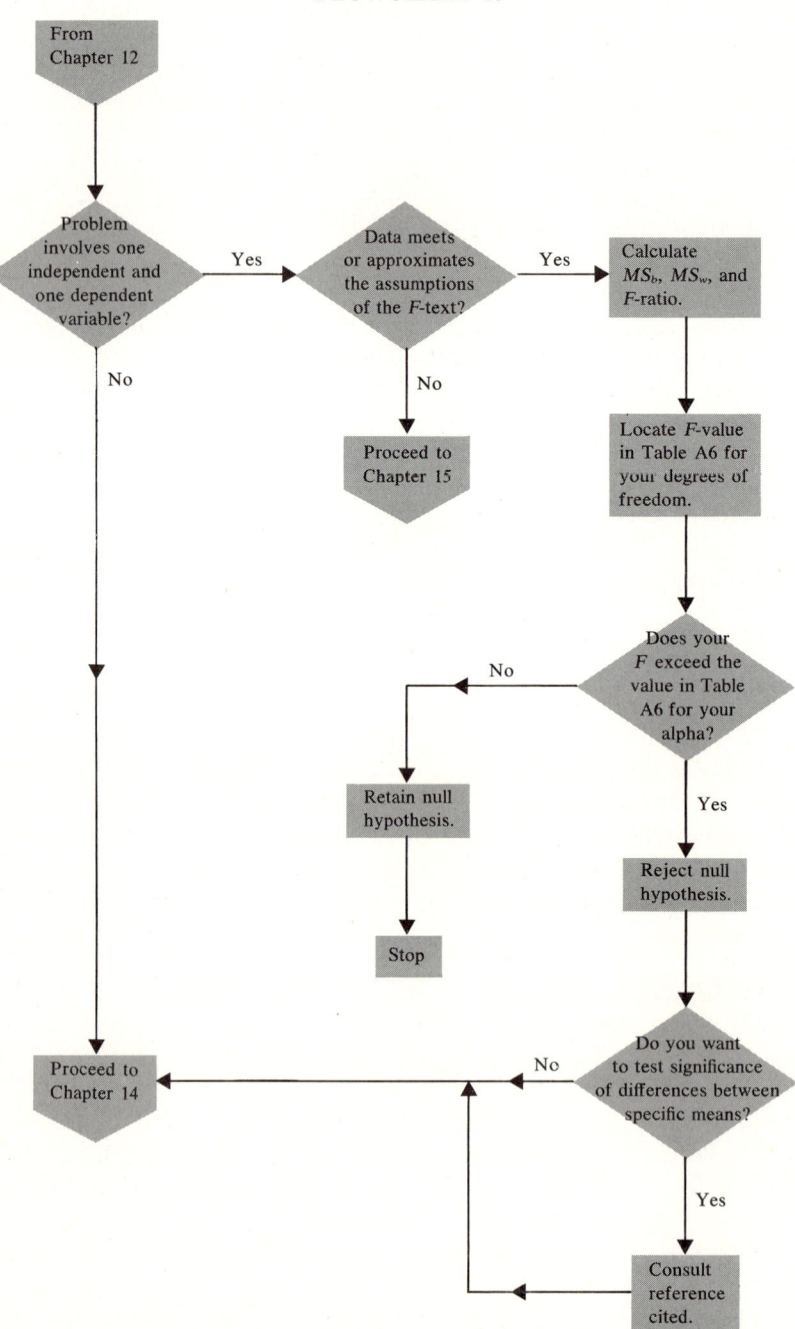

```
From Chapter 12
      │
      ▼
  Problem involves one independent and one dependent variable? ──Yes──▶ Data meets or approximates the assumptions of the F-text? ──Yes──▶ Calculate $MS_b$, $MS_w$, and F-ratio.
      │                                                                      │                                                              │
      No                                                                     No                                                             ▼
      │                                                                      ▼                                                      Locate F-value in Table A6 for your degrees of freedom.
      │                                                              Proceed to Chapter 15                                                   │
      │                                                                                                                                     ▼
      │                                                                                                                          Does your F exceed the value in Table A6 for your alpha? ──No──▶ Retain null hypothesis. ──▶ Stop
      │                                                                                                                                     │
      │                                                                                                                                     Yes
      │                                                                                                                                     ▼
      │                                                                                                                             Reject null hypothesis.
      │                                                                                                                                     │
      │                                                                                                                                     ▼
      ▼                                                                                                                          Do you want to test significance of differences between specific means? ──No──▶ Proceed to Chapter 14
  Proceed to Chapter 14  ◀────────────────────────────────────────────────────────────────────────────────────────────────────────────────── │
                                                                                                                                              Yes
                                                                                                                                              ▼
                                                                                                                                     Consult reference cited.
```

variance between groups, which represents the variance of the sample means around the total mean; and the variance within groups, which represents the variance of scores around their own group mean. The null hypothesis is that the samples are drawn from the same population. H_0: $\mu_1 = \mu_2 = \mu_3 \cdots = \mu_k$.

An estimate of the population variance derived from differences between groups (MS_b) is divided by an estimate of the population variance estimate derived from differences within groups (MS_w) to form the F-ratio. MS_b serves as a measure of experimental effect while MS_w serves as the error term. If the F-ratio is larger than the tabled value of F, the null hypothesis of equal means is rejected.

If there are no significant differences, the ratio of mean square between groups to mean square within groups is near 1, indicating that the differences between groups are no greater than could be expected by chance. The null hypothesis of equal means is retained.

The analysis of variance is a widely used statistical test because it is appropriate for testing hypotheses about means in a variety of experimental situations. It can be used with only two groups or with a number of groups and thus is much more versatile than the t-test.

EXERCISES

1. Complete the following summary table:

SOURCE	SUM OF SQUARES	d.f.	MEAN SQUARE	F
Between Groups	40	2	—	—
Within Groups	234	117	—	
Total	274	—		

2. (a) What ratio is used to test hypotheses in ANOVA?
   (b) What happens to the F-ratio as the difference between the groups increases?
3. When a sum of squares is divided by its degrees of freedom, an unbiased estimate of the population variance is obtained. What is this estimate of the population variance called?
4. Under which of the following conditions would the null hypothesis be rejected?
   (a) when the mean square between groups is about the same as the mean square within groups
   (b) when the mean square between groups is much larger than the mean square within groups

(c) when the mean square between groups is much smaller than the mean square within groups

5. Complete this summary table for the analysis of variance and check for significance at the .05 level.

SOURCE OF VARIATION	SS	d.f.	MS	F
Between Groups	60	—	—	—
Within Groups	—	56	—	
Total	450	59		

6. Explain the source of the variation measured by:
   (a) the total sum of squares
   (b) the between-groups sum of squares
   (c) the within-groups sum of squares

7. The following data were obtained for three groups of subjects:

	I	II	III
n	10	10	10
$\bar{X}$	7.5	9.0	10.50
ΣX^2	650	865	1,150

Apply the analysis of variance to test the null hypothesis H_0.

8. Forty students were randomly divided into four groups, each group receiving a different method of instruction in reading. The four groups, which did not differ initially in reading achievement, were tested at the end of the school year on a reading achievement test. The following data were obtained:

GROUP I	GROUP II	GROUP III	GROUP IV
11	10	15	14
9	10	13	11
6	10	12	8
7	10	10	7
6	6	7	6
5	6	6	4
3	5	6	3
2	5	5	2
2	5	3	1
1	5	2	1

Use the analysis of variance to determine whether the treatment means differ significantly.

ANSWERS

1.

SOURCE	SUM OF SQUARES	d.f.	MEAN SQUARE	F
Between Groups	40	2	20	10
Within Groups	234	117	2	
Total	274	119		

With 2 and 117 d.f. this is significant at the .01 level.

2. (a) the F-ratio
   (b) the F-ratio increases
3. mean square
4. (b)
5.

SOURCE OF VARIATION	SS	d.f.	MS	F
Between Groups	60	3	20	2.87
Within Groups	390	56	6.96	
Total	450	59		

F is significant at the .05 level

6. (a) The total sum of squares is the variation of the individual scores from the overall or total mean. It is a measure of the total variability present in the data.
   (b) The between-groups sum of squares is the variation of the means of the separate groups from the grand mean weighted to adjust for number in each group.
   (c) The within-groups sum of squares is the variation of individual scores about the mean of the group within which they are found.

7.

	GROUP I	GROUP II	GROUP III
	$n = 10$	$n = 10$	$n = 10$
	$\Sigma X_1 = 75$	$\Sigma X_2 = 90$	$\Sigma X_3 = 105$
	$\Sigma X_1^2 = 650$	$\Sigma X_2^2 = 865$	$\Sigma X_3^2 = 1,150$

$$SS_t = 2,665 - \frac{(270)^2}{30} = 2,665 - 2,430 = 235$$

$$SS_b = \left(\frac{(75)^2}{10} + \frac{(90)^2}{10} + \frac{(105)^2}{10}\right) - \frac{(270)^2}{30}$$
$$= (562.5 + 810 + 1,102.5) - 2,430$$
$$= 2,475 - 2,430$$
$$= 45$$

$$SS_w = \left\{\begin{array}{r} 650 - 562.5 = 87.5 \\ 865 - 810 = 55 \\ 1{,}150 - 1{,}102.5 = 47.5 \end{array}\right\} = 190 \text{ or } 235 - 45 = 190$$

SOURCE	SS	d.f.	MS	F
Between Groups	45	2	22.5	3.196
Within Groups	190	27	7.04	
Total	235	29		

F not significant at the .05 level

8.
GROUP I	GROUP II	GROUP III	GROUP IV
$n = 10$	$n = 10$	$n = 10$	$n = 10$
$\Sigma X_1 = 52$	$\Sigma X_2 = 72$	$\Sigma X_3 = 79$	$\Sigma X_4 = 57$
$\Sigma X_1^2 = 366$	$\Sigma X_2^2 = 572$	$\Sigma X_3^2 = 797$	$\Sigma X_4^2 = 497$
$(\Sigma X_1)^2 = 2{,}704$	$(\Sigma X_2)^2 = 5{,}184$	$(\Sigma X_3)^2 = 6{,}241$	$(\Sigma X_4)^2 = 3{,}249$

$$SS_t = 2{,}232 - \frac{(260)^2}{40} = 2{,}232 - 1{,}690 = 542$$

$$SS_b = \left(\frac{(52)^2}{10} + \frac{(72)^2}{10} + \frac{(79)^2}{10} + \frac{(57)^2}{10}\right) - \frac{(260)^2}{40}$$
$$= (270.4 + 518.4 + 624.1 + 324.9) - 1{,}690$$
$$= 1{,}737.8 - 1690$$
$$= 47.8$$

$$SS_w = 366 - \frac{(52)^2}{10} = 366 - 270.4 = 95.6 \left.\begin{array}{l}\\ \\ \\ \\ \\ \\ \end{array}\right\}$$
$$= 572 - \frac{(72)^2}{10} = 572 - 518.4 = 53.6$$
$$= 797 - \frac{(79)^2}{10} = 797 - 624.1 = 172.9 \quad 494.2$$
$$= 497 - \frac{(57)^2}{10} = 497 - 324.9 = 172.1$$

SOURCE	SS	d.f.	MS	F
Between Groups	47.8	3	15.93	1.16
Within Groups	494.2	36	13.73	
Total	542	39		

F is not significant at the .05 level.

CHAPTER FOURTEEN
TWO-WAY ANALYSIS OF VARIANCE

OBJECTIVES

After completing this chapter the reader should be able to:

1. Describe the advantages of two-way ANOVA designs.
2. Identify situations in which a two-way ANOVA design would be appropriate.
3. Calculate F-ratios in a two-way ANOVA problem.
4. Set up a summary table for a two-way ANOVA problem.
5. Distinguish between main effects and interaction.
6. Interpret interaction.

Factorial ANOVA

In Chapter Thirteen, we considered the one-way analysis of variance which is used to investigate the effect of one independent variable on a dependent variable. However, it is known that the effect of a single independent variable alone may not be the same as its effect when interacting with another independent variable. For example, the effectiveness of a given method of instruction may depend on a number of other variables, such as the ability level, motivation or sex of students, the size of the class, or the personality of the teacher using the method. In other words, two independent variables in combination may have an effect which cannot be accounted for by the effect of the two independent variables taken separately. The effect produced by the combined influence of two independent variables is called *interaction*.

In many experiments two or more independent variables are manipulated at the same time in order to determine not only the independent effects of each variable on the dependent variable, but also the interaction effects of the variables in combination. The advantages of such experimental designs are obvious. A single experiment allows one to investigate independent variables which would ordinarily require two or more separate studies and in addition to obtain a measure of the interaction of the variables, which would not be possible if the variables were analyzed in separate experiments. For example, instead of investigating the effect of a certain method of teaching on achievement in one experiment and the effect of the size of class on achievement in a second experiment, one may economize on time and subjects and investigate both of the variables at the same time in a single experiment. Questions could be answered concerning whether method, class size, or the interaction of method and class size affect achievement.

In a design with two independent variables, both independent variables may be manipulated experimentally or only one may be manipulated and the other used as a control variable. If there are variables such as sex, age, home background, family position, intelligence, and so on, that are believed to influence the dependent variable, then such variables may be controlled simply by incorporating them into the study as one of the independent variables. For example, a study may classify students at two levels of intelligence and then assess the effect of the main independent variable at each of these two levels of intelligence. Thus the variable of intelligence has not only been controlled, but also information has been gained about its independent effect on the dependent variable as well as its effect in interaction with the other independent variable of the study. This control function is another advantage of the two-variable design.

The independent variables in such experimental designs are referred to as *factors*. The statistical technique which makes possible an analysis of the independent and interaction effects of these independent variables or factors is thus called factorial analysis of variance or multifactor analysis of variance. When there are two factors, the technique is known as two-way analysis of variance.

More complicated designs have more than two independent variables or factors, with each variable having two or more possible values. Each possible value of an independent variable is called a *level*. For example, if we have three methods of instruction and two learning environments we have three levels for the first factor and two levels for the second for a 3×2 design. If the subjects in this study are classified boys and girls, we have a $3 \times 2 \times 2$ design.

Let us begin by considering a simple design in which there are two independent variables (factors) with each factor being varied in two ways. That is, each factor exists at two levels. Such a design is referred to as a 2 × 2 (two by two) factorial design. The statistical method to be used is a two-way analysis of variance.

2 × 2 Factorial Design: An Example

Assume an experimenter wishes to investigate the relative effectiveness of two methods of instruction: directed learning and discovery learning. He believes that the effectiveness of the method may depend on the cultural background of the student with whom it is being used. Thus he wishes to investigate the two methods with two different cultural groups—Spanish speaking and English speaking. The two independent variables under investigation are (1) method of instruction, and (2) cultural background. Each factor is varied in two ways; that is, two methods and two cultural groups are being considered. The dependent variable of the study is achievement.

Data from such a factorial design are generally organized into a double-entry table with one of the independent variables listed in columns and the other listed in rows. Scores on the dependent variable of the study are listed in the appropriate cells, with each cell representing the intersection of one level of the first variable with one level of the second variable. The paradigm for a 2 × 2 factorial design is shown in Figure 14.1. All the possible

	Variable A: Method	
Variable B: language	Directed A_1	Discovery A_2
B_1: English	A_1B_1	A_2B_1
B_2: Spanish	A_1B_2	A_2B_2

Figure 14.1 Paradigm for Two-Way Analysis of Variance

combinations of the levels of the independent variables, symbolized A and B, are indicated. The two levels of methods are indicated by A_1 and A_2; the two language groups are symbolized B_1 and B_2.

It can be seen that four groups of subjects would be required for this study: English speakers B_1 receiving Method A_1; Spanish speakers B_2 receiving Method A_1; English Speakers B_1 with Method A_2; and Spanish speakers B_2 with Method A_2. Note that all subjects appearing in the first column are taught by Method A_1 and all appearing in the second column are taught by Method A_2. All subjects appearing in the first row are English speakers while all those appearing in the second row are Spanish speakers. Thus, with this design one may determine the effect of Method A_1 on achievement as compared with the effect of Method A_2 by comparing column 1 scores with column 2 scores. By comparing row 1 scores with row 2 scores the effect of the native language on achievement is revealed.

We see that two questions are being investigated at the same time with the same group of subjects. One question investigates methods; the second investigates language. The effects associated with the two independent variables of the study are called *main effects*. In addition to the main effects, one can determine whether or not the two independent variables in combination have an effect on the dependent variable. This is called the *interaction effect* of the variables and is symbolized $A \times B$. In this example, interaction would be indicated if the relative effectiveness of Methods A_1 and A_2 was different for the English speakers and the Spanish speakers. If there is no interaction, the effects of method and language are independent of each other. That is, the same relative position of the two variables will be evident on the dependent variable, regardless of the level being considered.

Computation for Two-way Analysis of Variance[1]

The computational procedure is similar to that for one-way analysis of variance except that the sum of squares between groups is divided into component parts. The sum of squares within groups and the total sum of squares are the same as they are in one-way ANOVA.

In the two-way ANOVA the variation between-group means, the component of SS_b, is divided into three parts: (1) the sum of squares between columns SS_c (variation associated with the column independent variable), (2) the sum of squares between rows SS_r (variation associated with the row independent variable), and (3) the sum of squares for row by column interaction SS_{rc} (variation associated with the interaction of the row and column variables). Since sum of squares between SS_b plus sum of squares within SS_w equal sum of squares total ($SS_b + SS_w = SS_t$) when we break down the SS_b into its component parts we have $(SS_c + SS_r + SS_{rc}) + SS_w = SS_t$.

---

[1]*The computational procedures included in this chapter are those appropriate for cases where the number of subjects in all the cells are equal. The computational procedures for unequal n's are much more complicated. For procedures appropriate for unequal n's see G. V. Glass and J. C. Stanley, Statistical Methods in Education and Psychology (Englewood Cliffs, N.J.: Prentice Hall, 1970), Chapter 17.*

Each of these sums of squares is divided by its associated degrees of freedom in order to obtain a mean square for each. The degrees of freedom for the various sums of squares are:

$$\text{d.f. for between columns} = c - 1$$
$$\text{d.f. for between-rows} = r - 1$$
$$\text{d.f. for interaction} = (c - 1)(r - 1)$$
$$\text{d.f. for within-groups} = N - k$$
$$\text{d.f. for total sum of squares} = N - 1$$

Thus the total number of degrees of freedom has been partitioned:

$$(N - 1) = (c - 1) + (r - 1) + (c - 1)(r - 1) + (N - k)$$

where: c = number of columns
r = number of rows
N = number of subjects in all groups
k = number of groups

The two-way ANOVA yields three F-ratios; one for each of the independent variables and one for their interaction. The mean square for each of these three sources of between-groups variation is divided by the mean square within groups to form the F-ratios needed for hypothesis testing. (See Table 14.1.)

TABLE 14.1 Outline of Procedure for Two-Way Analysis of Variance

SOURCE OF VARIATION	DEGREES OF FREEDOM	SUM OF SQUARES	MEAN SQUARE	F
Variable A (columns)	$c - 1$	SS_c	$MS_c = \dfrac{SS_c}{c - 1}$	$\dfrac{MS_c}{MS_w}$
Variable B (rows)	$r - 1$	SS_r	$MS_r = \dfrac{SS_r}{r - 1}$	$\dfrac{MS_r}{MS_w}$
Interaction between A and B (rc)	$(c - 1)(r - 1)$	SS_{rc}	$MS_{rc} = \dfrac{SS_{rc}}{(c - 1)(r - 1)}$	$\dfrac{MS_{rc}}{MS_w}$
Within-groups	$N - rc$	SS_w	$MS_w = \dfrac{SS_w}{N - rc}$	

The F-ratio for columns ($F = MS_c/MS_w$) provides a test for the main effects of the *column* independent variable. If this F-ratio is significant, we conclude that the difference between the means for A_1 and A_2 columns is greater than would be expected through chance alone. This would be described as a significant main effect for columns. In the same manner the

F-ratio for rows ($F = MS_r/MS_w$) tests the significance of the difference between row means. The F-ratio for interaction ($F = MS_{rc}/MS_w$) tests the significance of the interaction.

Let us work a hypothetical example. An investigator wants to compare the effectiveness of directed learning and discovery learning (Variable A) among Spanish-speaking and English-speaking students (Variable B). Five Spanish speakers and five English speakers are assigned to each of the two instructional methods. At the end of the experiment, a measure of achievement is administered to the subjects in all four groups. The three null hypotheses to be tested are: (1) Main effect A, the mean of a population with directed learning is equal to the mean of a population with discovery learning, that is, $\mu_{A_1} = \mu_{A_2}$. (2) Main effect B, the mean of the English-speaking population is equal to the mean of the Spanish-speaking population, that is, $\mu_{B_1} = \mu_{B_2}$. (3) Interaction A × B, the difference between directed learning and discovery learning subpopulations is the same in the English-speaking population as it is in the Spanish-speaking population, that is, $(\mu_{A_1B_1} - \mu_{A_2B_1}) = (\mu_{A_1B_2} - \mu_{A_2B_2})$. The interaction null hypothesis could also be stated: "The difference between English-speaking and Spanish-speaking subpopulations is the same in the directed learning population as it is in the discovery learning population, that is, $(\mu_{A_1B_1} - \mu_{A_1B_2}) = (\mu_{A_2B_1} - \mu_{A_2B_2})$. Which of the two ways to state the interaction null hypothesis is a matter of choice. The results are summarized in Table 14.2.

Total Sum of Squares 1. The first step in the computation is to calculate the total sum of squares. This is done in the same way as it was for the simple analysis of variance, using the raw score procedure and Formula (13.6), $SS_t = \Sigma X^2 - [(\Sigma X)^2/N]$. That is, each score value in the total group is squared and the sum of the squared values found: $\Sigma X^2 = 10^2 + 11^2 + \cdots + 8^2 + 9^2 + \cdots + 12^2 + 15^2 + \cdots + 15^2 + 16^2 + \cdots = 4{,}410$. Next we find the sum of all the scores ΣX (total) $= 63 + 56 + 78 + 91 = 288$, square that sum, and divide by N, the total number of subjects:

$$\frac{(\Sigma X)^2}{N} = \frac{(288)^2}{20} = \frac{82{,}944}{20} = 4{,}147.2$$

To obtain the total sum of squares we substitute in the formula and subtract the second value from the first:

$$SS_t = \Sigma X^2 - \frac{(\Sigma X)^2}{N}$$
$$= 4{,}410 - \frac{(288)^2}{20}$$
$$= 4{,}410 - 4{,}147.2$$
$$= 262.8$$

Two-Way Analysis of Variance 377

TABLE 14.2 Achievement Measures for Four Groups Using Two Methods of Instruction and Two Language Groups

		METHOD OF TEACHING (VARIABLE A)	
		Directed Learning A_1	Discovery Learning A_2
LANGUAGE SPOKEN (VARIABLE B)	English Speaking B_1	10 11 12 14 16 $\Sigma = \overline{63}$	8 9 11 13 15 $\Sigma = \overline{56}$	$\Sigma X_{r1} = 119$
	Spanish Speaking B_2	12 15 16 17 18 $\Sigma = \overline{78}$	15 16 18 19 23 $\Sigma = \overline{91}$	$\Sigma X_{r2} = 169$
		$\Sigma X_{c1} = 141$	$\Sigma X_{c2} = 147$	ΣX (total) = 288

$\Sigma X_{r1} = 119 \quad \bar{X}_{r1} = 11.9$
$\Sigma X_{r2} = 169 \quad \bar{X}_{r2} = 16.9$
$\Sigma X_{c1} = 141 \quad \bar{X}_{c1} = 14.1$
$\Sigma X_{c2} = 147 \quad \bar{X}_{c2} = 14.7$
ΣX (total) = 288

Between-Groups Sum of Squares 2. The between-groups sum of squares is found using the same raw score procedure, Formula (13.7), we used in one-way ANOVA. For each cell the sum of the scores is squared and this square is divided by the number of subjects in the cell. These quotients are then summed and the quantity $(\Sigma X)^2/N$ is subtracted from this sum:

$$SS_b = \frac{(\Sigma X_1)^2}{n_1} + \frac{(\Sigma X_2)^2}{n_2} + \frac{(\Sigma X_3)^2}{n_3} + \frac{(\Sigma X_4)^2}{n_4} - \frac{(\Sigma X)^2}{N}$$

$$= \frac{(63)^2}{5} + \frac{(56)^2}{5} + \frac{(78)^2}{5} + \frac{(91)^2}{5} - \frac{(288)^2}{20}$$

$$= 4{,}294 - 4{,}147.2$$

$$= 146.8$$

Within-Groups Sum of Squares 3. The within-groups sum of squares is found by subtracting the between-groups sum of squares from the total sum of squares with Formula (13.10):

$$SS_w = SS_t - SS_b$$
$$= 262.8 - 146.8$$
$$= 116$$

Partitioning the Between-Groups Sum of Squares 4. The between-groups sum of squares is broken down into three sums of squares each of which is calculated separately.

(a) The first component of the between-groups sum of squares is that related to the method of instruction. We compare A_1 and A_2 or directed versus discovery learning in their effect on achievement; therefore, we consider the total data listed in each of the two columns. We find the sum of scores for each column, square each column sum, and divide each of the squares by the total number of scores in each column. The sum of these two quotients is found and from it is subtracted $(\Sigma X)^2/N$:

$$SS_{bc} = \frac{(\Sigma X_{c1})^2}{n_{c1}} + \frac{(\Sigma X_{c2})^2}{n_{c2}} - \frac{(\Sigma X)^2}{N}$$

where: SS_{bc} = sum of squares between columns
ΣX_{c1} = sum of scores in column 1
ΣX_{c2} = sum of scores in column 2
n_{c1} = number of scores in column 1
n_{c2} = number of scores in column 2
ΣX = total sum of scores
N = total number of scores

Formula (14.1)
Sum of Squares Between Columns

In our example;

$$SS_{bc} = \frac{(141)^2}{10} + \frac{(147)^2}{10} - \frac{(288)^2}{20}$$
$$= \frac{19,881}{10} + \frac{21,609}{10} - 4,147.2$$
$$= 4,149 - 4,147.2$$
$$= 1.8$$

(b) The second component of the between-groups sum of squares is that related to the language. One is comparing B_1 and B_2, or Spanish- and English-speaking subjects. So we consider the data listed in the two rows. To find the sum of squares between rows, first square the total value of scores in each and divide by the number in that row. Finally, sum these quotients for all rows and subtract, $(\Sigma X)^2/N$. That is:

$$SS_{br} = \frac{(\Sigma X_{r1})^2}{n_{r1}} + \frac{(\Sigma X_{r2})^2}{n_{r2}} - \frac{(\Sigma X)^2}{N}$$

where: ΣX_{r1} = sum of scores in row 1

ΣX_{r2} = sum of scores in row 2
n_{r1} = number of scores in row 1
n_{r2} = number of scores in row 2
ΣX = total sum of scores
N = total number of scores
(Formula (14.2)
Sum of Squares Between Rows

In our example:

$$SS_{br} = \frac{(119)^2}{10} + \frac{(169)^2}{10} - \frac{(288)^2}{20}$$

$$= \frac{14{,}161}{10} + \frac{28{,}561}{10} - 4{,}147.2$$

$$= 125$$

(c) The third component of the between-groups sum of squares is that due to the interaction between method and language. Since we have partitioned the between-groups sum of squares into three parts and have already calculated two of these, the third component can be found by subtraction. That is, one can find the interaction sum of squares by adding the between-columns sum of squares and the between-rows sum of squares and subtracting their sum from the between-groups sum of squares. Interaction of method × language = between-all-groups − (between methods + between language):

$$SS_{rc} = SS_b - SS_{bc} + SS_{br}$$

Formula (14.3)
Row by Column Interaction
Sum of Squares

In our example:

$$SS_{rc} = 146.8 - (1.8 + 125)$$
$$= 146.8 - 126.8$$
$$= 20.0$$

Mean Squares 5. Each of the above three sums of squares is divided by its associated degrees of freedom in order to arrive at a mean square for each. The mean square for columns MS_c is $1.8/1 = 1.8$; for rows, $MS_r = 125/1 = 125$; for interaction, $MS_{rc} = 20/1 = 20$. The sum of squares within groups is divided by its associated degrees of freedom $(N - k)$ to obtain the mean square within groups, $MS_w = 116/16 = 7.25$.

F-Ratio 6. The F-ratios for the main effects of the two independent variables and for the interaction effect are found by dividing the mean squares for each of the between-groups components by the within-groups mean square. Columns: $F = 1.8/7.25 = .248$. Rows: $F = 125/7.25 = 17.24$. Interaction: $F = 20/7.25 = 2.758$. All this is summarized in Table 14.3.

The significance of the F-ratios is determined by consulting the F-table (Table A6) in the Appendix. As before, one enters the table with the number

TABLE 14.3 Summary Table for 2 × 2 Analysis of Variance of Data in Table 14.2

SOURCE OF VARIATION	SUM OF SQUARES	d.f.	MS	F
Between columns (methods)	1.8	1	1.8	.248
Between rows (languages)	125.0	1	125.0	17.241
Columns by rows (interaction)	20	1	20	2.758
Within groups	116	16	7.25	
Total	262.8	19		

of degrees of freedom associated with the numerator and the denominator of the *F*-ratio; in this example each ratio has 1 and 16. Entering the table with 1 and 16 d.f., one finds that an *F*-ratio of 4.49 is needed for significance at the .05 level. Only the *F*-ratio for language (17.24) exceeds the tabled *F*-value at the .05 level of significance. The other two *F*-ratios are not significant.

We conclude that the method of instruction (directed learning versus discovery learning) had no significant effect on achievement and that there was no significant interaction between the method and the language of the subjects. A nonsignificant interaction effect means that the difference between the two methods of instruction is independent of the language variable. That is, we find approximately the same difference between the two language groups regardless of the method of instruction. The relationship between language and achievement was significant. An examination of the scores in Table 14.2 reveals a significant difference in the two row means with mean achievement higher in the Spanish group. Therefore, we reject the null hypothesis of no difference in the row means.

The Interpretation of Interaction

One of the important features of factorial ANOVA is that it enables us to observe interaction. The presence of interaction indicates that the dependent variable means cannot be explained by the separate effects of the two independent variables alone. The effect of one independent variable is different under different values of the other independent variable. Interaction may occur with or without main effects.

Let us take a single example and look at various possible outcomes. An instructor has 50 boys and 50 girls who are taught handwriting by Method *A* and another 50 boys and 50 girls who are taught handwriting by Method *B*. After treatment a handwriting specimen is collected from each subject and scored. The numbers in the cells represent the means of each of the four groups, the row totals give the means for all boys and all girls, the column totals give the means for all Method *A* subjects and all Method *B* subjects. We will illustrate four out of various possible sets of results.

CASE A:

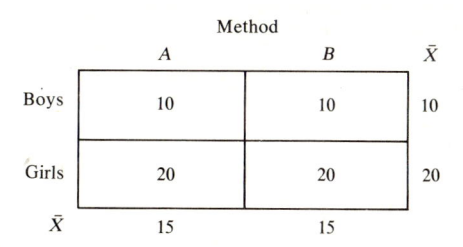

	Method A	B	$\bar{X}$
Boys	10	10	10
Girls	20	20	20
$\bar{X}$	15	15	

Diagram Showing the Means of the Four Groups

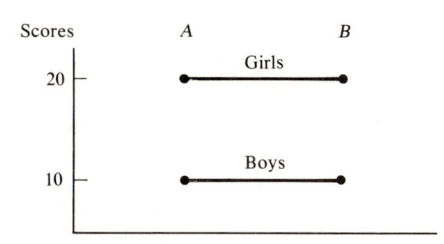

```
Scores     A              B
  20  ├─   ●──── Girls ────●
  10  ├─   ●──── Boys ─────●
```

Above we have the cell means, column means, and row means followed by a diagram showing the cell means. On the ordinate we have values for the dependent variable. Under A we have dots for the two A means. Under B we have dots for the two B means. The two boys' means are connected by a line and so are the two girls' means.

Comparing column means we see no main effect for method as shown by the fact that the means for both columns are the same. Comparing rows we see there is a sex difference. There is no interaction; comparing cell means we see that the difference between boys and girls is the same under both methods. The girls' line and the boys' line are parallel in our graph.

CASE B:

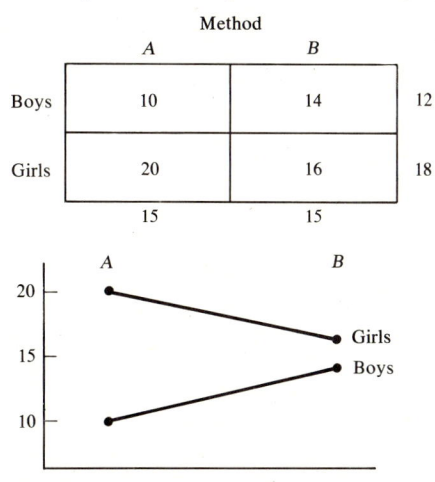

	Method A	B	
Boys	10	14	12
Girls	20	16	18
	15	15	

```
          A              B
  20  ├─  ●
                         ● Girls
  15  ├─                 ● Boys
  10  ├─  ●
```

382 INFERENTIAL STATISTICS

Case B shows a main effect for sex, the girls' mean is higher than the boys' mean; there is no main effect for method, both column means are the same. However, the girls are scoring higher with Method A than with Method B while the reverse is true for boys, showing that there is interaction.

CASE C:

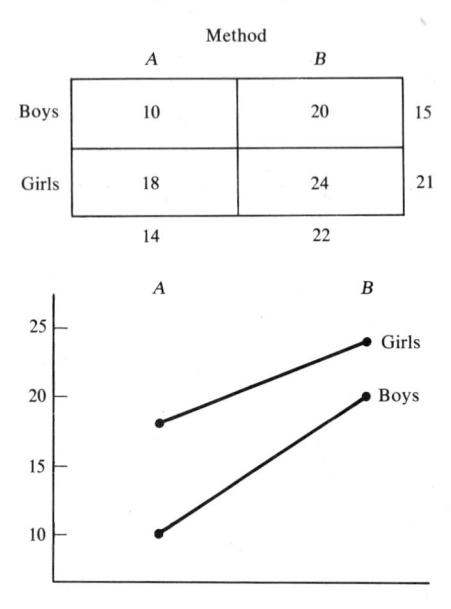

	Method A	Method B	
Boys	10	20	15
Girls	18	24	21
	14	22	

Case C shows a main effect for sex, the girls' mean is higher than the boys' mean; there is a main effect for method, the B mean is higher than the A mean. There is interaction, although both boys and girls do better with Method B than with Method A, there is more difference between the means of the two methods for boys than there is for girls.

CASE D:

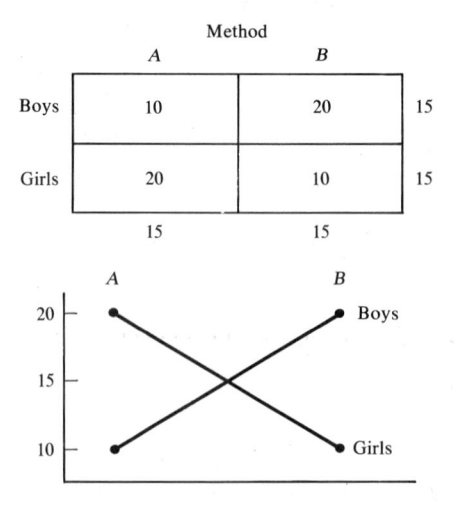

	Method A	Method B	
Boys	10	20	15
Girls	20	10	15
	15	15	

In case *D* the boys' mean is equal to the girls' mean and the *A* and *B* means are equal, so there is neither sex nor method main effect but there is interaction as the boys' mean is higher with Method *B* and the girls' mean is higher with Method *A*.

In those cases where the lines in our graph are not parallel we have apparent interaction. The lines may depart from parallel through sampling error or there may be a genuine interaction. The *F*-ratio for interaction tells us whether the difference in the slopes of the lines is statistically significant or not.

An Example with Significant Interaction Assume an experiment which investigates the effect of programmed instruction compared with conventional instruction on the skill scores of students with high achievement motivation and those with low achievement motivation. The two independent variables under consideration are method of instruction and achievement motivation with each variable existing at two levels, giving us a 2×2 factorial design. The scores on the criterion for the four groups of students are summarized in Table 14.4.

TABLE 14.4 Skill Scores for Four Groups Using Two Methods of Instruction and Two Levels of Achievement Motivation

METHOD OF INSTRUCTION (VARIABLE *A*)

		A_1 Conventional Instruction	A_2 Programmed Instruction	
LEVEL OF ACHIEVEMENT MOTIVATION (VARIABLE *B*)	High B_1	22 20 18 17 16 15 — 108	26 22 20 19 18 15 — 120	$\Sigma X_{r1} = 228$ $\bar{X}_{r1} = 19$
	Low B_2	25 20 19 18 17 15 — 114	18 16 15 13 12 10 — 84	$\Sigma X_{r2} = 198$ $\bar{X}_{r2} = 16.5$
		$\Sigma X_{c1} = 222$ $\bar{X}_{c1} = 18.5$	$\Sigma X_{c2} = 204$ $\bar{X}_{c2} = 17$	

384 INFERENTIAL STATISTICS

Again we find the total sum of squares, the between-groups sum of squares and the within-groups sum of squares:

$$SS_t = \Sigma X^2 - \frac{(\Sigma X)^2}{N}$$

$$= (22)^2 + (20)^2 + \cdots + (10^2) - \frac{(426)^2}{24}$$

$$= 7{,}890 - 7{,}561.5$$

$$= 328.5$$

$$SS_b = \frac{(\Sigma X_1)^2}{n_1} + \frac{(\Sigma X_2)^2}{n_2} + \frac{(\Sigma X_3)^2}{n_3} + \frac{(\Sigma X_4)^2}{n_4} - \frac{(\Sigma X)^2}{N}$$

$$= \frac{(108)^2}{6} + \frac{(120)^2}{6} + \frac{(114)^2}{6} + \frac{(84)^2}{6} - \frac{(426)^2}{24}$$

$$= (1{,}944 + 2{,}400 + 2{,}166 + 1{,}176) - 7{,}561.5$$

$$= 7{,}686 - 7{,}561.5$$

$$= 124.5$$

$$SS_w = \Sigma x_t^2 - \Sigma x_b^2$$

$$= 328.5 - 124.5$$

$$= 204$$

Partitioning the between-groups sum of squares:
Between-columns (method):

$$SS_{bc} = \frac{(\Sigma X_{c1})^2}{n_{c1}} + \frac{(\Sigma X_{c2})^2}{n_{c2}} - \frac{(\Sigma X)^2}{N}$$

$$= \frac{(222)^2}{12} + \frac{(204)^2}{12} - \frac{(426)^2}{24}$$

$$= (4{,}107 + 3{,}468) - 7{,}561.5$$

$$= 7{,}575 - 7{,}561.5$$

$$= 13.5$$

Between-rows (achievement motivation):

$$SS_{br} = \frac{(\Sigma X_{r1})^2}{n_{r1}} + \frac{(\Sigma X_{r2})^2}{n_{r2}} - \frac{(\Sigma X)^2}{N}$$

$$= \frac{(228)^2}{12} + \frac{(198)^2}{12} - \frac{(426)^2}{24}$$

$$= (4{,}332 + 3{,}267) - 7{,}561.5$$

$$= 7{,}599.0 - 7{,}561.5$$

$$= 37.5$$

Interaction (method × level of achievement motivation);

$$SS_{rc} = \Sigma x_{bg}^2 - (\Sigma x_{bc}^2 + \Sigma x_{br}^2)$$
$$= 124.5 - (13.5 + 37.5)$$
$$= 124.5 - 51$$
$$= 73.5.$$

The results of the computations are summarized in Table 14.5.

TABLE 14.5 Summary Table for 2 × 2 Analysis of Variance of Data in Table 14.4

SOURCE OF VARIATION	SUM OF SQUARES	d.f.	MS	F
Between-columns (methods)	13.5	1	13.5	1.32
Between-rows (achievement motivation)	37.5	1	37.5	3.68
Columns by rows (interaction)	73.5	1	73.5	7.21
Within groups	204	20	10.2	
Total	328.5	23		

Entering the F-table with 1 and 20 d.f., we find that an F-ratio of 4.35 is needed for significance at the .05 level. The F-values for method of instruction and for level of achievement motivation, main effects, do not reach this value and are therefore not significant. We conclude that the two methods of instruction do not differ significantly from each other in their effect on the achievement of subjects in the experiment. Furthermore, there is no significant difference in the performance of those subjects with high achievement motivation and those with low achievement motivation. However, the F-ratio for the interaction of method and level of achievement motivation (7.21) does exceed the tabled F-value at the .05 level of significance. A significant interaction indicates that the effect of the method is dependent on the level of achievement motivation of the students.

Let us interpret this interaction effect by comparing the means for each of the four groups. See Table 14.6.

TABLE 14.6 Mean Scores for Four Groups Using Two Methods of Instruction and Two Levels of Achievement Motivation

	CONVENTIONAL A_1	PROGRAMMED A_2	
HIGH B_1	18	20	19
LOW B_2	19	14	16.5
	18.5	17	

Looking at the row means shown in Table 14.6, we see that the difference between the means of A_1 and A_2 at the B_1 level is $18 - 20 = -2$. The difference between A_1 and A_2 means at the B_2 level is $19 - 14 = 5$. Since these two differences are not equal we know that there is interaction between the variables.

Our F-ratio for interaction showed that method of instruction and level of achievement motivation in combination produce a significant effect on achievement. The cell means show that Method A_1 is superior to Method A_2, at motivation level B_2, but that Method A_2 is superior to A_1 at motivation level B_1. Students with high achievement motivation perform at a higher level under programmed instruction, while those of low achievement motivation perform at a higher level under conventional instruction.

We can diagram these relationships as shown in Figure 14.2. Original data are given in Table 14.4 with A representing method of instruction and B the level of achievement motivation.

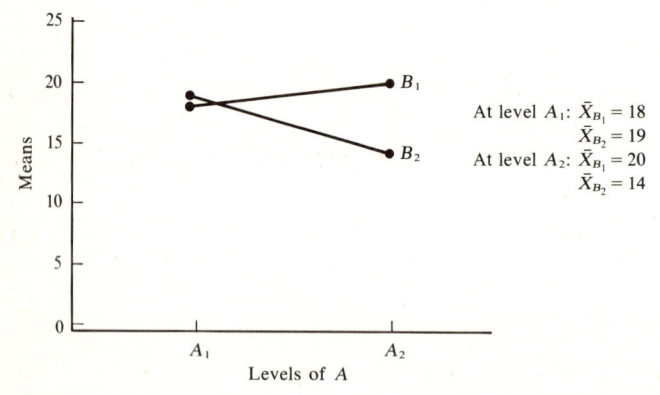

Figure 14.2 Means for the Levels of B_1 and B_2 for Each Level of A

At level A_1: $\bar{X}_{B_1} = 18$
$\bar{X}_{B_2} = 19$
At level A_2: $\bar{X}_{B_1} = 20$
$\bar{X}_{B_2} = 14$

The different slopes of the lines in Figure 14.2 indicate interaction. It can be seen that the effect of A on the dependent variable of achievement depends on the level of the B variable. Whenever there is interaction between the variables, the two lines for B_1 and B_2 are nonparallel. However, when there is *no* interaction between the variables, the two lines will be parallel or nearly parallel; the effect of A is the same at each level of B.

K-Factor Analysis of Variance

Factorial designs are not limited to two factors. They may include three, four, or more independent variables with each variable classified at two or more levels. For example, a study may be designed to investigate two methods of instruction in classes of two different sizes with students classified at three ability levels. This would be a $2 \times 2 \times 3$ design with three independent variables—two of them at two levels and one at three levels. The paradigm for a $2 \times 2 \times 3$ design is illustrated in Figure 14.3. The number

Method

		A_1 (Programmed)		A_2 (Conventional)	
		Class Size		Class Size	
		B_1 (Large)	B_2 (Small)	B_1 (Large)	B_2 (Small)
Ability Level	(Low) C_1	$A_1B_1C_1$	$A_1B_2C_1$	$A_2B_1C_1$	$A_2B_2C_1$
	(Average) C_2	$A_1B_1C_2$	$A_1B_2C_2$	$A_2B_1C_2$	$A_2B_2C_2$
	(High) C_3	$A_1B_1C_3$	$A_1B_2C_3$	$A_2B_1C_3$	$A_2B_2C_3$

Figure 14.3 Paradigm for a $2 \times 2 \times 3$ Factorial Design

of digits used indicates the number of factors or independent variables involved; the numerical values of the digits indicate the number of levels for the factors. For example, in a $2 \times 3 \times 4$ factorial design, there are three independent variables with two, three and four levels respectively.

In a $2 \times 2 \times 3$ factorial design, one can test for the main effects of each of three variables: A (method), B (class size), and C (ability level) on the dependent variable. In addition there are four interaction effects to be tested: $A \times B$, $A \times C$, $B \times C$, and $A \times B \times C$. Seven F-ratios would be computed and tested for significance.

The two-factor interactions such as $A \times B$ are called first-order interactions. The three-factor interactions such as $A \times B \times C$ are called second-order interactions, while a four-factor interaction such as $A \times B \times C \times D$ is a third-order interaction. The latter two are examples of the high-order interactions which arise when additional independent variables are added to a factorial design.

Of course, as variables are added to a factorial design, the computation and interpretation become more complex. In addition, there is the practical problem of getting sufficient subjects to fill in the different treatment groups. In the $2 \times 2 \times 3$ design illustrated in Figure 14.3, 12 groups of subjects would be required. If there were only 10 subjects in each treatment group, 120 subjects would be needed for the study. A $2 \times 2 \times 2 \times 2$ factorial design would require 16 groups and a $2 \times 2 \times 2 \times 2 \times 2$ would require 32 groups. For this reason designs with more than four variables are not often used.

A consideration of the computation and interpretation in a higher-order classification analysis of variance is beyond the scope of this text.[2]

SUMMARY

Two-way analysis of variance is an extension of simple analysis of variance to an experimental situation in which two independent variables are manipulated simultaneously. This technique enables one to determine the main effects of each of these independent variables on a dependent variable as well as the effect of interaction between the variables. Thus, in a single experiment one can accomplish what would ordinarily require two separate studies and, in addition, learn the effect of the variables in combination. Such information may have much practical and theoretical importance especially in research dealing with human subjects. To restrict research to one variable is to oversimplify a complex situation because the effect of a variable may depend on the presence of another variable or variables. For this reason factorial designs involving two or more factors are very appropriate for research in education and the social sciences. Tremendous advances have been made in educational research as a result of the development and use of multifactor analysis of variance.

The computation of two-way analysis of variance is similar to that for one-way ANOVA except there are more sources of variation to consider. One computes the total sum of squares, which is partitioned into four parts: a sum of squares for each of the two independent variables, a sum of squares for interaction, and a sum of squares within. Each of these sums of squares is divided by the appropriate number of degrees of freedom. Three F-ratios are computed by comparing the mean square for: (1) variable A, (2) variable B, and (3) interaction between $A \times B$, to the within mean square in each case. The F-ratios thus test the significance of the two main effects and the interaction effect. A significant interaction indicates that the effect of one independent variable is not constant at the various levels of the second variable. The effect depends on the different combinations of levels of the two independent variables.

---

[2]*For a discussion of the higher-order classification ANOVA see: A. L. Edwards, Experimental Design in Psychological Research, 3rd ed. (New York: Holt, Rinehart and Winston, 1968), Chapters 11 and 12.*

FLOWCHART 14

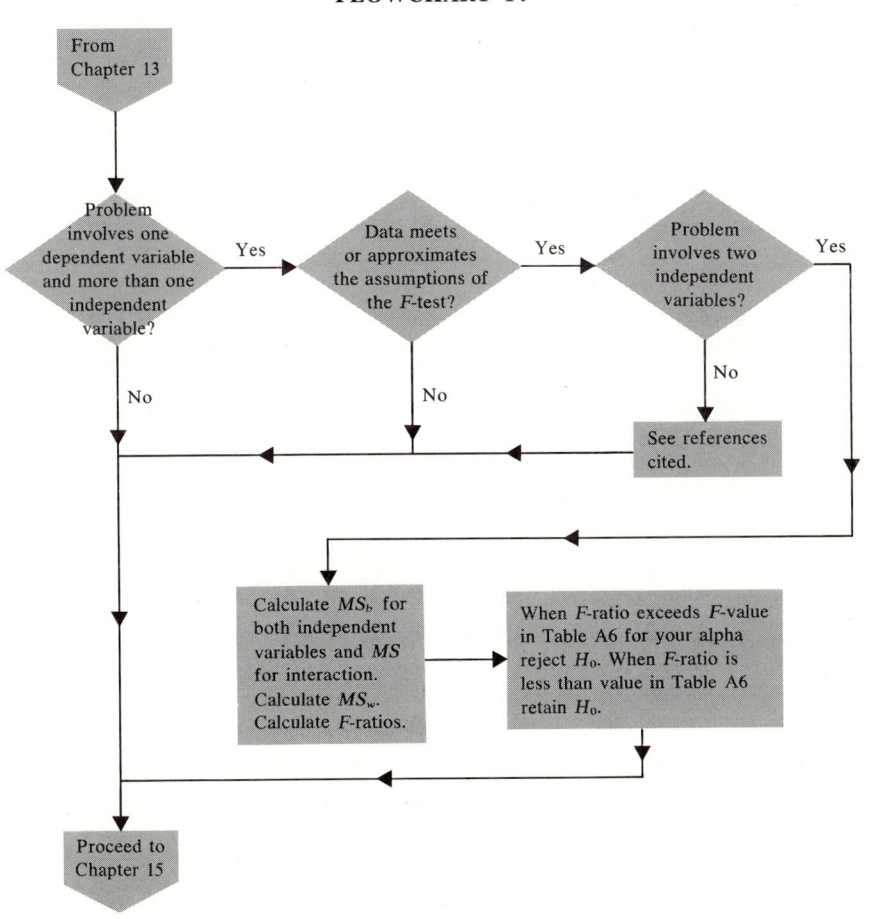

```
From Chapter 13
      │
      ▼
Problem involves one dependent variable and more than one independent variable?
   │ Yes →  Data meets or approximates the assumptions of the F-test?
   │            │ Yes →  Problem involves two independent variables?
   │            │              │ Yes ─────────────────────────┐
   │            │              │ No                            │
   │            │              ▼                               │
   │            │         See references cited.                │
   │            │ No ──────────────┐                           │
   │ No ──────────────────────────┐│                           │
                                   ▼▼                          ▼
                          Calculate $MS_b$ for both independent variables and $MS$ for interaction. Calculate $MS_w$. Calculate F-ratios.  →  When F-ratio exceeds F-value in Table A6 for your alpha reject $H_0$. When F-ratio is less than value in Table A6 retain $H_0$.
                                   │
                                   ▼
                          Proceed to Chapter 15
```

EXERCISES

1. Define the following terms:
   factor
   levels of a factor
   two-way analysis of variance
   main effect
   interaction
   first-order interaction
2. Give an example of an experiment in which two independent variables might be expected to interact significantly.
3. A study is conducted to ascertain the effect of the level of anxiety of the learner and the meaningfulness of the material on the speed of learning lists of words. Each independent variable is classified at two levels: high and low.
   (a) Identify the independent variables and the dependent variable of the study. Show a paradigm for the design using A and B to represent the independent variables.
   (b) State the null hypotheses to be tested.
   (c) If a total of 60 subjects are equally distributed among the groups, what F-values are necessary for the tests to be significant at the .05 level?
4. A 2×2 factorial design with 10 scores per cell is presented below. Complete the summary table.

	A_1	A_2
B_1		
B_2		

SOURCE OF VARIATION	SUM OF SQUARES	d.f.	MS	F	SIGNIFICANCE LEVEL
A	60				
B	48				
$A \times B$	48				
Within	144				
Total	300				

5. Given the following data, complete the analysis of variance and set up the summary table. Test the null hypotheses at the .05 level.

	A_1	A_2		A_1	A_2
	7	3		11	5
	6	2		10	4
B_1	4	2	B_2	9	3
	3	1		8	2
	2	1		4	1

6. Graph the means of the data in Problem 5. Describe the appearance of the lines and give an interpretation.
7. The effect produced by the combined influence of two independent variables is called a(n):
   (a) factor
   (b) level
   (c) interaction
   (d) main effect
8. In the two-way ANOVA, the variation among group means, the component of SS_b is divided into three parts. Which of the following symbolizes this division?
   (a) $SS_b = SS_r + SS_w + SS_t$
   (b) $SS_b = SS_c + SS_r + SS_{rc}$
   (c) $SS_b = SS_{rc} + SS_r + SS_w$
   (d) $SS_b = SS_r + SS_w + SS_c$
9. In the two-way ANOVA the F-ratio for columns ($F = MS_c/MS_w$) provides a test for the _____ of the column independent variable.
   (a) interaction
   (b) error variance
   (c) main effects
   (d) none of the above

The following graphs pertain to Questions 10–13:

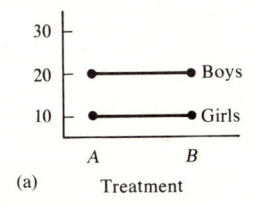

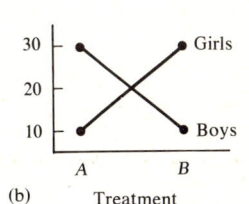

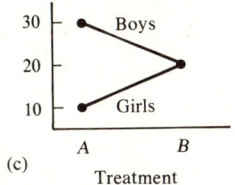

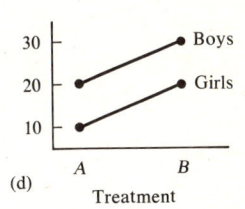

10. Which of the graphs shows a main effect for sex, no main effect for treatment, and no interaction?
11. Which of the graphs shows a main effect for sex, no main effect for treatment, and interaction?
12. Which of the graphs shows a main effect for sex, a main effect for treatment, and no interaction?
13. Which of the graphs shows no main effect for sex, no main effect for treatment, and interaction?

14. Three-factor interactions such as $A \times B \times C$ are called:
    (a) zero-order interactions
    (b) first-order interactions
    (c) second-order interactions
    (d) third-order interactions
15. In a $2 \times 2 \times 2 \times 3$ factorial design with 10 subjects in each treatment group how many subjects would be needed for the study?
    (a) 60
    (b) 120
    (c) 180
    (d) 240
16. Harrington wishes to experiment with the sequencing of instructional material and "organizers" that structure the material for the learner. A group of 90 persons were randomly split into three groups of 30 each. Group I received organizing material before studying instructional materials on mathematics; Group II received the "organizer" after studying the mathematics; Group III received no organizing material in connection with studying the mathematics instructional materials. A 50-item test of the mathematics covered in the instructional materials served as a measure of the dependent variable. What method should Harrington use for the analysis of the data, and what statistical tests of significance should be used?
    (a) one-way ANOVA, t-test
    (b) one-way ANOVA, F-test
    (c) two-way ANOVA, t-test
    (d) two-way ANOVA, F-test
17. A researcher plans to do an experiment involving the Forced Syntksis, Free Syntksis, and No Syntksis methods of instruction. The researcher is also interested in the possible effects of the instructional methods on students with above-average and below-average intelligence. The dependent variable will be scores on an achievement test. What method should the researcher use for the analysis of the data, and what statistical tests of significance should be used?
    (a) one-way ANOVA, t-test
    (b) one-way ANOVA, F-test
    (c) two-way ANOVA, t-test
    (d) two-way ANOVA, F-test

ANSWERS

3. (a) The independent variables are level of anxiety and meaningfulness of material. The dependent variable is speed of learning. The paradigm is:

A_1B_1	A_2B_1
A_1B_2	A_2B_2

(b) The null hypotheses are: 1. $\mu_{A_1} = \mu_{A_2}$; 2. $\mu_{B_1} = \mu_{B_2}$; 3. $(\mu_{A_1B_1} - \mu_{A_1B_2}) = (\mu_{A_2B_1} - \mu_{A_2B_2})$ or $(\mu_{A_1B_1} - \mu_{A_2B_1}) = (\mu_{A_1B_2} - \mu_{A_2B_2})$
(c) 4.02 (1 and 56 d.f.)

4.

SOURCE OF VARIATION	SUM OF SQUARES	d.f.	MS	F	SIGNIFICANCE LEVEL
A	60	1	60	15	.01
B	48	1	48	12	.01
A × B	48	1	48	12	.01
Within	144	36	4		
Total	300	39			

5.

SOURCE OF VARIATION	SUM OF SQUARES	d.f.	MS	F	SIGNIFICANCE LEVEL
Columns (A)	80	1	80	21.62	.01
Rows (B)	33.8	1	33.8	9.14	.01
(A × B)	9.8	1	9.8	2.65	ns
Within	59.2	16	3.7		
Total	182.8	19			

Significant difference between the column means; reject the null hypothesis.
Significant difference between the row means; reject the null hypothesis.
The interaction between A and B is not significant; retain the null hypothesis.

6. There is observed interaction since the lines are not parallel. However, since the F-ratio for interaction is not significant the apparent interaction could easily be a function of chance.

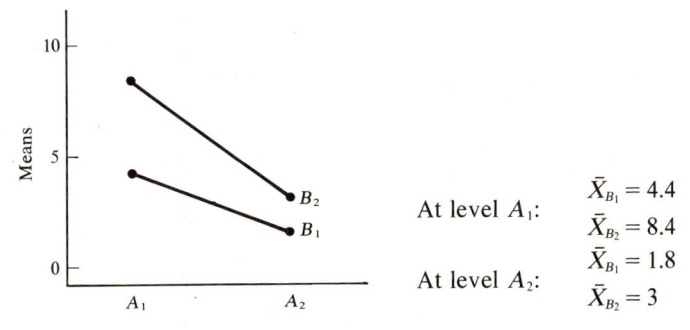

At level A_1: $\bar{X}_{B_1} = 4.4$, $\bar{X}_{B_2} = 8.4$
At level A_2: $\bar{X}_{B_1} = 1.8$, $\bar{X}_{B_2} = 3$

7. (c) 8. (b) 9. (c) 10. (a) 11. (c) 12. (d) 13. (b) 14. (c)
15. (d) 16. (b) 17. (d)

CHAPTER FIFTEEN
NONPARAMETRIC STATISTICS

OBJECTIVES

After completing this chapter the reader should be able to:
1. Distinguish between data that meet the parametric assumptions and data that do not.
2. Identify situations where a nonparametric test would be used instead of a parametric test.
3. Recognize situations where the chi square goodness-of-fit test should be used.
4. Recognize situations where the chi square test of the independence of categorical values should be used.
5. Compute chi square for both goodness-of-fit and independence of categories problems.
6. Determine degrees of freedom for chi square tests.
7. Convert a chi square from a two by two table into a phi coefficient.
8. Identify situations where it would be appropriate to convert a chi square to a contingency coefficient.
9. Calculate a contingency coefficient from a chi square.
10. Identify situations where the sign test should be used.
11. Compute a sign test.
12. Identify situations where the median test should be used.
13. Compute a median test.

Introduction

In the previous chapters on inferential statistics we have been concerned with using the statistics computed from representative samples as estimates of the corresponding population parameters. The statistical tests designed for this purpose are known therefore as *parametric* tests of significance. You recall that these parametric tests, such as the t-test and the F-ratio were based on certain assumptions regarding the parameters. Some of the assumptions implicit in the use of the parametric tests are: (1) that the data represent population characteristics that are continuous and symmetrical. These characteristics are usually only found in data that are interval or ratio, so typically parametric tests are restricted to such data; (2) that the variable has a normal or near normal distribution in the population; and (3) that the sample statistic provides an estimate of the population parameter.

Often there is a need to test hypotheses with data that are ordinal or nominal or with interval data that are not normally distributed or that fail to meet other assumptions necessary for the use of parametric tests. For this purpose *nonparametric* tests have been developed. Nonparametric tests require fewer assumptions about data than parametric tests and therefore can be used in a wide variety of situations where parametric tests would be inappropriate. The nonparametric tests are used to test hypotheses involving nominal or ordinal data.

In an introductory book such as this it is impossible to introduce all the nonparametric methods available. In this chapter we introduce three widely used nonparametric procedures: chi square, sign test, and median test. We have already discussed, in Chapter Nine, the Spearman rho and other nonparametric correlation methods. Other nonparametric procedures can be found in Siegel[1] and other books on the subject.

Chi Square

The *chi square* (χ^2) procedure is used to test hypotheses about the independence of frequency counts in various categories. For example, the data may be the proportions of smokers preferring each one of three brands of low-tar cigarettes or the proportion of people favoring a repeal of the ban on phosphate detergents along with the proportion opposed to a repeal. Categories of responses are set up, such as Brands A, B, and C of cigarettes or Approve-Disapprove of a certain issue, and the number of individuals or events falling into each category is recorded. In such a situation one can obtain nothing more than the frequency, or number of times, that a particular category is chosen, which constitutes nominal data. With such data the only analysis possible is to determine whether the frequencies observed in the sample differ significantly from hypothesized frequencies.

[1] Sidney Siegel, *Nonparametric Statistics for the Behavioral Sciences* (New York: McGraw-Hill, 1956).

Such hypothesized frequencies may be completely theoretical, for example, one blindly guesses that half of an incoming freshman class will identify themselves as liberals and half as conservatives, or may be based on previous information, for example, 60 percent of last year's freshmen identified themselves as liberal and 40 percent classified themselves conservative, so one hypothesizes that the same proportions are true for this year's freshmen. These two uses of the chi square are often called "goodness-of-fit" tests because they tell us how well an observed distribution fits a hypothesized or theoretical distribution within a single population.[2] A third use of the χ^2 is with the hypothesis that the same proportions will be found in two or more populations, for example, the proportions of liberals and conservatives among female students will be the same as among male students. In each case one compares the observed proportions in a sample with the hypothesized proportions and applies the chi square test to determine whether a difference between observed and hypothesized proportions is likely to be a function of sampling error (nonsignificant) or unlikely to be a function of sampling error (significant).

Goodness-of-Fit Chi Square

As an example of the first use of the chi square (single sample, goodness-of-fit), let us picture a food processor who has developed three different blends of instant coffee, A, B, and C. He wants to determine whether any one of them will be a better prospect for marketing than the others. His null hypothesis would be that among prospective customers the proportion preferring each of the blends would be equal. From among prospective customers he selects a single sample of 90 and asks each of them to try the three blends and indicate a preference. The results are shown in Table 15.1.

TABLE 15.1 Distribution of Preferences for Each of Three Blends of Instant Coffee in a Sample of 90 Prospective Customers

BRAND	FREQUENCY
A	26
B	38
C	26

Blend B has been chosen more often than the other two blends. Is this good evidence that among prospective customers from whom our sample was drawn Blend B would be preferred? Or could the apparent preference

---
[2] Some texts reserve the term "goodness-of-fit" for tests of a hypothesis that sample data are derived from a population with a normal distribution. Other texts use the term to describe a test concerning any hypothetical distribution, which is the way we are using this term.

easily be a function of sampling error? To arrive at an estimate of the probability that the observed frequency distribution is due to chance he would apply the χ^2 test. The chi square test permits him to estimate the probability that observed frequencies differ from expected frequencies through chance alone.

In this example, his null hypothesis is that no one blend of coffee would be preferred over any other. In this case, he would expect that each of the three blends would be selected by one-third of the prospective customers, giving expected frequencies of 30:30:30. If the null hypothesis is true, any departure from these frequencies would be the result of pure chance.

In order to compare his expected and observed frequencies using the χ^2 test he first enters them in a table as in the first two columns of Table 15.2.

TABLE 15.2 Calculation of χ^2 for Data on Preferences for Three Blends of Instant Coffee in a Sample of 90

BLEND	OBSERVED O	EXPECTED E	$O - E$	$(O - E)^2$	$\dfrac{(O - E)^2}{E}$
A	26	30	−4	16	.53
B	38	30	8	64	2.1
C	26	30	−4	16	.53
Sum	90	90	0		$3.16 = \chi^2$

Note that the sum of the expected frequencies is equal to the sum of the observed frequencies. This will always be the case. The first column shows how the 90 subjects in his sample were actually distributed. The second column shows how they were expected to be distributed under the null hypothesis. The third column shows the differences between these two values. He now has the information needed to calculate the value of χ^2 for his experiment:

$$\chi^2 = \sum \frac{(O - E)^2}{E}$$

where: O = observed frequency in a given category
E = expected frequency in that category
Σ = sum of the ratios over all categories

Formula (15.1)
Chi Square

The formula indicates that to calculate χ^2, one finds for *each* category the difference between the observed and expected values, squares the difference, and divides the squared difference by the expected value. Then these quotients are summed. We see that χ^2 will always have a positive value because of the squaring of the difference between O and E. In our example the value of χ^2 is 3.16.

Degrees of Freedom

The next step is to determine the degrees of freedom. Recall that the degrees of freedom (d.f.) of a statistic represent the number of observations free to vary while producing that statistic. In our example the food processor has a single sample of subjects who have selected from three categories. Once he has established the frequency of any two of these categories the frequency of the third category is fixed. His first two frequencies (26 and 38) add up to 64; the last frequency, therefore, has to be 26 in order to make the total 90. The same principle would hold true for any combination of categories. For any one sample χ^2 test, the degrees of freedom are the number of categories (k) minus one because, as we have seen, the frequency in one category is not free to vary. In the present problem, the number of d.f. = 3 − 1 = 2.

The final step is to refer to Table A7 in the Appendix in order to determine whether the obtained χ^2 value is statistically significant or not. Note that the chi square table has rows for the different degrees of freedom and columns for indicating the approximate probability of obtaining a χ^2 value equal to or greater than any given value for a certain degrees of freedom when the null hypothesis is true. To use the table one must enter it with the appropriate degrees of freedom. In our example it is entered with 2 d.f.

The essential question one asks is: "How often would a χ^2 of a certain value or greater occur by chance when the null hypothesis is true and there are $k - 1$ degrees of freedom?" As is true with the t-table and the F-table, if the χ^2 value one has calculated is equal to or greater than the value required for significance at a predetermined probability level for the d.f., then the null hypothesis of no real difference between the observed and expected frequencies is rejected at that level of significance. If the calculated χ^2 is smaller than the tabled value required for significance at that level, then one may *not* reject the null hypothesis. Table A7 tells us that with 2 d.f. a χ^2 value of at least 5.99 is necessary to reject the null hypothesis at the .05 level. The value of χ^2 in our example is only 3.16. Since the food processor's 3.16 is less than the value required for significance, he cannot reject his null hypothesis. Thus, he would retain his null hypothesis and conclude that the observed frequencies do not differ significantly from the expected frequencies. The χ^2 test shows that he has insufficient evidence of a systematic preference for one blend of coffee over another.

At this point it is important to note the effect of the degrees of freedom upon the significance of any calculated χ^2. In Table A7 it can be seen that the χ^2 required for significance at any given level gets larger as the number of degrees of freedom gets larger. We saw that for 2 d.f. a χ^2 of 5.99 was needed for significance at the 5 percent level. With 3 d.f. this χ^2 value increases to 7.82, to 11.07 for 5 d.f., and to 43.77 for 30 d.f. Thus, it is

Goodness-of-Fit Employing Prior Information

Let us consider an example of the use of the χ^2 procedure where the hypothesized distribution is based on previous information. A principal suspects the band program is not attracting participation from the black students in her school as well as it is attracting participation from the white students. The enrollment in her school is 60 percent white and 40 percent black. *Having access to this information*, the principal has a *basis* for establishing the hypothetical frequency distribution or the expected frequencies. She will *not* assume, as in the previous problem, that the total expected observations will be evenly distributed among the categories. Based on her information, she would reason that she could expect that 40 percent of the 350 students enrolled in band would be black, that is, 140 would be the expected frequency for blacks. Similarly, 60 percent of the 350, or 210, of the students enrolled in band could be expected to be white. When the principal checked the enrollment figures for band, she found that there were 120 blacks and 230 whites actually enrolled. These are the observed frequencies for the chi square analysis. The question is whether these observed frequencies indicate a significantly greater participation on the part of white students. (See Table 15.3.)

TABLE 15.3 Observed and Expected Frequencies of Blacks and Whites in Middle School Band

	OBSERVED FREQUENCY	EXPECTED FREQUENCY
Black	120	140
White	230	210
Total	350	350

In this problem, there is $k - 1$ or $2 - 1 = 1$ degree of freedom. Once it is known that the number of blacks is 120, the number of whites must be 230 since the total is 350. Therefore, only one observation is free to vary; therefore, d.f. = 1.

Special Correction for 1 Degree of Freedom Chi Square

The sampling distribution of χ^2 as represented in Table A7 is a continuous theoretical frequency curve. For situations where the number of degrees of freedom is one, this continuous curve somewhat underestimates the actual probabilities. An adjustment is necessary to make the χ^2 probabilities in Table A7 more accurately approximate the actual probabilities of the discrete events. The appropriate adjustment is known as Yates' correction. It consists of subtracting .5 from the *absolute difference* between O and E for each category and then squaring the difference. (The absolute difference between two numbers is the difference recorded as a positive number regardless of whether that difference is actually positive or negative.) The formula for χ^2 then becomes:

$$\chi^2 = \sum \frac{[|O - E| - .5]^2}{E}$$

Formula (15.2)
Chi Square with Yates' Correction

Table 15.4 shows the calculation of χ^2 for the data in Table 15.3 using Yates' correction.

TABLE 15.4 Calculation of χ^2 for Data in Table 15.3

	O	E	$O - E$	$|O - E| - .5$	$[(O - E) - .5]^2$	$\frac{[|O - E| - .5]^2}{E}$
Blacks	120	140	20	19.5	380.25	2.72
Whites	230	210	20	19.5	380.25	1.81
						$4.53 = \chi^2$

Now the principal enters Table A7 with 1 d.f. The values required for significance at the 5 and 1 percent levels for 1 d.f. are 3.84 and 6.64, respectively. The calculated χ^2 of 4.5 is significant at the .05 level, which means that there are fewer than 5 chances in 100 of observing such a difference through chance alone. Thus the principal's null hypothesis of no difference in the proportion of blacks and whites enrolled can be rejected at the .05 level of significance and the conclusion drawn that the proportion of whites enrolled in band at the middle school is significantly greater than that of blacks. The principal would infer that the difference between observed and hypothetical frequencies is not likely to be a function of chance and that there is evidence for a systematic tendency for a larger proportion of whites than blacks to participate in the band program.

Summary, Goodness-of-Fit Chi Square

Let us summarize the steps to be followed in applying the chi square test to a single sample with multiple categories or, as it is called, a chi square test for goodness of fit.

1. The first step is to set up categories for responses that are mutually exclusive and exhaustive of all possible responses.
2. Next, expected frequencies are proposed for each of the categories either on the basis of a theoretical model or on the basis of whatever information one may already have available.
3. A random sample of subjects is drawn and their responses recorded in the appropriate categories. This provides us with the obtained frequencies which we will compare with the theoretical or expected chance frequencies.
4. The null hypothesis, which states that no actual difference exists between the observed and expected frequencies, is assumed. The χ^2 is calculated. If the number of degrees of freedom is one, Yates correction is applied in the calculation.
5. At a predetermined level of significance, the calculated χ^2 is compared with the tabled value for the appropriate degrees of freedom $(k-1)$. If the calculated χ^2 is equal to or greater than the tabled value, then the null hypothesis of no difference can be rejected. The investigator may conclude that the difference between the observed and expected frequencies is significant at the predetermined level and is unlikely to be due to chance factors. If the calculated χ^2 is smaller than the tabled value, the null hypothesis cannot be rejected; that is, the null hypothesis is retained and it is concluded that any difference between the observed and expected distributions is not significant and therefore could easily be due to chance.

Exercises

1. A random sample of taxpayers in a school district were asked "Should the school economize by eliminating intramural sports?" The results were as follows: Agree 12, Indifferent 12, Disagree 24.
   (a) What is the null hypothesis?
   (b) What is the expected value for each category under the null hypothesis?
   (c) What is the value of chi square?
   (d) What are the degrees of freedom?
   (e) Is the value significant at the .05 level?
   (f) What conclusion would be reached?

2. All the soldiers in an Army corps have been classified according to religious preference. The population proportions are: Catholic 30 percent, Jewish 10 percent, Protestant 40 percent, and Other 20 percent. A random sample of 50 soldiers who have volunteered for overseas duty show the following preferences: Catholic 17, Jewish 8, Protestant 15, and Other 10. Is there a significant difference between these proportions and the proportions in the total population of the corps?

3. An investigator studying sex stereotypes in basic reading texts finds that one entitled *Happy Times with the Smith Twins* has 16 episodes in which Joe plays the role of leader and four in which Jane plays the role of leader. Could these proportions easily be a function of chance?

Solutions

1. (a) In the population of taxpayers there is an even division among the three opinions.
   (b) Agree 16, Indifferent 16, Disagree 16
   (c) $\dfrac{(12-16)^2}{16} + \dfrac{(12-16)^2}{16} + \dfrac{(24-16)^2}{16} = 1 + 1 + 4 = 6$
   (d) d.f. $= k - 1 = 3 - 1 = 2$
   (e) yes
   (f) There is sufficient evidence to conclude that among all taxpayers in the district there is not an even division among the three opinions.

2. $\chi^2 = \dfrac{(17-15)^2}{15} + \dfrac{(8-5)^2}{5} + \dfrac{(15-20)^2}{20} + \dfrac{(10-10)^2}{10}$
   $= .27 + 1.80 + 1.25 + 0 = 3.32$

   With 3 d.f. a χ^2 of 3.32 is not statistically significant at the .05 level. There is no evidence that the volunteer proportions differ from the corps proportions.

3. Applying Yates' correction because there is 1 d.f. we have:
   $\chi^2 = \dfrac{(|16-10|-5)^2}{10} + \dfrac{(|4-10|-.5)^2}{10} = 3.025 + 3.025 = 6.05$

   With 1 d.f. a χ^2 of 6.05 is statistically significant at the .05 and .02 levels but not at the .01 level. The proportions are not likely to be a function of chance.

Chi Square Test of the Independence of Categorical Variables

In the previous section we have been dealing with categories divided on the basis of a single variable. However, another widely used application of the χ^2 procedure involves its use with data that are in the form of paired observations on two variables. That is, a sample of subjects is classified into categories on two variables and the question concerns the presence or absence of a relationship between the variables. For example, one might ask: Is there a relationship between the socioeconomic background of a child and his preference among extracurricular activities at school? Is there a relationship between income level and attitudes toward some current issue? Is there a difference in the extent of reported drug use among adolescents coming from different socioeconomic backgrounds? The independence or the association of these variables can be determined by means of the chi square test.

When data of these types are gathered, they are recorded in what is called a *contingency table.* The paired responses are categorized into cells organized by rows and columns.

Let us consider an example: A pollster has drawn a random sample of 200 individuals whom he asked to indicate their political preference: Republican, Democratic, or Independent, and also whether they approve or disapprove of a national health insurance plan. Thus one has a single random sample classified on two different nominal variables: political preference and approval-disapproval of national health insurance. The frequency data are tabulated and placed in the appropriate cells of the table. The pollster wants to know whether the proportions of Republicans, Democrats, and Independents who favor the health insurance plan differ significantly, indicating a relationship between political preference and attitude toward a national health insurance plan. Since the data are in the form of frequencies in categories, the χ^2 test is the appropriate statistical procedure to use in analyzing the data. The null hypothesis to be tested in such cases will state that the variables are independent or unrelated; that is, knowledge of one variable tells us nothing about the other variable. In our example the null hypothesis would be: "There is no relationship between political preference and attitude toward a national health insurance plan." This null hypothesis will be retained if the χ^2 value is found to be *not* statistically significant. If the obtained χ^2 is statistically significant, the null hypothesis of independence is rejected. One would therefore conclude that the two variables are dependent and that knowledge of one would tell us something about the other.

Table 15.5 is a contingency table showing the results of the poll. The *observed* frequencies were obtained by asking the subjects their political preference and how they felt about the health plan. Note that the cells are arranged in rows according to political preference and in columns according to attitude toward the health plan. This creates six cells plus marginal totals for the rows and columns.

It is now necessary to determine the *expected* frequency values for each of the cells in the contingency table. When the chi square procedure is applied to a contingency table to test for independence of the row and column variables, there is no a priori basis for hypothesizing the expected frequency distribution as there was in the χ^2 goodness-of-fit test. *The expected frequencies in each cell are derived from the data themselves.* These expected cell frequencies are those one would expect to get if the two variables were completely independent of each other and are derived from the row and column totals. The first step in calculating the expected frequencies is to consider each variable separately. Thus in the case of

404 INFERENTIAL STATISTICS

TABLE 15.5 The Observed Frequencies of Responses of Democrats, Republicans, and Independents as to Their Approval-Disapproval of a National Health Insurance Plan

POLITICAL PREFERENCE	RESPONSE TO QUESTION APPROVE	DISAPPROVE	TOTAL (ROWS)
Republican	10	50	60
Democrat	60	40	100
Independent	20	20	40
Total (Columns)	90	110	200

attitude toward the health insurance plan, we find that 90 of our sample of 200 approve and 110 out of the 200 disapprove; that is, 45 percent approve and 55 percent disapprove. Now we apply these proportions to the separate categories of Republican, Democrat, and Independent. That is, we would expect to find 45 percent of each political category in the approve group and 55 percent in the disapprove group. Using the actual numbers from our sample, this would be 45 percent of 60 Republicans (27), 45 percent of 100 Democrats (45), and 45 percent of the 40 Independents (18) in the approve column. Similarly, we would expect 55 percent of 60 Republicans (33), 55 percent of 100 Democrats (55), and 55 percent of the 40 Independents (22) in the disapprove column. We now have the necessary information to add the expected values to our contingency table as shown in Table 15.6. The expected values are shown in parentheses. Note that the row and column totals of the expected frequencies will always equal the row and column totals of the observed frequencies.

In calculating expected values in our example we first found the observed

TABLE 15.6 Observed and Expected Responses Classified by Political Preference and Approval-Disapproval of a National Health Insurance Plan

POLITICAL PREFERENCE	RESPONSE TO QUESTION APPROVE O E	DISAPPROVE O E	TOTAL
Republican	10 (27)	50 (33)	60 (60)
Democrat	60 (45)	40 (55)	100 (100)
Independent	20 (18)	20 (22)	40 (40)
Total	90 (90)	110 (110)	200 (200)

percentages with reference to one of the variables and then multiplied these percentages by the observed frequencies for the other variable in order to illustrate the logic involved in determining expected frequencies. However, there is a more efficient procedure that also has the advantage of minimizing rounding error. To calculate the expected frequency for each cell multiply the row total associated with that cell by the column total associated with that cell and then divide this product by N. In our example the row and column totals associated with the first cell (Republican, approve) in Table 15.6 are 60 and 90. When the product of these two numbers is divided by N (200), the result is 27, the figure we had already obtained by the longer method. The computation of the next cell (Republican, disapprove) is $60 \times 110/200$. This method of determining expected cell frequencies is expressed by:

$$\text{Expected frequency} = \frac{\text{(Row total)(Column total)}}{\text{Grand total}} \quad \text{Formula (15.3)}$$

Expected Cell Frequencies in a Chi Square Contingency Table

Having determined the expected value for each cell we are ready to compute the chi square value for our example:

$$\chi^2 = \sum \frac{(O-E)^2}{E}$$

$$= \frac{(10-27)^2}{27} + \frac{(50-33)^2}{33} + \frac{(60-45)^2}{45} + \frac{(40-55)^2}{55} + \frac{(20-18)^2}{18} + \frac{(20-22)^2}{22}$$

$$= \frac{289}{27} + \frac{289}{33} + \frac{225}{45} + \frac{225}{55} + \frac{4}{18} + \frac{4}{22}$$

$$= 10.70 + 8.76 + 5.00 + 4.09 + .22 + .18$$

$$= 28.95$$

Now we must determine whether our obtained chi square value of 28.95 is statistically significant. In order to do this the degrees of freedom must first be determined.

Degrees of Freedom for Contingency Table

Recall that in any statistic the degrees of freedom are the number of values that are free to vary. The general rule is that the degrees of freedom for a contingency table equal the number of rows minus one multiplied by the number of columns minus one:

$$\text{d.f.} = (\text{rows} - 1)(\text{columns} - 1) \quad \text{Formula (15.4)}$$

Degrees of Freedom for Contingency Table

In our example with three rows and two columns we have $(3-1)(2-1) = 2$ d.f. Let us use our example to show that only two values are free to vary in a 2×3 contingency table.

406 INFERENTIAL STATISTICS

	Approve	Disapprove	Total
Republican	10		60
Democrat			100
Independent	20		40
	90	110	

If we know, for example, that 10 Republicans approve, then the other 50 Republicans must disapprove. Now if we know the number in any one of the other four cells, we can determine the number in the remainder through subtraction. For example, if we also know that 20 Independents approve, the other 60 approve responses must have come from the Democrats. Knowing the 20 Independents approve, we know that the other 20 Independents disapprove. If 60 Democrats approve, the other 40 Democrats must disapprove. Only two of the six cells are free to vary. In a two by two contingency table only one cell $(2-1)(2-1)$ is free to vary. In a three by four table $(3-1)(4-1) = 6$ cells are free to vary.

Returning to our example, we are ready to determine the significance of the calculated chi square value of 28.95. Entering Table A7 with 2 d.f. we find that the value of χ^2 required for significance at the 5 percent level is 5.99 and at the 1 percent level is 9.21. The obtained χ^2 value of 28.95 is therefore highly significant. The pollster can reject the null hypothesis of independence. In other words, he would conclude that there is a relationship between the variables of political preference and attitude toward the national health insurance plan. Whether the respondent approves of the plan seems to be in part dependent upon his political preference. Thus, when the pollster rejects the hypothesis of independence, he accepts the alternate hypothesis of dependence. Or, if one had stated the null hypothesis in the form of no difference between Republicans, Democrats, and Independents in their attitudes, then this hypothesis of no difference would have been rejected. Rather, one would conclude that Republicans, Democrats, and Independents do differ significantly in their approval of a national health insurance plan.

2 × 2 Contingency Table

One of the most common uses of the chi square test of the independence of categorical variables is with the 2 × 2 contingency table, in which there are two variables each divided into two categories. For example, an investigator hypothesizes that high school guidance counselors tend to advise boys to prepare for professional careers more often than they advise girls to prepare

TABLE 15.7 Example of 2×2 Chi Square, Careers Suggested to Boys and Girls

	GIRLS	BOYS	
Professional Career Suggested	8	22	30
Nonprofessional Career Suggested	42	28	70
Total	50	50	

for professional careers. He randomly draws a sample of 100 students who have had career counseling, classifies the career suggested by the counselor into professional and nonprofessional categories, and classifies the subjects as girls and boys. The results are shown in Table 15.7.

The degrees of freedom for the 2 × 2 contingency table are (2 − 1)(2 − 1) = 1. Since there is only 1 d.f. in this case, we need to apply Yates' correction. We could calculate the expected values for each cell as we did in the other example and apply Formula (15.2) to compute the value of chi square. However, a special formula is available for computing chi square values with a 2 × 2 contingency table. This formula [Formula (15.5)] is mathematically equivalent to Formula (15.2) but does not require the calculation of expected values for each cell. It also minimizes rounding error:

$$\chi^2 = \frac{(|ad - bc| - N/2)^2 N}{(a + b)(c + d)(a + c)(b + d)} \qquad \text{Formula (15.5)}$$

Computation of 2×2 Chi Square with Yates' Correction

In order to apply Formula (15.5) we first label the cells as follows:

a	b	$a + b$
c	d	$c + d$

$a + c \quad b + d$

This schema has been applied to Table 15.7 data and appears in Table 15.8.

TABLE 15.8 Schematic Representation of Frequencies in Table 15.7.

	GIRLS	BOYS	TOTAL
Professional Career Suggested	a_8	b_{22}	$a + b$ (30)
Nonprofessional Career Suggested	c_{42}	d_{28}	$c + d$ (70)
	$a + c$ (50)	$b + d$ (50)	N (100)

Note that $|ad - bc|$ in Formula (15.5) represents an absolute value; the difference between ad and bc will always be expressed as a positive number. Applying Formula (15.5) to the data in Table 15.7 we have:

$$\chi^2 = \frac{(|(8)(28) - (22)(42)| - 100/2)^2 100}{(30)(70)(50)(50)}$$

$$= \frac{(|224 - 924| - 50)^2 100}{5,250,000}$$

$$= \frac{(650)^2 100}{5,250,000}$$

$$= \frac{42,250,000}{5,250,000}$$

$$= 8.048$$

Consulting Table A7 the investigator finds that with 1 d.f. a chi square value beyond 6.635 is statistically significant at the .01 level. He therefore concludes that in the population from which his sample was drawn there is a systematic tendency for counselors to suggest professional careers for boys more often than for girls.

Let us summarize the procedure to be followed in applying the chi square test of independence:

1. The null hypothesis states that the two variables are independent of each other; that is, knowledge of an individual's classification on one variable would indicate nothing of his classification on the other variable.
2. A random sample is drawn and the subjects classified on two or more variables.
3. A contingency table is set up with the variables indicated in the rows and columns of the table.
4. The observed frequencies are recorded in the proper cells and the marginal totals determined.
5. The expected frequencies are derived from the observed data themselves. The sum of the expected frequencies will equal the sum of the observed frequencies.
6. Since the data consist of frequencies in categories, the χ^2 test of independence is the appropriate statistical test. The calculated χ^2 value is compared with the tabled value at a predetermined level of significance and with (rows − 1)(columns − 1) degrees of freedom. In a 2 × 2 table where the number of degrees of freedom is one, a continuity correction should be used in calculating χ^2 [Formula (15.2)] or the alternative formula (15.5) may be used.
7. If the calculated χ^2 value equals or exceeds the tabled value, the finding is significant and the null hypothesis of independence is rejected; one concludes that the two variables are dependent or related at the given level of significance. If the calculated χ^2 value is smaller than the tabled

value, the null hypothesis of independence may *not* be rejected; one retains the null hypothesis and concludes that there is not sufficient evidence of relationship between two variables.

Restrictions in the Use of the Chi Square

It is important to remember that χ^2 is most appropriate for the analysis of data that are classified as frequency of occurrence within categories (nominal data). When the data are ordinal or interval other tests of significance are usually preferred.

When nominal data are arranged into categories for a chi square analysis, these categories must be mutually exclusive categories, which means that each response can be classified only once. This is true because a fundamental assumption in the use of the chi square test is that each observation or frequency is independent of all others. Thus one could not obtain several responses or observations from the same individual and classify each as if it were independent of the others. If this is done, the same individual is placed in several categories, which inflates the size of N, and may lead to the rejection of the null hypothesis when it should not be rejected.

Another restriction in the use of the chi square test is that when there are multiple categories, larger samples are needed. If N is small and consequently the expected frequency in any cell is small, the sample statistic may not approximate the theoretical χ^2 distribution very closely. A rule-of-thumb which one may follow is that in a χ^2 analysis with 1 d.f., the expected frequency in all cells should at least equal or be greater than 5. When the number of degrees of freedom is greater than one, the expected frequency should be equal to or greater than 5 in at least 80 percent of the cells. If the data do not meet these restrictions, a useful (or a possible) remedy is to combine some of the categories so that the expected frequencies will be raised to an acceptable size. However, there must be some logical basis for the combination of categories, otherwise one is not justified in combining them and should not proceed with a χ^2 analysis.

Suppose one conducted an opinion poll among elementary school children concerning their attitude toward some new school policy. The frequencies are:

GRADE LEVEL

	Kdgn.	1	2	3	4	5	6	Total
Agree	2	4	5	7	5	10	12	45
Disagree	1	4	5	6	8	9	12	45

410 INFERENTIAL STATISTICS

The expected frequency in each of the kindergarten cells is 1.5 and in each of the first grade cells it is 4. The investigator could logically combine the kindergarten and first grade categories and then none of the categories would have fewer than five. The expected frequencies would then become:

	Kdgn.–1	2	3	4	5	6	Total
Agree	5.5	5	6.5	6.5	9.5	12	45
Disagree	5.5	5	6.5	6.5	9.5	12	45

Then one could proceed with the chi square analysis.

When categories cannot be logically combined in order to meet the restrictions of the chi square test, an alternate procedure known as Fisher's exact test may be used.[3]

Exercises

1. A random sample of teachers was classified as elementary, middle school, secondary, and junior college teachers and they were asked if they would favor organizing a teacher's union. The results were as follows:

	ELEMENTARY	MIDDLE SCHOOL	SECONDARY	JUNIOR COLLEGE
Yes	4	20	26	30
No	36	20	14	10

Are the differences in the proportions statistically significant at the .01 level?

2. In a school with a merit system for pay raises a random sample of the faculty were asked if they wished that system to be continued. When these teachers were also classified according to whether they had had above-average or below-average merit raises, the results were:

	CONTINUE	DON'T CONTINUE
ABOVE AVERAGE	7	3
BELOW AVERAGE	1	9

Is the difference in proportions statistically significant at the .05 level?

---

[3] See Sidney Siegel, Nonparametric Statistics for the Behavior Sciences (New York: McGraw-Hill, 1956), pp. 96–104.

Solutions

1. $\chi^2 = \frac{(4-20)^2}{20} + \frac{(36-20)^2}{20} + \frac{(20-20)^2}{20} + \frac{(26-20)^2}{20} + \frac{(14-20)^2}{20} + \frac{(30-20)^2}{20}$
$+ \frac{(10-20)^2}{20}$
$= 12.8 + 12.8 + 0 + 0 + 1.8 + 1.8 + 5.0 + 5.0$
$= 39.2$ With 3 d.f. this χ^2 is statistically significant at the .01 level.

2. $\chi^2 = \frac{(|ad - bc| - N/2)^2 N}{(a+b)(c+d)(a+c)(b+d)}$ (with Yates' correction)
$= \frac{(|63-3| - 10)^2 (20)}{(10)(10)(8)(12)}$
$= \frac{(50)^2 (20)}{9,600}$
$= 5.2$ With 1 d.f. this is statistically significant at the .05 level.

Correlation Coefficients Derived from Chi Square

Two correlation coefficients can be calculated directly from a χ^2 derived from a contingency table. One of these, the phi coefficient, ϕ, has already been considered in Chapter Nine; the other, the contingency coefficient, C, was mentioned but not described in that chapter. When a χ^2 is statistically significant an investigator may wish to calculate a correlation coefficient in order to assess the strength of a relationship. If the χ^2 is not statistically significant, the frequencies in the cells of the contingency table could easily be a result of chance and there is no point in converting the χ^2 to a correlation coefficient.

Phi Coefficient

A chi square derived from a two by two contingency table can be transformed into the phi coefficient, ϕ, which was described in Chapter 9. The phi coefficient is derived by taking the square root of the quotient obtained by dividing the chi square by the total number of subjects. Yates' correction should *not* be used in calculating a chi square which is converted into a phi coefficient:

$$\phi = \sqrt{\frac{\chi^2}{N}}$$

Formula (15.6)[4]

Phi Coefficient Derived from Chi Square

---

[4] *Only the absolute value of ϕ can be determined by this formula. In order to know the sign of the coefficient ϕ one must examine the $bc - ad$ in the formula introduced in Chapter 9.*

412 INFERENTIAL STATISTICS

In our example involving professional and nonprofessional careers suggested to boys and girls, the χ^2 without Yates' correction is 9.33. Dividing this by $N(100)$ and taking the square root, we have a phi coefficient of .31. The phi coefficient is interpreted in the same manner as the other correlation coefficients. Whereas the χ^2 only tells us whether a relationship is statistically significant or not, the phi coefficient provides an index of the strength of the relationship.

The Contingency Coefficient

When the contingency table has more than 2 × 2 cells, the phi coefficient cannot be used. The contingency coefficient, C, is used in such cases. C is a measure of the relationship between two nominal variables when the data have been organized into a contingency table. The contingency coefficient is calculated by dividing χ^2 by $\chi^2 + N$ and taking the square root of the result:

$$C = \sqrt{\frac{\chi^2}{\chi^2 + N}}$$ where: N = total number of observations   Formula (15.7)

Contingency Coefficient

Example: The dean of a school of nursing has classified her seniors according to her perception of whether their dedication to the profession is high, average, or low. When the results of the state licensure examination become available she also classifies them into pass-fail categories and computes a χ^2 to see if there is a relationship between the variables.

LICENSURE EXAM RESULT	DEDICATION TO PROFESSION			
	High	Average	Low	
Pass	17	10	3	30
Fail	3	10	7	20
	20	20	10	

$$\chi^2 = \frac{(17-12)^2}{12} + \frac{(10-12)^2}{12} + \frac{(3-6)^2}{6} + \frac{(3-8)^2}{8} + \frac{(10-8)^2}{8} + \frac{(7-4)^2}{4}$$

$$= 2.08 + .33 + 1.5 + 3.12 + .5 + 2.25$$

$$= 9.78$$

With 2 d.f. the chi square of 9.78 is statistically significant at the .01 level so the dean can conclude that there is a relationship between the two variables. She would also like a measure of the strength of the relationship. Since the categories can be ordered in a meaningful manner a contingency coefficient would assist in the interpretation of the data:

$$C = \sqrt{\frac{\chi^2}{\chi^2 + N}}$$

$$= \sqrt{\frac{9.78}{9.78 + 50}}$$

$$= \sqrt{.1636}$$

$$= .404$$

The dean would conclude that there is a moderate relationship between the two variables. The minimum value of C is zero, which is obtained when two variables are independent. The maximum value is a function of the number of cells in the table. It never reaches 1.0 even though a perfect relationship exists. The maximum value of C for a 2×2 table is .707; for a 3×3 table, .816; for a 4×4 table, .866, and so on. Since the value of C depends on the number of cells in the table, one cannot compare different values of C directly unless they are based on tables having the same number of rows and columns. It is not a widely used statistic.

Sign Test

The sign test is the nonparametric alternative to the t-test for correlated samples. The sign test is so named because it counts plus and minus signs to test hypotheses. The only assumptions required to utilize the sign test are (1) that the dependent variable has an underlying continuous distribution; (2) that the subjects can be matched in a meaningful manner; and (3) that it is possible to rank the two members of each resulting pair on the dependent variables. Neither normal distribution nor interval data is required for the sign test.

The sign test is appropriate in two types of research situations: (1) when there is a sample consisting of matched pairs and one wishes to test the hypothesis that they are from the same population or (2) when the sample is composed of one group of subjects and one wishes to test a hypothesis about change in that group over time. In the latter case, data may consist of pre- and posttest scores for an experimental group. In either situation, the data need only be ordinal.

For example, an investigator hypothesizes that fathers and mothers of school children will differ in their attitudes toward the desirability of sex education in the schools. He asks a random sample of 20 pairs of parents to indicate which of the following statements most nearly represents their opinion.

414 INFERENTIAL STATISTICS

1. Teaching sex education in the schools should be prohibited.
2. Schools have more important things to do than teach sex education.
3. The advantages and disadvantages of sex education are about equal.
4. Sex education is a good idea but it should be an elective rather than a required course.
5. It is essential that all school children have instruction in sex education.

The investigator believes it is reasonable to assume that the responses represent a rank order along a hypothetically continuous score, that is, 5 is a more positive response than 4, 4 is a more positive response than 3, and so on. However, he has no evidence for assuming that the differences between the responses are equal; therefore, it would be inappropriate to treat the data as interval data. Therefore, he chooses to test his hypothesis with the sign test. He chooses an alpha of .05. (See Table 15.9.)

TABLE 15.9 Parental Attitudes Toward Sex Education

FAMILY	MOTHER	FATHER	SIGN
1. Adams	1	3	−
2. Baker	4	3	+
3. Brown	4	3	+
4. Carr	2	2	0
5. Cline	2	1	+
6. Denver	1	1	0
7. Dunn	4	3	+
8. Edwards	1	2	−
9. Fox	5	4	+
10. Green	5	5	0
11. Hughes	5	4	+
12. Jacobs	2	1	+
13. Katz	5	4	+
14. Lane	4	3	+
15. McCoy	4	4	0
16. Neilson	3	2	+
17. Petty	2	1	+
18. Ross	3	2	+
19. Smith	1	2	−
20. Washington	2	1	+

There are 13 families where the mother has a more positive attitude than the father, 3 cases where the father has a more positive attitude than the mother, and 4 cases where both parents gave the same response.

In the sign test, pairs with equal scores are dropped from the analysis because only those showing differences can be used. The null hypothesis is that the median difference between the pairs is zero; that is, if the fathers and mothers *do not differ* in their attitudes, then half the differences between the pairs will be positive and the other half negative so that the median difference equals zero. In our example the question is phrased: "How likely is it that by chance 3 of 16 pairs would have one sign and 13 would have the opposite sign?"

Under the null hypothesis, the probability of having a plus is equal to the probability of having a minus which in turn is equal to .50. Thus by chance we would expect 8 pluses and 8 minuses. The binomial expansion is the model for determining the probability of obtaining 13 pluses where the probability is .50. (The use of the binomial expansion was explained in Chapter 8.) For numbers less than 25, Table A3 is used. In this table we find that the one-tailed probability of a 3-13 split is .011. Doubling that figure for a two-tailed test we have .022. This is less than the investigator's alpha of .05. Since the probability of obtaining 13 plus signs by chance is only .02, the investigator can reject the null hypothesis and conclude that there is sufficient evidence to support the assumption that in his district the mothers have more positive attitudes than fathers on the issue of sex education in the schools.

When the number of signed pairs exceeds 25, the z-score method can be used to approximate the binomial probability. For example, a principal has a hunch that the teachers in her school tend to raise pupils' conduct grades from the first to the second six weeks more often than they lower them. She has her secretary check conduct grades of 100 randomly selected pupils for the two reporting periods. She reports:

Second period higher than first:	26
Second period equal to first:	64
Second period lower than first:	10

The formula for finding the z in the sign test is:

$$z = \frac{(X \pm .5) - N/2}{\frac{1}{2}\sqrt{N}}$$

where: X = number of pluses

N = number of pluses, plus number of minuses = number of nontied pairs

$\pm$ = a correction for continuity. Recall that where whole numbers are used, the number represents real limits from .5 less to .5 more than that number. In the sign test the limit nearest $\frac{1}{2}N$ is used. If the X (number of pluses) is *more* than $N/2$, use $X - .5$; if X is *less* than $N/2$, use $X + .5$ in the formula

Formula (15.8)
z-Score for Sign Test

In the example:

$$z = \frac{(26 - .5) - \frac{36}{2}}{\frac{1}{2}\sqrt{36}}$$

$$= \frac{25.5 - 18}{\frac{1}{2}(6)}$$

$$= \frac{7.5}{3}$$

$$= 2.5$$

Consulting the normal curve table (Table A2) we find a z of $+2.5$ or more would occur by chance only six times out of 1,000. The principal has convincing evidence that raising conduct grades is more common than lowering them.

Median Test

A nonparametric test which serves a function similar to the independent t-test or the one-way ANOVA, is the median test. When using the median test, the null hypothesis is that two groups have the same median. The median test is used when scores are ordinal. It is also used when the data are interval or ratio, but do not meet parametric assumptions.

The data consist of two independent samples of $n_1 + n_2$ subjects. The basic strategy of the median test is to first combine the groups and compute the median for the total number of subjects. Then determine the number of subjects in each of the original groups which fall above and below this median. A contingency table is set up with these two categories as rows and with as many columns as there are groups in the study. Finally, the chi square test is used to find whether the proportions of subjects above and below the median differ significantly from group to group.

Example: A seventh grade English teacher has developed a unit which she thinks will encourage independent reading and use of the library. She has two sections of average ability students who have been assigned to their sections in a manner approximating randomness. She uses her unit with one section (experimental) while the other section (control) does a unit on punctuation. She hypothesizes that the experimental group will check out more books from the school library than the control group. The null hypothesis to be tested is that the median number of books checked out of the library will be the same for each group. She records the number of books each pupil has checked out, and tabulates the results as shown in Table 15.10.

TABLE 15.10 Numbers of Books Checked Out by Experimental and Control Groups

NUMBER OF BOOKS	EXPERIMENTAL f	CONTROL f	TOTAL f	cf
15	1		1	50
14		1	1	49
12			0	48
11	1	2	3	48
10	1		1	45
9	2		2	44
8	1	1	2	42
7	2		2	40
6	3	1	4	38
5	3	2	5	34
4	2	4	6	29
3	2	3	5	23
2	1	3	4	18
1	1	4	5	14
0	4	5	9	9
	$n = 24$	$n = 26$	$n = 50$	

Computing the median for all 50 subjects we have:

$$\text{Mdn} = L + \left(\frac{N/2 - f_b}{f_w}\right) i$$

$$= 3.5 + \left(\frac{50/2 - 23}{6}\right) 1$$

$$= 3.5 + \frac{25 - 23}{6}$$

$$= 3.5 + .33$$

$$= 3.83$$

Now we count the cases above and below the median in each group. The experimental group has 16 above the median and 8 below. The control group has 11 above and 15 below. These numbers are recorded as observed frequencies in Table 15.11.

418 INFERENTIAL STATISTICS

TABLE 15.11 Observed and Expected Frequencies Above and Below the Median for Experimental and Control Groups

	EXPERIMENTAL		CONTROL		TOTAL
	O	E	O	E	
ABOVE MEDIAN	16	12.96	11	14.04	27
BELOW MEDIAN	8	11.04	15	11.96	23
	24		26		50

The expected frequencies shown in Table 15.11 are found in the way that was explained earlier in the χ^2 test. That is, one multiplies the respective row and column totals associated with each cell and then divides that product by N. For example, the expected frequency for the experimental group above the median would be $(24 \times 27)/50 = 12.96$.

Then chi square is computed by applying Formula (15.2) with a correction for continuity since the d.f. = 1:

$$\chi^2 = \sum \frac{(|O - E| - .5)^2}{E}$$

$$= \frac{(|16 - 12.96| - .5)^2}{12.96} + \frac{(|11 - 14.04| - .5)^2}{14.04} + \frac{(|8 - 11.04| - .5)^2}{11.04} + \frac{(|15 - 11.96| - .5)^2}{11.96}$$

$$= \frac{2.54^2}{12.96} + \frac{2.54^2}{14.04} + \frac{2.54^2}{11.04} + \frac{2.54^2}{11.96}$$

$$= .50 + .46 + .58 + .54$$

$$= 2.08$$

[We could have obtained the same result using Formula (15.5).] The chi square of 2.08 is not significant at the .05 level. The teacher would conclude that there is not sufficient evidence for predicting that other similar classes would check out more library books after studying her unit than after studying the unit on punctuation.

The median test can also be used with more than two groups. In this case, it functions as a one-word alternative to one-way ANOVA. Let us consider an example of the median test involving more than two groups. A school psychologist wants to compare the relative effectiveness of three methods of counseling. Students referred to the counseling center by their teachers are randomly assigned to the three treatment groups. After the treatment each student's teacher is asked to rate the change in adjustment by indicating a point on a scale ranging from nine, for very much improved, to five, for no change, to one, for very much worse. The null hypothesis is that the three samples represent populations with identical medians.

A	B	C
9	8	7
9	8	6
8	7	5
8	7	5
6	5	4
5	4	3
5	3	3
4	2	3
2	2	2
2	2	1
2	1	

The grand median is 4.5. Counting the scores above and below the median in each of the three treatment groups we have the following contingency table:

	A	B	C	
Above	7	5	4	16
Below	4	6	6	16
	11	11	10	

The data in the contingency table again constitute the observed frequencies for the x^2 analysis. The expected frequencies are determined in the usual way. Computing the chi square we have:

$$\chi^2 = \frac{(7-5.5)^2}{5.5} + \frac{(5-5.5)^2}{5.5} + \frac{(4-5)^2}{5} + \frac{(4-5.5)^2}{5.5} + \frac{(6-5.5)^2}{5.5} + \frac{(6-5)^2}{5}$$
$$= .41 + .04 + .20 + .41 + .04 + .20$$
$$= 1.3$$

With $(r-1)(c-1)$, d.f. = 2. Consulting the chi square table for 2 d.f. we find a value of 4.605 is required to reject the null hypothesis at the .05 level. The null hypothesis that the three counseling methods do not differ must be retained. It is concluded that the medians for the three treatment groups do not differ significantly.

Comparison of Parametric and Nonparametric Statistics

As we have seen, nonparametric statistics make fewer demands on data than parametric statistics. They can be applied in a wide variety of situations including many in which parametric statistics are entirely inappropriate. Why, then, are parametric statistics more widely used than nonparametric? The reason for choosing a parametric test over a nonparametric test is that when the parametric assumptions are met the parametric test has more *power*. Power in statistics is defined as the probability that an inferential statistical test will detect when a null hypothesis is false and should be rejected. For any given N, a parametric test is more powerful than the parallel nonparametric test when the parametric assumptions are true. The parametric test involves less risk of a Type II error because it uses interval or ratio data which furnishes a maximum amount of information. For example, we have more information when we work with raw scores themselves, such as 70, 55, 30, 21, 10, than if we convert them to ordinal data in the forms of ranks 1, 2, 3, 4, 5 because the raw scores indicate distance as well as rank. The nonparametric procedure using the ordinal data involves a loss of information because it would consider only the *order* of measurements. Information concerning the magnitude of the scores and how far apart they are is lost in the computations. Thus, since the nonparametric tests use less of the available information, they are less powerful statistically.

Therefore, when given the choice between a parametric and a nonparametric test, one should choose the parametric test *provided that the data can satisfy the underlying assumptions* of the parametric test. Unfortunately, there have been many cases where researchers have chosen parametric tests when their data did not meet the assumptions of parametric tests. When this is done it can lead to conclusions that are really not justified by the data (Type I errors). Both investigators and readers of research need to be wary of this hazard. Perhaps parametric tests have been more widely used than they should be because the nonparametric tests are not as well known as the parametric tests.

SUMMARY

When the data in a study do not meet the parametric assumptions a variety of descriptive and inferential nonparametric procedures are available. Among the more widely used nonparametric procedures are the nonparametric correlations described in Chapter Nine, the chi square test, the sign test, and the median test.

The chi square test is used with nominal and ordinal data. The goodness-of-fit chi square is used to establish whether observed proportions differ significantly from a hypothesized distribution. When subjects are categorized on the basis of two different classification variables, the chi square test of independence tests the null hypothesis that the two variables are independent of each other. A chi square computed to test the independence of categories can be converted into a correlation coefficient. If the contingency table is 2×2, a phi coefficient is calculated. For other contingency tables a contingency coefficient can be derived if the categories can be ordered meaningfully. When subjects can be ranked on the dependent variable the sign test serves a role similar to that of the nonindependent t-test. With ranked data, the median test serves as a nonparametric alternative to the independent t-test or one-way analysis of variance. In general, parametric tests are more powerful than nonparametric tests and are preferred when the parametric assumptions are met.

EXERCISES

1. One hundred and twenty college students were interviewed and asked their preference among methods of class presentation. It was found that 60 preferred lecture, 35 preferred group discussion, and 25 preferred individualized instruction. Test the null hypothesis that the three types of presentation are equally preferred.

2. A new brand of low tar and nicotine cigarette has been developed. In order to test its acceptance, a sample of 300 smokers were asked to try the cigarette and to indicate whether or not they liked it. The data obtained from the sample are listed below:

YES	NO
170	130

    Compute the chi square and evaluate its significance at the .01 level. Would you accept or reject the hypothesis that these subjects are a random sample from a population where "Yes" responses equal "No" responses? What conclusion would you reach?

3. In the 1976 election the mayor received 75 percent of the vote in a certain large city. Suppose that in the summer of 1977 a poll of 10,000 people selected at random in that same city showed that 6,500 said they would vote for the mayor again. Is there a significant difference between the 1976 election results and the 1977 poll?

4. What is the number of degrees of freedom in the following types of contingency tables?
    (a) 2×2   (b) 3×5   (c) 2×3

5. As an economy measure, the school board has recommended that art and music programs be eliminated from the elementary school. A survey was taken to determine whether there was a relationship between income level and attitudes toward this proposal. Results are listed below.

	ATTITUDE	
	APPROVE	DISAPPROVE
HIGH	46	54
MIDDLE	30	70
LOW	65	35

FLOWCHART 15

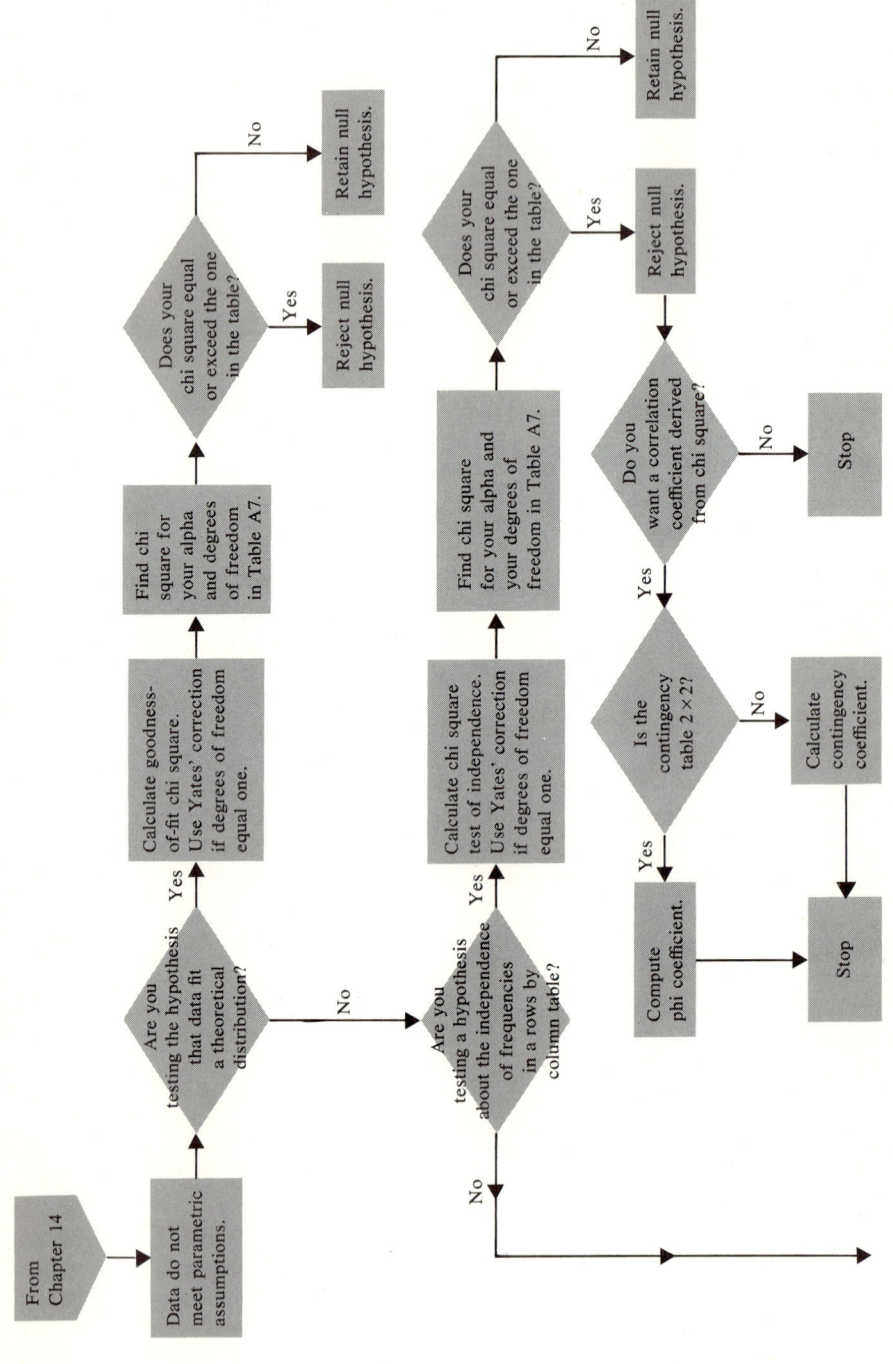

```
From Chapter 14 → Data do not meet parametric assumptions. →

Are you testing the hypothesis that data fit a theoretical distribution?
  — Yes → Calculate goodness-of-fit chi square. Use Yates' correction if degrees of freedom equal one. → Find chi square for your alpha and degrees of freedom in Table A7. → Does your chi square equal or exceed the one in the table?
    — Yes → Reject null hypothesis.
    — No → Retain null hypothesis.
  — No ↓

Are you testing a hypothesis about the independence of frequencies in a rows by column table?
  — Yes → Calculate chi square test of independence. Use Yates' correction if degrees of freedom equal one. → Find chi square for your alpha and your degrees of freedom in Table A7. → Does your chi square equal or exceed the one in the table?
    — Yes → Reject null hypothesis. → Do you want a correlation coefficient derived from chi square?
        — No → Stop
        — Yes → Is the contingency table 2 × 2?
            — Yes → Compute phi coefficient. → Stop
            — No → Calculate contingency coefficient. → Stop
    — No → Retain null hypothesis.
  — No ↓
```

422

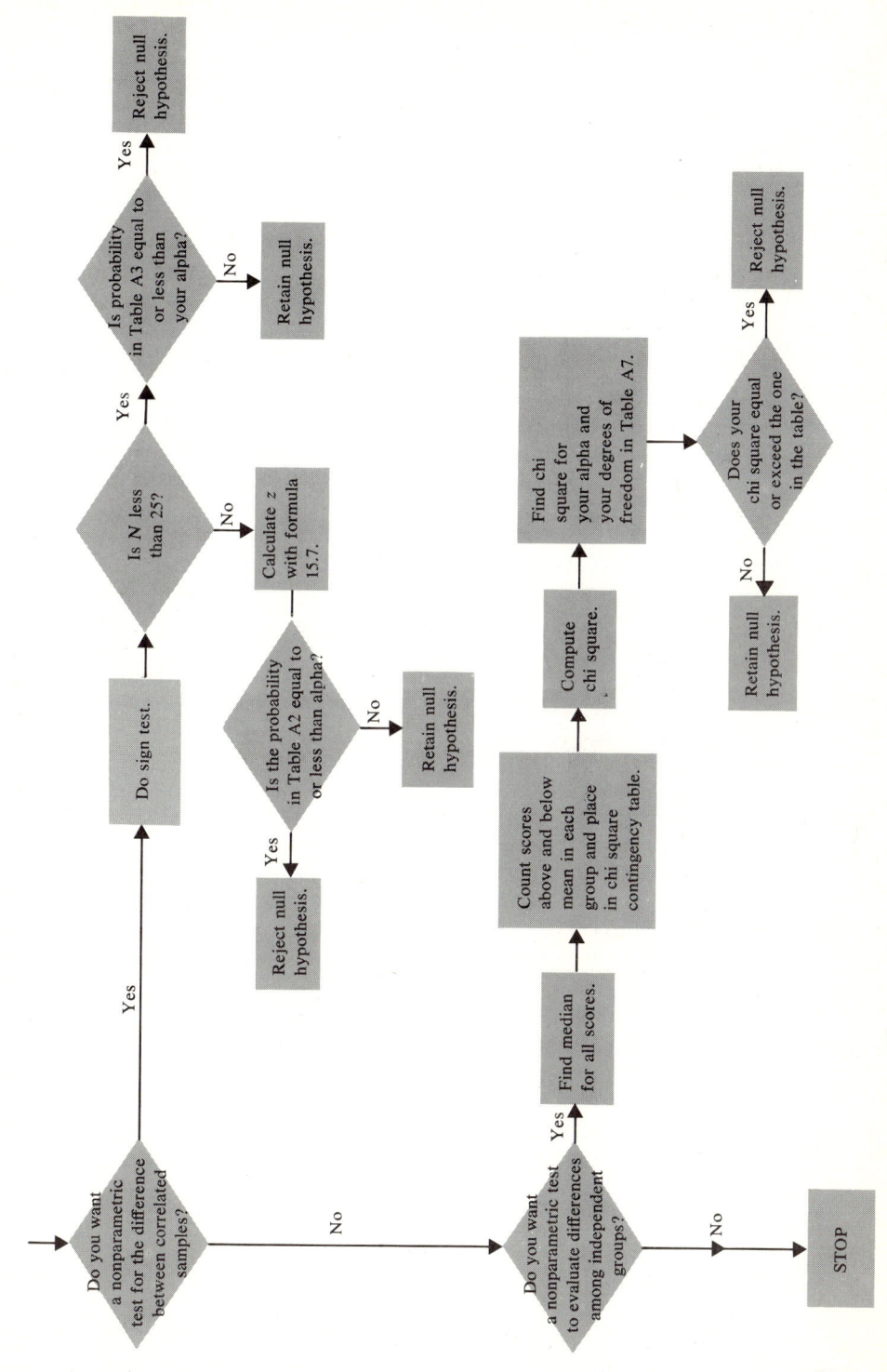

```
                                              ┌─────────────┐
                                              │ Reject null │
                                              │ hypothesis. │
                                              └─────────────┘
                                                     ▲
                                                   Yes
                                                     │
                                              ╱ Is probability ╲
                                             ╱  in Table A3 equal ╲      ┌─────────────┐
                                             ╲  to or less than   ╱─No──▶│ Retain null │
                                              ╲   your alpha?    ╱       │ hypothesis. │
                                               ╲               ╱         └─────────────┘
                                                     ▲
                                                    Yes
                                                     │
                                          ╱ Is N less ╲        ┌───────────┐
                                         ╱            ╲──No───▶│Calculate z│
                                         ╲  than 25?  ╱        │with formula│
                                          ╲          ╱         │   15.7.   │
                                                ▲              └───────────┘
                                                │                    │
                                         ┌──────────┐                ▼
                                         │Do sign   │        ╱ Is the probability ╲
                                         │ test.    │       ╱  in Table A2 equal   ╲──No──▶┌─────────────┐
                                         └──────────┘       ╲  to or less than     ╱       │ Retain null │
                                                ▲            ╲     alpha?         ╱        │ hypothesis. │
                                                │                  ▲                       └─────────────┘
                                                │                 Yes
                                                │                  │
                                                │           ┌─────────────┐
                                                │           │ Reject null │
                                                │           │ hypothesis. │
                                                │           └─────────────┘
                                                │
                                              Yes
                                                │
╱ Do you want     ╲                                                                    ┌──────────────┐
╱ a nonparametric ╲                                                                    │Find chi      │
╲ test for the    ╱                                                                    │square for    │
 ╲ difference    ╱                                                                     │your alpha and│
  ╲ between     ╱                                                                      │your degrees  │
   ╲correlated ╱                                                                       │of freedom in │
    ╲samples? ╱                                                                        │ Table A7.    │
         │                                                                             └──────────────┘
        No                                                                                     │
         │                                                                                     ▼
         ▼                                                                            ┌──────────────┐         ┌─────────────┐
╱ Do you want         ╲                            ┌──────────┐                       │   Compute    │         │ Reject null │
╱ a nonparametric test ╲                Yes        │Find median│                      │ chi square.  │         │ hypothesis. │
╲ to evaluate differ-  ╱───────────────────────────▶│for all   │                      └──────────────┘         └─────────────┘
 ╲ ences among inde-  ╱                            │scores.   │                              ▲                        ▲
  ╲pendent groups?   ╱                             └──────────┘                              │                      Yes
         │                                               │                                   │                        │
         No                                              ▼                                   │           ╱ Does your chi ╲
         │                                       ┌──────────────┐                            │          ╱ square equal or ╲
         ▼                                       │Count scores  │                            │          ╲ exceed the one  ╱
   ┌──────────┐                                  │above and below│                           │           ╲ in the table? ╱
   │   STOP   │                                  │mean in each   │───────────────────────────┘                 │
   └──────────┘                                  │group and place│                                            No
                                                 │in chi square  │                                             │
                                                 │contingency    │                                             ▼
                                                 │table.         │                                    ┌─────────────┐
                                                 └──────────────┘                                     │ Retain null │
                                                                                                      │ hypothesis. │
                                                                                                      └─────────────┘
```

424   INFERENTIAL STATISTICS

Test the null hypothesis that attitude toward elimination of art and music programs is independent of the income level of the respondent.

6. A researcher selected 60 secondary school teachers and 60 elementary school teachers at random from a large school system. He asked them to check "Yes" or "No" to the following question: "Do you like the present method of teacher evaluation being employed in this school system?" The following results were obtained:

	YES	NO
Elementary Teachers	36	24
Secondary Teachers	23	37

State the null hypothesis to be tested. Compute the chi square and test its significance at the .05 level. What conclusions would you reach?

7. A doctoral student in administration was investigating the hypothesis that school administrators are less likely to recommend the removal of disruptive elementary school children from regular classrooms than are these children's teachers. The doctoral student surveyed 15 teachers and principals of students referred for psychological sessions due to disruptive behavior in the classroom sitting. The survey consisted of both the teacher and principal checking one of the following alternatives:

1. The child should be removed from the regular class on a permanent basis and placed in a special education class for behavior disordered children.
2. The child should be kept in the regular classroom for one-half of the school day and receive special help the other half of the day.
3. The child should be kept in a regular classroom.

CHILD	TEACHER	PRINCIPAL
1	2	3
2	1	3
3	1	2
4	3	1
5	2	3
6	3	2
7	2	1
8	1	2
9	1	3
10	1	2
11	2	1
12	1	3
13	3	2
14	1	3
15	3	1

Test the null hypothesis for this problem.

8. A professor wanted to test the hypothesis "Blonds have more fun." He had random samples of 20 blonds and 30 nonblonds. On a questionnaire he asked

each subject to indicate either:

> 5. I have lots of fun.
> 4. I have quite a bit of fun.
> 3. I have a moderate amount of fun.
> 2. I have very little fun.
> 1. I have no fun.

The professor decided it would be inappropriate to treat his scores as parametric data. His results were:

BLONDS		NONBLONDS	
X	f	X	f
5	2	5	3
4	8	4	4
3	6	3	14
2	3	2	7
1	1	1	2

Test the null hypothesis (nonblonds have as much fun as blonds).

9. A classroom observer thought that the effects of fatigue and weariness make *problem behavior* less frequent in the afternoon as compared to the morning of the school day. The observer collected data from 10 randomly selected fifth grade rooms. Is there support for the hypothesis?

CLASSROOM	NUMBER OF PROBLEM BEHAVIORS	
	A.M.	P.M.
1	10	6
2	2	6
3	12	8
4	8	16
5	5	5
6	6	8
7	9	11
8	3	4
9	1	6
10	0	3

ANSWERS

1. $$\chi^2 = \frac{(60-40)^2}{40} + \frac{(35-40)^2}{40} + \frac{(25-40)^2}{40}$$

$$= \frac{400}{40} + \frac{225}{40} + \frac{25}{40}$$

$$= 10 + 5.625 + .625$$

$$= 16.25$$

With 2 d.f. the chi square of 16.25 is statistically significant at the .01 level.

2. $$\frac{(|170-150|-.5)^2}{150}+\frac{(|130-150|-.5)^2}{150}=\frac{(19.5)^2}{150}+\frac{(19.5)^2}{150}$$
$$=\frac{380.25}{150}+\frac{380.25}{150}=5.07$$

With 1 d.f. the chi square of 5.07 is statistically significant at the .05 level but not at the .01 level. Retain null hypothesis.

3. $$\frac{999.5^2}{7500}+\frac{999.5^2}{2500}=133.2+399.6=532.8$$

The poll indicates a highly significant drop in the mayor's support.

4. (a) 1   (b) 8   (c) 2

5.

	APPROVE		DISAPPROVE		TOTAL
	OBSERVED	EXPECTED	OBSERVED	EXPECTED	
HIGH	46	47	54	53	100
MIDDLE	30	47	70	53	100
LOW	65	47	35	53	100
TOTAL	141		159		

$\chi^2 = 24.64$ with 2 d.f. is statistically significant at .01 level. The two variables are not independent.

6. The null hypothesis is that teacher status (elementary or secondary) is unrelated to attitude toward present evaluation system.

$$\chi^2 = \frac{(|(36)(37)-(23)(24)|-120/2)^2\,120}{(60)(60)(59)(61)} = \frac{(|1,332-552|-60)^2\,120}{12,956,400}$$
$$=\frac{(720)^2(120)}{12,956,400}=\frac{62,208,000}{12,956,400}=4.80$$

With 1 d.f. of 4.80 it is statistically significant at .05 level so one would conclude that the populations from which the samples were drawn differ in their attitude toward the teacher evaluation method.

7. There are 6 cases where the teacher's score is higher than the principal's and 9 cases where the principal's score is higher than the teachers. In Table A3 we find that the one-tailed probability of a 6-9 split is .304. The null hypothesis is retained.

8. The median for the entire group is 3.1. Among blonds 10 are above the median and 10 are below it. Among nonblonds 7 are above the median and 23 are below it. The chi square for these data, 2.70, is not statistically significant at the .05 level. The evidence is inconclusive.

9. There are two classes with more problem behavior in the morning, seven with more in the afternoon, and one class with a tie. In Table A3 we find that the one-tailed probability of a 2-7 split in 9 cases is .02. The two-tailed probability is .04. The results are statistically significant at the .05 level.

APPENDIX

Table A1: Squares and Square Roots of Numbers from 1 To 1000

Table A1 may be expanded to permit one to find the square root of numbers larger than 1,000. This may be done by either one of two methods:

1. The student should note the relationship between column 1 (Number) and column 2 (Square) in the table. If column 2 contains the square of the number, then column 1 contains the number one started with, which would be the square root of the product listed in column 2. If 6 squared is 36 (column 2), then the square root of this number (36) is 6 (column 1). If we look at the Square column and see 2,500, then we can look back at the Number column and see that the square root of 2,500 is 50.

   Thus, if one wishes to know the square root of a number greater than 1,000, he can find it by working backward from the Square column.

For example, assume that one wants the square root of 1,700. Begin by moving down the Square column until one finds the number as close as possible to 1,700. It is 1,681; then looking back at the Number column, one would see that 41 is its square root. To approximate the square root of 1,700 more closely multiply 41 by 10 and turn to 410 in column 1. Proceeding along this column, we find that the square of 412 is equal to 169,744; so 41.2 is a reasonable approximation of the square root of 1,700.

2. An alternative approach is to change the decimal point in the Number and in the Square root columns. If one wants the square root of 1,700, then 1,700 is equal to 17×100, and its square root is equal to $\sqrt{17} \times \sqrt{100}$ or $\sqrt{17} \times 10$. The table gives the square root of 17 as 4.1231. So $4.1231 \times 10 = 41.231$, which is the square root of 1,700.

428 APPENDIX

Similarly, the square root of 55,100 (same as 551×100) is equal to $\sqrt{551} \times \sqrt{100}$, or $23.4734 \times 10 = 234.734$.

Let us assume that one wants the square root of 6.50. This number is equivalent to $650 \times 1/100$ and its square root is $\sqrt{650} \times 1/\sqrt{100}$ or $25.4951 \times 1/10$, which equals 2.54951 or about 2.55.

N	N^2	$\sqrt{N}$	N	N^2	$\sqrt{N}$	N	N^2	$\sqrt{N}$	N	N^2	$\sqrt{N}$
1	1	1.0000	41	1681	6.4031	81	6561	9.0000	121	14641	11.0000
2	4	1.4142	42	1764	6.4807	82	6724	9.0554	122	14884	11.0454
3	9	1.7321	43	1849	6.5574	83	6889	9.1104	123	15129	11.0905
4	16	2.0000	44	1936	6.6332	84	7056	9.1652	124	15376	11.1355
5	25	2.2361	45	2025	6.7082	85	7225	9.2195	125	15625	11.1803
6	36	2.4495	46	2116	6.7823	86	7396	9.2736	126	15876	11.2250
7	49	2.6458	47	2209	6.8557	87	7569	9.3274	127	16129	11.2694
8	64	2.8284	48	2304	6.9282	88	7744	9.3808	128	16384	11.3137
9	81	3.0000	49	2401	7.0000	89	7921	9.4340	129	16641	11.3578
10	100	3.1623	50	2500	7.0711	90	8100	9.4868	130	16900	11.4018
11	121	3.3166	51	2601	7.1414	91	8281	9.5394	131	17161	11.4455
12	144	3.4641	52	2704	7.2111	92	8464	9.5917	132	17424	11.4891
13	169	3.6056	53	2809	7.2801	93	8649	9.6437	133	17689	11.5326
14	196	3.7417	54	2916	7.3485	94	8836	9.6954	134	17956	11.5758
15	225	3.8730	55	3025	7.4162	95	9025	9.7468	135	18225	11.6190
16	256	4.0000	56	3136	7.4833	96	9216	9.7980	136	18496	11.6619
17	289	4.1231	57	3249	7.5498	97	9409	9.8489	137	18769	11.7047
18	324	4.2426	58	3364	7.6158	98	9604	9.8995	138	19044	11.7473
19	361	4.3589	59	3481	7.6811	99	9801	9.9499	139	19321	11.7898
20	400	4.4721	60	3600	7.7460	100	10000	10.0000	140	19600	11.8322
21	441	4.5826	61	3721	7.8102	101	10201	10.0499	141	19881	11.8743
22	484	4.6904	62	3844	7.8740	102	10404	10.0995	142	20164	11.9164
23	529	4.7958	63	3969	7.9373	103	10609	10.1489	143	20449	11.9583
24	576	4.8990	64	4096	8.0000	104	10816	10.1980	144	20736	12.0000
25	625	5.0000	65	4225	8.0623	105	11025	10.2470	145	21025	12.0416
26	676	5.0990	66	4356	8.1240	106	11236	10.2956	146	21316	12.0830
27	729	5.1962	67	4489	8.1854	107	11449	10.3441	147	21609	12.1244
28	784	5.2915	68	4624	8.2462	108	11664	10.3923	148	21904	12.1655
29	841	5.3852	69	4761	8.3066	109	11881	10.4403	149	22201	12.2066
30	900	5.4772	70	4900	8.3666	110	12100	10.4881	150	22500	12.2474
31	961	5.5678	71	5041	8.4261	111	12321	10.5357	151	22801	12.2882
32	1024	5.6569	72	5184	8.4853	112	12544	10.5830	152	23104	12.3288
33	1089	5.7446	73	5329	8.5440	113	12769	10.6301	153	23409	12.3693
34	1156	5.8310	74	5476	8.6023	114	12996	10.6771	154	23716	12.4097
35	1225	5.9161	75	5625	8.6603	115	13225	10.7238	155	24025	12.4499
36	1296	6.0000	76	5776	8.7178	116	13456	10.7703	156	24336	12.4900
37	1369	6.0828	77	5929	8.7750	117	13689	10.8167	157	24649	12.5300
38	1444	6.1644	78	6084	8.8318	118	13924	10.8628	158	24964	12.5698
39	1521	6.2450	79	6241	8.8882	119	14161	10.9087	159	25281	12.6095
40	1600	6.3246	80	6400	8.9443	120	14400	10.9545	160	25600	12.6491

Portions of Table A1 have been reproduced from J. W. Dunlap and A. K. Kurtz, Handbook of Statistical Nomographs, Tables, and Formulas, World Book Company, New York (1932).

Table A1 Continued

N	N^2	$\sqrt{N}$	N	N^2	$\sqrt{N}$	N	N^2	$\sqrt{N}$	N	N^2	$\sqrt{N}$
161	25921	12.6886	201	40401	14.1774	241	58081	15.5242	281	78961	16.7631
162	26244	12.7279	202	40804	14.2127	242	58564	15.5563	282	79524	16.7929
163	26569	12.7671	203	41209	14.2478	243	59049	15.5885	283	80089	16.8226
164	26896	12.8062	204	41616	14.2829	244	59536	15.6205	284	80656	16.8523
165	27225	12.8452	205	42025	14.3178	245	60025	15.6525	285	81225	16.8819
166	27556	12.8841	206	42436	14.3527	246	60516	15.6844	286	81796	16.9115
167	27889	12.9228	207	42849	14.3875	247	61009	15.7162	287	82369	16.9411
168	28224	12.9615	208	43264	14.4222	248	61504	15.7480	288	82944	16.9706
169	28561	13.0000	209	43681	14.4568	249	62001	15.7797	289	83521	17.0000
170	28900	13.0384	210	44100	14.4914	250	62500	15.8114	290	84100	17.0294
171	29241	13.0767	211	44521	14.5258	251	63001	15.8430	291	84681	17.0587
172	29584	13.1149	212	44944	14.5602	252	63504	15.8745	292	85264	17.0880
173	29929	13.1529	213	45369	14.5945	253	64009	15.9060	293	85849	17.1172
174	30276	13.1909	214	45796	14.6287	254	64516	15.9374	294	86436	17.1464
175	30625	13.2288	215	46225	14.6629	255	65025	15.9687	295	87025	17.1756
176	30976	13.2665	216	46656	14.6969	256	65536	16.0000	296	87616	17.2047
177	31329	13.3041	217	47089	14.7309	257	66049	16.0312	297	88209	17.2337
178	31684	13.3417	218	47524	14.7648	258	66564	16.0624	298	88804	17.2627
179	32041	13.3791	219	47961	14.7986	259	67081	16.0935	299	89401	17.2916
180	32400	13.4164	220	48400	14.8324	260	67600	16.1245	300	90000	17.3205
181	32761	13.4536	221	48841	14.8661	261	68121	16.1555	301	90601	17.3494
182	33124	13.4907	222	49284	14.8997	262	68644	16.1864	302	91204	17.3781
183	33489	13.5277	223	49729	14.9332	263	69169	16.2173	303	91809	17.4069
184	33856	13.5647	224	50176	14.9666	264	69696	16.2481	304	92416	17.4356
185	34225	13.6015	225	50625	15.0000	265	70225	16.2788	305	93025	17.4642
186	34596	13.6382	226	51076	15.0333	266	70756	16.3095	306	93636	17.4929
187	34969	13.6748	227	51529	15.0665	267	71289	16.3401	307	94249	17.5214
188	35344	13.7113	228	51984	15.0997	268	71824	16.3707	308	94864	17.5499
189	35721	13.7477	229	52441	15.1327	269	72361	16.4012	309	95481	17.5784
190	36100	13.7840	230	52900	15.1658	270	72900	16.4317	310	96100	17.6068
191	36481	13.8203	231	53361	15.1987	271	73441	16.4621	311	96721	17.6352
192	36864	13.8564	232	53824	15.2315	272	73984	16.4924	312	97344	17.6635
193	37249	13.8924	233	54289	15.2643	273	74529	16.5227	313	97969	17.6918
194	37636	13.9284	234	54756	15.2971	274	75076	16.5529	314	98596	17.7200
195	38025	13.9642	235	55225	15.3297	275	75625	16.5831	315	99225	17.7482
196	38416	14.0000	236	55696	15.3623	276	76176	16.6132	316	99856	17.7764
197	38809	14.0357	237	56169	15.3948	277	76729	16.6433	317	100489	17.8045
198	39204	14.0712	238	56644	15.4272	278	77284	16.6733	318	101124	17.8326
199	39601	14.1067	239	57121	15.4596	279	77841	16.7033	319	101761	17.8606
200	40000	14.1421	240	57600	15.4919	280	78400	16.7332	320	102400	17.8885

Table A1 Continued

N	N^2	$\sqrt{N}$	N	N^2	$\sqrt{N}$	N	N^2	$\sqrt{N}$	N	N^2	$\sqrt{N}$
321	103041	17.9165	361	130321	19.0000	401	160801	20.0250	441	194481	21.0000
322	103684	17.9444	362	131044	19.0263	402	161604	20.0499	442	195364	21.0238
323	104329	17.9722	363	131769	19.0526	403	162409	20.0749	443	196249	21.0476
324	104976	18.0000	364	132496	19.0788	404	163216	20.0998	444	197136	21.0713
325	105625	18.0278	365	133225	19.1050	405	164025	20.1246	445	198025	21.0950
326	106276	18.0555	366	133956	19.1311	406	164836	20.1494	446	198916	21.1187
327	106929	18.0831	367	134689	19.1572	407	165649	20.1742	447	199809	21.1424
328	107584	18.1108	368	135424	19.1833	408	166464	20.1990	448	200704	21.1660
329	108241	18.1384	369	136161	19.2094	409	167281	20.2237	449	201601	21.1896
330	108900	18.1659	370	136900	19.2354	410	168100	20.2485	450	202500	21.2132
331	109561	18.1934	371	137641	19.2614	411	168921	20.2731	451	203401	21.2368
332	110224	18.2209	372	138384	19.2873	412	169744	20.2978	452	204304	21.2603
333	110889	18.2483	373	139129	19.3132	413	170569	20.3224	453	205209	21.2838
334	111556	18.2757	374	139876	19.3391	414	171396	20.3470	454	206116	21.3073
335	112225	18.3030	375	140625	19.3649	415	172225	20.3715	455	207025	21.3307
336	112896	18.3303	376	141376	19.3907	416	173056	20.3961	456	207936	21.3542
337	113569	18.3576	377	142129	19.4165	417	173889	20.4206	457	208849	21.3776
338	114244	18.3848	378	142884	19.4422	418	174724	20.4450	458	209764	21.4009
339	114921	18.4120	379	143641	19.4679	419	175561	20.4695	459	210681	21.4243
340	115600	18.4391	380	144400	19.4936	420	176400	20.4939	460	211600	21.4476
341	116281	18.4662	381	145161	19.5192	421	177241	20.5183	461	212521	21.4709
342	116964	18.4932	382	145924	19.5448	422	178084	20.5426	462	213444	21.4942
343	117649	18.5203	383	146689	19.5704	423	178929	20.5670	463	214369	21.5174
344	118336	18.5472	384	147456	19.5959	424	179776	20.5913	464	215296	21.5407
345	119025	18.5742	385	148225	19.6214	425	180625	20.6155	465	216225	21.5639
346	119716	18.6011	386	148996	19.6469	426	181476	20.6398	466	217156	21.5870
347	120409	18.6279	387	149769	19.6723	427	182329	20.6640	467	218089	21.6102
348	121104	18.6548	388	150544	19.6977	428	183184	20.6882	468	219024	21.6333
349	121801	18.6815	389	151321	19.7231	429	184041	20.7123	469	219961	21.6564
350	122500	18.7083	390	152100	19.7484	430	184900	20.7364	470	220900	21.6795
351	123201	18.7350	391	152881	19.7737	431	185761	20.7605	471	221841	21.7025
352	123904	18.7617	392	153664	19.7990	432	186624	20.7846	472	222784	21.7256
353	124609	18.7883	393	154449	19.8242	433	187489	20.8087	473	223729	21.7486
354	125316	18.8149	394	155236	19.8494	434	188356	20.8327	474	224676	21.7715
355	126025	18.8414	395	156025	19.8746	435	189225	20.8567	475	225625	21.7945
356	126736	18.8680	396	156816	19.8997	436	190096	20.8806	476	226576	21.8174
357	127449	18.8944	397	157609	19.9249	437	190969	20.9045	477	227529	21.8403
358	128164	18.9209	398	158404	19.9499	438	191844	20.9284	478	228484	21.8632
359	128881	18.9473	399	159201	19.9750	439	192721	20.9523	479	229441	21.8861
360	129600	18.9737	400	160000	20.0000	440	193600	20.9762	480	230400	21.9089

Table A1 Continued

N	N^2	$\sqrt{N}$	N	N^2	$\sqrt{N}$	N	N^2	$\sqrt{N}$	N	N^2	$\sqrt{N}$
481	231361	21.9317	521	271441	22.8254	561	314721	23.6854	601	361201	24.5153
482	232324	21.9545	522	272484	22.8473	562	315844	23.7065	602	362404	24.5357
483	233289	21.9773	523	273529	22.8692	563	316969	23.7276	603	363609	24.5561
484	234256	22.0000	524	274576	22.8910	564	318096	23.7487	604	364816	24.5764
485	235225	22.0227	525	275625	22.9129	565	319225	23.7697	605	366025	24.5967
486	236196	22.0454	526	276676	22.9347	566	320356	23.7908	606	367236	24.6171
487	237169	22.0681	527	277729	22.9565	567	321489	23.8118	607	368449	24.6374
488	238144	22.0907	528	278784	22.9783	568	322624	23.8328	608	369664	24.6577
489	239121	22.1133	529	279841	23.0000	569	323761	23.8537	609	370881	24.6779
490	240100	22.1359	530	280900	23.0217	570	324900	23.8747	610	372100	24.6982
491	241081	22.1585	531	281961	23.0434	571	326041	23.8956	611	373321	24.7184
492	242064	22.1811	532	283024	23.0651	572	327184	23.9165	612	374544	24.7386
493	243049	22.2036	533	284089	23.0868	573	328329	23.9374	613	375769	24.7588
494	244036	22.2261	534	285156	23.1084	574	329476	23.9583	614	376996	24.7790
495	245025	22.2486	535	286225	23.1301	575	330625	23.9792	615	378225	24.7992
496	246016	22.2711	536	287296	23.1517	576	331776	24.0000	616	379456	24.8193
497	247009	22.2935	537	288369	23.1733	577	332929	24.0208	617	380689	24.8395
498	248004	22.3159	538	289444	23.1948	578	334084	24.0416	618	381924	24.8596
499	249001	22.3383	539	290521	23.2164	579	335241	24.0624	619	383161	24.8797
500	250000	22.3607	540	291600	23.2379	580	336400	24.0832	620	384400	24.8998
501	251001	22.3830	541	292681	23.2594	581	337561	24.1039	621	385641	24.9199
502	252004	22.4054	542	293764	23.2809	582	338724	24.1247	622	386884	24.9399
503	253009	22.4277	543	294849	23.3024	583	339889	24.1454	623	388129	24.9600
504	254016	22.4499	544	295936	23.3238	584	341056	24.1661	624	389376	24.9800
505	255025	22.4722	545	297025	23.3452	585	342225	24.1868	625	390625	25.0000
506	256036	22.4944	546	298116	23.3666	586	343396	24.2074	626	391876	25.0200
507	257049	22.5167	547	299209	23.3880	587	344569	24.2281	627	393129	25.0400
508	258064	22.5389	548	300304	23.4094	588	345744	24.2487	628	394384	25.0599
509	259081	22.5610	549	301401	23.4307	589	346921	24.2693	629	395641	25.0799
510	260100	22.5832	550	302500	23.4521	590	348100	24.2899	630	396900	25.0998
511	261121	22.6053	551	303601	23.4734	591	349281	24.3105	631	398161	25.1197
512	262144	22.6274	552	304704	23.4947	592	350464	24.3311	632	399424	25.1396
513	263169	22.6495	553	305809	23.5160	593	351649	24.3516	633	400689	25.1595
514	264196	22.6716	554	306916	23.5372	594	352836	24.3721	634	401956	25.1794
515	265225	22.6936	555	308025	23.5584	595	354025	24.3926	635	403225	25.1992
516	266256	22.7156	556	309136	23.5797	596	355216	24.4131	636	404496	25.2190
517	267289	22.7376	557	310249	23.6008	597	356409	24.4336	637	405769	25.2389
518	268324	22.7596	558	311364	23.6220	598	357604	24.4540	638	407044	25.2587
519	269361	22.7816	559	312481	23.6432	599	358801	24.4745	639	408321	25.2784
520	270400	22.8035	560	313600	23.6643	600	360000	24.4949	640	409600	25.2982

Table A1 Continued

N	N^2	$\sqrt{N}$	N	N^2	$\sqrt{N}$	N	N^2	$\sqrt{N}$	N	N^2	$\sqrt{N}$
641	410881	25.3180	681	463761	26.0960	721	519841	26.8514	761	579121	27.5862
642	412164	25.3377	682	465124	26.1151	722	521284	26.8701	762	580644	27.6043
643	413449	25.3574	683	466489	26.1343	723	522729	26.8887	763	582169	27.6225
644	414736	25.3772	684	467856	26.1534	724	524176	26.9072	764	583696	27.6405
645	416025	25.3969	685	469225	26.1725	725	525625	26.9258	765	585225	27.6586
646	417316	25.4165	686	470596	26.1916	726	527076	26.9444	766	586756	27.6767
647	418609	25.4362	687	471969	26.2107	727	528529	26.9629	767	588289	27.6948
648	419904	25.4558	688	473344	26.2298	728	529984	26.9815	768	589824	27.7128
649	421201	25.4755	689	474721	26.2488	729	531441	27.0000	769	591361	27.7308
650	422500	25.4951	690	476100	26.2679	730	532900	27.0185	770	592900	27.7489
651	423801	25.5147	691	477481	26.2869	731	534361	27.0370	771	594441	27.7669
652	425104	25.5343	692	478864	26.3059	732	535824	27.0555	772	595984	27.7849
653	426409	25.5539	693	480249	26.3249	733	537289	27.0740	773	597529	27.8029
654	427716	25.5734	694	481636	26.3439	734	538756	27.0924	774	599076	27.8209
655	429025	25.5930	695	483025	26.3629	735	540225	27.1109	775	600625	27.8388
656	430336	25.6125	696	484416	26.3818	736	541696	27.1293	776	602176	27.8568
657	431649	25.6320	697	485809	26.4008	737	543169	27.1477	777	603729	27.8747
658	432964	25.6515	698	487204	26.4197	738	544644	27.1662	778	605284	27.8927
659	434281	25.6710	699	488601	26.4386	739	546121	27.1846	779	606841	27.9106
660	435600	25.6905	700	490000	26.4575	740	547600	27.2029	780	608400	27.9285
661	436921	25.7099	701	491401	26.4764	741	549081	27.2213	781	609961	27.9464
662	438244	25.7294	702	492804	26.4953	742	550564	27.2397	782	611524	27.9643
663	439569	25.7488	703	494209	26.5141	743	552049	27.2580	783	613089	27.9821
664	440896	25.7682	704	495616	26.5330	744	553536	27.2764	784	614656	28.0000
665	442225	25.7876	705	497025	26.5518	745	555025	27.2947	785	616225	28.0179
666	443556	25.8070	706	498436	26.5707	746	556516	27.3130	786	617796	28.0357
667	444889	25.8263	707	499849	26.5895	747	558009	27.3313	787	619369	28.0535
668	446224	25.8457	708	501264	26.6083	748	559504	27.3496	788	620944	28.0713
669	447561	25.8650	709	502681	26.6271	749	561001	27.3679	789	622521	28.0891
670	448900	25.8844	710	504100	26.6458	750	562500	27.3861	790	624100	28.1069
671	450241	25.9037	711	505521	26.6646	751	564001	27.4044	791	625681	28.1247
672	451584	25.9230	712	506944	26.6833	752	565504	27.4226	792	627264	28.1425
673	452929	25.9422	713	508369	26.7021	753	567009	27.4408	793	628849	28.1603
674	454276	25.9615	714	509796	26.7208	754	568516	27.4591	794	630436	28.1780
675	455625	25.9808	715	511225	26.7395	755	570025	27.4773	795	632025	28.1957
676	456976	26.0000	716	512656	26.7582	756	571536	27.4955	796	633616	28.2135
677	458329	26.0192	717	514089	26.7769	757	573049	27.5136	797	635209	28.2312
678	459684	26.0384	718	515524	26.7955	758	574564	27.5318	798	636804	28.2489
679	461041	26.0576	719	516961	26.8142	759	576081	27.5500	799	638401	28.2666
680	462400	26.0768	720	518400	26.8328	760	577600	27.5681	800	640000	28.2843

Table A1 Continued

N	N^2	$\sqrt{N}$	N	N^2	$\sqrt{N}$	N	N^2	$\sqrt{N}$	N	N^2	$\sqrt{N}$
801	641601	28.3019	841	707281	29.0000	881	776161	29.6816	921	848241	30.3480
802	643204	28.3196	842	708964	29.0172	882	777924	29.6985	922	850084	30.3645
803	644809	28.3373	843	710649	29.0345	883	779689	29.7153	923	851929	30.3809
804	646416	28.3549	844	712336	29.0517	884	781456	29.7321	924	853776	30.3974
805	648025	28.3725	845	714025	29.0689	885	783225	29.7489	925	855625	30.4138
806	649636	28.3901	846	715716	29.0861	886	784996	29.7658	926	857476	30.4302
807	651249	28.4077	847	717409	29.1033	887	786769	29.7825	927	859329	30.4467
808	652864	28.4253	848	719104	29.1204	888	788544	29.7993	928	861184	30.4631
809	654481	28.4429	849	720801	29.1376	889	790321	29.8161	929	863041	30.4795
810	656100	28.4605	850	722500	29.1548	890	792100	29.8329	930	864900	30.4959
811	657721	28.4781	851	724201	29.1719	891	793881	29.8496	931	866761	30.5123
812	659344	28.4956	852	725904	29.1890	892	795664	29.8664	932	868624	30.5287
813	660969	28.5132	853	727609	29.2062	893	797449	29.8831	933	870489	30.5450
814	662596	28.5307	854	729316	29.2233	894	799236	29.8998	934	872356	30.5614
815	664225	28.5482	855	731025	29.2404	895	801025	29.9166	935	874225	30.5778
816	665856	28.5657	856	732736	29.2575	896	802816	29.9333	936	876096	30.5941
817	667489	28.5832	857	734449	29.2746	897	804609	29.9500	937	877969	30.6105
818	669124	28.6007	858	736164	29.2916	898	806404	29.9666	938	879844	30.6268
819	670761	28.6182	859	737881	29.3087	899	808201	29.9833	939	881721	30.6431
820	672400	28.6356	860	739600	29.3258	900	810000	30.0000	940	883600	30.6594
821	674041	28.6531	861	741321	29.3428	901	811801	30.0167	941	885481	30.6757
822	675684	28.6705	862	743044	29.3598	902	813604	30.0333	942	887364	30.6920
823	677329	28.6880	863	744769	29.3769	903	815409	30.0500	943	889249	30.7083
824	678976	28.7054	864	746496	29.3939	904	817216	30.0666	944	891136	30.7246
825	680625	28.7228	865	748225	29.4109	905	819025	30.0832	945	893025	30.7409
826	682276	28.7402	866	749956	29.4279	906	820836	30.0998	946	894916	30.7571
827	683929	28.7576	867	751689	29.4449	907	822649	30.1164	947	896809	30.7734
828	685584	28.7750	868	753424	29.4618	908	824464	30.1330	948	898704	30.7896
829	687241	28.7924	869	755161	29.4788	909	826281	30.1496	949	900601	30.8058
830	688900	28.8097	870	756900	29.4958	910	828100	30.1662	950	902500	30.8221
831	690561	28.8271	871	758641	29.5127	911	829921	30.1828	951	904401	30.8383
832	692224	28.8444	872	760384	29.5296	912	831744	30.1993	952	906304	30.8545
833	693889	28.8617	873	762129	29.5466	913	833569	30.2159	953	908209	30.8707
834	695556	28.8791	874	763876	29.5635	914	835396	30.2324	954	910116	30.8869
835	697225	28.8964	875	765625	29.5804	915	837225	30.2490	955	912025	30.9031
836	698896	28.9137	876	767376	29.5973	916	839056	30.2655	956	913936	30.9192
837	700569	28.9310	877	769129	29.6142	917	840889	30.2820	957	915849	30.9354
838	702244	28.9482	878	770884	29.6311	918	842724	30.2985	958	917764	30.9516
839	703921	28.9655	879	772641	29.6479	919	844561	30.3150	959	919681	30.9677
840	705600	28.9828	880	774400	29.6648	920	846400	30.3315	960	921600	30.9839

Table A1 Continued

N	N^2	$\sqrt{N}$	N	N^2	$\sqrt{N}$
961	923521	31.0000	981	962361	31.3209
962	925444	31.0161	982	964324	31.3369
963	927369	31.0322	983	966289	31.3528
964	929296	31.0483	984	968256	31.3688
965	931225	31.0644	985	970225	31.3847
966	933156	31.0805	986	972196	31.4006
967	935089	31.0966	987	974169	31.4166
968	937024	31.1127	988	976144	31.4325
969	938961	31.1288	989	978121	31.4484
970	940900	31.1448	990	980100	31.4643
971	942841	31.1609	991	982081	31.4802
972	944784	31.1769	992	984064	31.4960
973	946729	31.1929	993	986049	31.5119
974	948676	31.2090	994	988036	31.5278
975	950625	31.2250	995	990025	31.5436
976	952576	31.2410	996	992016	31.5595
977	954529	31.2570	997	994009	31.5753
978	956484	31.2730	998	996004	31.5911
979	958441	31.2890	999	998001	31.6070
980	960400	31.3050	1000	1000000	31.6228

Table A2 Areas and Ordinates of the Standard Normal Curve

(1) z	(2) A Area from μ to z	(3) B Area in Larger Portion	(4) C Area in Smaller Portion	(5) y Ordinate at z
0.00	.0000	.5000	.5000	.3989
0.01	.0040	.5040	.4960	.3989
0.02	.0080	.5080	.4920	.3989
0.03	.0120	.5120	.4880	.3988
0.04	.0160	.5160	.4840	.3986
0.05	.0199	.5199	.4801	.3984
0.06	.0239	.5239	.4761	.3982
0.07	.0279	.5279	.4721	.3980
0.08	.0319	.5319	.4681	.3977
0.09	.0359	.5359	.4641	.3973
0.10	.0398	.5398	.4602	.3970
0.11	.0438	.5438	.4562	.3965
0.12	.0478	.5478	.4522	.3961
0.13	.0517	.5517	.4483	.3956
0.14	.0557	.5557	.4443	.3951
0.15	.0596	.5596	.4404	.3945
0.16	.0636	.5636	.4364	.3939
0.17	.0675	.5675	.4325	.3932
0.18	.0714	.5714	.4286	.3925
0.19	.0753	.5753	.4247	.3918
0.20	.0793	.5793	.4207	.3910
0.21	.0832	.5832	.4168	.3902
0.22	.0871	.5871	.4129	.3894
0.23	.0910	.5910	.4090	.3885
0.24	.0948	.5948	.4052	.3876
0.25	.0987	.5987	.4013	.3867
0.26	.1026	.6026	.3974	.3857
0.27	.1064	.6064	.3936	.3847
0.28	.1103	.6103	.3897	.3836
0.29	.1141	.6141	.3859	.3825
0.30	.1179	.6179	.3821	.3814
0.31	.1217	.6217	.3783	.3802
0.32	.1255	.6255	.3745	.3790
0.33	.1293	.6293	.3707	.3778
0.34	.1331	.6331	.3669	.3765

Table A2 Continued

(1) z	(2) A Area from μ to z	(3) B Area in Larger Portion	(4) C Area in Smaller Portion	(5) y Ordinate at z
0.35	.1368	.6368	.3632	.3752
0.36	.1406	.6406	.3594	.3739
0.37	.1443	.6443	.3557	.3725
0.38	.1480	.6480	.3520	.3712
0.39	.1517	.6517	.3483	.3697
0.40	.1554	.6554	.3446	.3683
0.41	.1591	.6591	.3409	.3668
0.42	.1628	.6628	.3372	.3653
0.43	.1664	.6664	.3336	.3637
0.44	.1700	.6700	.3300	.3621
0.45	.1736	.6736	.3264	.3605
0.46	.1772	.6772	.3228	.3589
0.47	.1808	.6808	.3192	.3572
0.48	.1844	.6844	.3156	.3555
0.49	.1879	.6879	.3121	.3538
0.50	.1915	.6915	.3085	.3521
0.51	.1950	.6950	.3050	.3503
0.52	.1985	.6985	.3015	.3485
0.53	.2019	.7019	.2981	.3467
0.54	.2054	.7054	.2946	.3448
0.55	.2088	.7088	.2912	.3429
0.56	.2123	.7123	.2877	.3410
0.57	.2157	.7157	.2843	.3391
0.58	.2190	.7190	.2810	.3372
0.59	.2224	.7224	.2776	.3352
0.60	.2257	.7257	.2743	.3332
0.61	.2291	.7291	.2709	.3312
0.62	.2324	.7324	.2676	.3292
0.63	.2357	.7357	.2643	.3271
0.64	.2389	.7389	.2611	.3251
0.65	.2422	.7422	.2578	.3230
0.66	.2454	.7454	.2546	.3209
0.67	.2486	.7486	.2514	.3187
0.68	.2517	.7517	.2483	.3166
0.69	.2549	.7549	.2451	.3144

Table A2 Continued

(1) z	(2) A AREA FROM μ TO z	(3) B AREA IN LARGER PORTION	(4) C AREA IN SMALLER PORTION	(5) y ORDINATE AT z
0.70	.2580	.7580	.2420	.3123
0.71	.2611	.7611	.2389	.3101
0.72	.2642	.7642	.2358	.3079
0.73	.2673	.7673	.2327	.3056
0.74	.2704	.7704	.2296	.3034
0.75	.2734	.7734	.2266	.3011
0.76	.2764	.7764	.2236	.2989
0.77	.2794	.7794	.2206	.2966
0.78	.2823	.7823	.2177	.2943
0.79	.2852	.7852	.2148	.2920
0.80	.2881	.7881	.2119	.2897
0.81	.2910	.7910	.2090	.2874
0.82	.2939	.7939	.2061	.2850
0.83	.2967	.7967	.2033	.2827
0.84	.2995	.7995	.2005	.2803
0.85	.3023	.8023	.1977	.2780
0.86	.3051	.8051	.1949	.2756
0.87	.3078	.8078	.1922	.2732
0.88	.3106	.8106	.1894	.2709
0.89	.3133	.8133	.1867	.2685
0.90	.3159	.8159	.1841	.2661
0.91	.3186	.8186	.1814	.2637
0.92	.3212	.8212	.1788	.2613
0.93	.3238	.8238	.1762	.2589
0.94	.3264	.8264	.1736	.2565
0.95	.3289	.8289	.1711	.2541
0.96	.3315	.8315	.1685	.2516
0.97	.3340	.8340	.1660	.2492
0.98	.3365	.8365	.1635	.2468
0.99	.3389	.8389	.1611	.2444
1.00	.3413	.8413	.1587	.2420
1.01	.3438	.8438	.1562	.2396
1.02	.3461	.8461	.1539	.2371
1.03	.3485	.8485	.1515	.2347
1.04	.3508	.8508	.1492	.2323

Table A2 Continued

(1) z	(2) A Area from μ to z	(3) B Area in Larger Portion	(4) C Area in Smaller Portion	(5) y Ordinate at z
1.05	.3531	.8531	.1469	.2299
1.06	.3554	.8554	.1446	.2275
1.07	.3577	.8577	.1423	.2251
1.08	.3599	.8599	.1401	.2227
1.09	.3621	.8621	.1379	.2203
1.10	.3643	.8643	.1357	.2179
1.11	.3665	.8665	.1335	.2155
1.12	.3686	.8686	.1314	.2131
1.13	.3708	.8708	.1292	.2107
1.14	.3729	.8729	.1271	.2083
1.15	.3749	.8749	.1251	.2059
1.16	.3770	.8770	.1230	.2036
1.17	.3790	.8790	.1210	.2012
1.18	.3810	.8810	.1190	.1989
1.19	.3830	.8830	.1170	.1965
1.20	.3849	.8849	.1151	.1942
1.21	.3869	.8869	.1131	.1919
1.22	.3888	.8888	.1112	.1895
1.23	.3907	.8907	.1093	.1872
1.24	.3925	.8925	.1075	.1849
1.25	.3944	.8944	.1056	.1826
1.26	.3962	.8962	.1038	.1804
1.27	.3980	.8980	.1020	.1781
1.28	.3997	.8997	.1003	.1758
1.29	.4015	.9015	.0985	.1736
1.30	.4032	.9032	.0968	.1714
1.31	.4049	.9049	.0951	.1691
1.32	.4066	.9066	.0934	.1669
1.33	.4082	.9082	.0918	.1647
1.34	.4099	.9099	.0901	.1626
1.35	.4115	.9115	.0885	.1604
1.36	.4131	.9131	.0869	.1582
1.37	.4147	.9147	.0853	.1561
1.38	.4162	.9162	.0838	.1539
1.39	.4177	.9177	.0823	.1518

Table A2 Continued

(1) z	(2) A Area from μ to z	(3) B Area in Larger Portion	(4) C Area in Smaller Portion	(5) y Ordinate at z
1.40	.4192	.9192	.0808	.1497
1.41	.4207	.9207	.0793	.1476
1.42	.4222	.9222	.0778	.1456
1.43	.4236	.9236	.0764	.1435
1.44	.4251	.9251	.0749	.1415
1.45	.4265	.9265	.0735	.1394
1.46	.4279	.9279	.0721	.1374
1.47	.4292	.9292	.0708	.1354
1.48	.4306	.9306	.0694	.1334
1.49	.4319	.9319	.0681	.1315
1.50	.4332	.9332	.0668	.1295
1.51	.4345	.9345	.0655	.1276
1.52	.4357	.9357	.0643	.1257
1.53	.4370	.9370	.0630	.1238
1.54	.4382	.9382	.0618	.1219
1.55	.4394	.9394	.0606	.1200
1.56	.4406	.9406	.0594	.1182
1.57	.4418	.9418	.0582	.1163
1.58	.4429	.9429	.0571	.1145
1.59	.4441	.9441	.0559	.1127
1.60	.4452	.9452	.0548	.1109
1.61	.4463	.9463	.0537	.1092
1.62	.4474	.9474	.0526	.1074
1.63	.4484	.9484	.0516	.1057
1.64	.4495	.9495	.0505	.1040
1.65	.4505	.9505	.0495	.1023
1.66	.4515	.9515	.0485	.1006
1.67	.4525	.9525	.0475	.0989
1.68	.4535	.9535	.0465	.0973
1.69	.4545	.9545	.0455	.0957
1.70	.4554	.9554	.0446	.0940
1.71	.4564	.9564	.0436	.0925
1.72	.4573	.9573	.0427	.0909
1.73	.4582	.9582	.0418	.0893
1.74	.4591	.9591	.0409	.0878

Table A2 Continued

(1) z	(2) A Area from μ to z	(3) B Area in Larger Portion	(4) C Area in Smaller Portion	(5) y Ordinate at z
1.75	.4599	.9599	.0401	.0863
1.76	.4608	.9608	.0392	.0848
1.77	.4616	.9616	.0384	.0833
1.78	.4625	.9625	.0375	.0818
1.79	.4633	.9633	.0367	.0804
1.80	.4641	.9641	.0359	.0790
1.81	.4649	.9649	.0351	.0775
1.82	.4656	.9656	.0344	.0761
1.83	.4664	.9664	.0336	.0748
1.84	.4671	.9671	.0329	.0734
1.85	.4678	.9678	.0322	.0721
1.86	.4686	.9686	.0314	.0707
1.87	.4693	.9693	.0307	.0694
1.88	.4699	.9699	.0301	.0681
1.89	.4706	.9706	.0294	.0669
1.90	.4713	.9713	.0287	.0656
1.91	.4719	.9719	.0281	.0644
1.92	.4726	.9726	.0274	.0632
1.93	.4732	.9732	.0268	.0620
1.94	.4738	.9738	.0262	.0608
1.95	.4744	.9744	.0256	.0596
1.96	.4750	.9750	.0250	.0584
1.97	.4756	.9756	.0244	.0573
1.98	.4761	.9761	.0239	.0562
1.99	.4767	.9767	.0233	.0551
2.00	.4772	.9772	.0228	.0540
2.01	.4778	.9778	.0222	.0529
2.02	.4783	.9783	.0217	.0519
2.03	.4788	.9788	.0212	.0508
2.04	.4793	.9793	.0207	.0498
2.05	.4798	.9798	.0202	.0488
2.06	.4803	.9803	.0197	.0478
2.07	.4808	.9808	.0192	.0468
2.08	.4812	.9812	.0188	.0459
2.09	.4817	.9817	.0183	.0449

Table A2 Continued

(1) z	(2) A Area from μ to z	(3) B Area in Larger Portion	(4) C Area in Smaller Portion	(5) y Ordinate at z
2.10	.4821	.9821	.0179	.0440
2.11	.4826	.9826	.0174	.0431
2.12	.4830	.9830	.0170	.0422
2.13	.4834	.9834	.0166	.0413
2.14	.4838	.9838	.0162	.0404
2.15	.4842	.9842	.0158	.0396
2.16	.4846	.9846	.0154	.0387
2.17	.4850	.9850	.0150	.0379
2.18	.4854	.9854	.0146	.0371
2.19	.4857	.9857	.0143	.0363
2.20	.4861	.9861	.0139	.0355
2.21	.4864	.9864	.0136	.0347
2.22	.4868	.9868	.0132	.0339
2.23	.4871	.9871	.0129	.0332
2.24	.4875	.9875	.0125	.0325
2.25	.4878	.9878	.0122	.0317
2.26	.4881	.9881	.0119	.0310
2.27	.4884	.9884	.0116	.0303
2.28	.4887	.9887	.0113	.0297
2.29	.4890	.9890	.0110	.0290
2.30	.4893	.9893	.0107	.0283
2.31	.4896	.9896	.0104	.0277
2.32	.4898	.9898	.0102	.0270
2.33	.4901	.9901	.0099	.0264
2.34	.4904	.9904	.0096	.0258
2.35	.4906	.9906	.0094	.0252
2.36	.4909	.9909	.0091	.0246
2.37	.4911	.9911	.0089	.0241
2.38	.4913	.9913	.0087	.0235
2.39	.4916	.9916	.0084	.0229
2.40	.4918	.9918	.0082	.0224
2.41	.4920	.9920	.0080	.0219
2.42	.4922	.9922	.0078	.0213
2.43	.4925	.9925	.0075	.0208
2.44	.4927	.9927	.0073	.0203

Table A2 Continued

(1) z	(2) A Area from μ to z	(3) B Area in Larger Portion	(4) C Area in Smaller Portion	(5) y Ordinate at z
2.45	.4929	.9929	.0071	.0198
2.46	.4931	.9931	.0069	.0194
2.47	.4932	.9932	.0068	.0189
2.48	.4934	.9934	.0066	.0184
2.49	.4936	.9936	.0064	.0180
2.50	.4938	.9938	.0062	.0175
2.51	.4940	.9940	.0060	.0171
2.52	.4941	.9941	.0059	.0167
2.53	.4943	.9943	.0057	.0163
2.54	.4945	.9945	.0055	.0158
2.55	.4946	.9946	.0054	.0154
2.56	.4948	.9948	.0052	.0151
2.57	.4949	.9949	.0051	.0147
2.58	.4951	.9951	.0049	.0143
2.59	.4952	.9952	.0048	.0139
2.60	.4953	.9953	.0047	.0136
2.61	.4955	.9955	.0045	.0132
2.62	.4956	.9956	.0044	.0129
2.63	.4957	.9957	.0043	.0126
2.64	.4959	.9959	.0041	.0122
2.65	.4960	.9960	.0040	.0119
2.66	.4961	.9961	.0039	.0116
2.67	.4962	.9962	.0038	.0113
2.68	.4963	.9963	.0037	.0110
2.69	.4964	.9964	.0036	.0107
2.70	.4965	.9965	.0035	.0104
2.71	.4966	.9966	.0034	.0101
2.72	.4967	.9967	.0033	.0099
2.73	.4968	.9968	.0032	.0096
2.74	.4969	.9969	.0031	.0093
2.75	.4970	.9970	.0030	.0091
2.76	.4971	.9971	.0029	.0088
2.77	.4972	.9972	.0028	.0086
2.78	.4973	.9973	.0027	.0084
2.79	.4974	.9974	.0026	.0081

Table A2 Continued

(1) z	(2) A Area from μ to z	(3) B Area in Larger Portion	(4) C Area in Smaller Portion	(5) y Ordinate at z
2.80	.4974	.9974	.0026	.0079
2.81	.4975	.9975	.0025	.0077
2.82	.4976	.9976	.0024	.0075
2.83	.4977	.9977	.0023	.0073
2.84	.4977	.9977	.0023	.0071
2.85	.4978	.9978	.0022	.0069
2.86	.4979	.9979	.0021	.0067
2.87	.4979	.9979	.0021	.0065
2.88	.4980	.9980	.0020	.0063
2.89	.4981	.9981	.0019	.0061
2.90	.4981	.9981	.0019	.0060
2.91	.4982	.9982	.0018	.0058
2.92	.4982	.9982	.0018	.0056
2.93	.4983	.9983	.0017	.0055
2.94	.4984	.9984	.0016	.0053
2.95	.4984	.9984	.0016	.0051
2.96	.4985	.9985	.0015	.0050
2.97	.4985	.9985	.0015	.0048
2.98	.4986	.9986	.0014	.0047
2.99	.4986	.9986	.0014	.0046
3.00	.4987	.9987	.0013	.0044
3.01	.4987	.9987	.0013	.0043
3.02	.4987	.9987	.0013	.0042
3.03	.4988	.9988	.0012	.0040
3.04	.4988	.9988	.0012	.0039
3.05	.4989	.9989	.0011	.0038
3.06	.4989	.9989	.0011	.0037
3.07	.4989	.9989	.0011	.0036
3.08	.4990	.9990	.0010	.0035
3.09	.4990	.9990	.0010	.0034
3.10	.4990	.9990	.0010	.0033
3.11	.4991	.9991	.0009	.0032
3.12	.4991	.9991	.0009	.0031
3.13	.4991	.9991	.0009	.0030
3.14	.4992	.9992	.0008	.0029

Table A2 Continued

(1) z	(2) A AREA FROM μ TO z	(3) B AREA IN LARGER PORTION	(4) C AREA IN SMALLER PORTION	(5) y ORDINATE AT z
3.15	.4992	.9992	.0008	.0028
3.16	.4992	.9992	.0008	.0027
3.17	.4992	.9992	.0008	.0026
3.18	.4993	.9993	.0007	.0025
3.19	.4993	.9993	.0007	.0025
3.20	.4993	.9993	.0007	.0024
3.21	.4993	.9993	.0007	.0023
3.22	.4994	.9994	.0006	.0022
3.23	.4994	.9994	.0006	.0022
3.24	.4994	.9994	.0006	.0021
3.30	.4995	.9995	.0005	.0017
3.40	.4997	.9997	.0003	.0012
3.50	.4998	.9998	.0002	.0009
3.60	.4998	.9998	.0002	.0006
3.70	.4999	.9999	.0001	.0004

Table A3 Table of Probabilities Associated with Values as Small as Observed Values of x in the Binomial Test [†]

Given in the body of this table are one-tailed probabilities under H_0 for the binomial test when $P = Q = \frac{1}{2}$. To save space, decimal points are omitted in the p's.

n \ r	0	1	2	3	4	5	6	7	8	9	10	11	12	13	14	15
5	031	188	500	812	969	*										
6	016	109	344	656	891	984	*									
7	008	062	227	500	773	938	992	*								
8	004	035	145	363	637	855	965	996	*							
9	002	020	090	254	500	746	910	980	998	*						
10	001	011	055	172	377	623	828	945	989	999	*					
11		006	033	113	274	500	726	887	967	994	*	*				
12		003	019	073	194	387	613	806	927	981	997	*	*			
13		002	011	046	133	291	500	709	867	954	989	998	*	*		
14		001	006	029	090	212	395	605	788	910	971	994	999	*	*	
15			004	018	059	151	304	500	696	849	941	982	996	*	*	*
16			002	011	038	105	227	402	598	773	895	962	989	998	*	*
17			001	006	025	072	166	315	500	685	834	928	975	994	999	*
18			001	004	015	048	119	240	407	593	760	881	952	985	996	999
19				002	010	032	084	180	324	500	676	820	916	968	990	998
20				001	006	021	058	132	252	412	588	748	868	942	979	994
21				001	004	013	039	095	192	332	500	668	808	905	961	987
22					002	008	026	067	143	262	416	584	738	857	933	974
23					001	005	017	047	105	202	339	500	661	798	895	953
24					001	003	011	032	076	154	271	419	581	729	846	924
25						002	007	022	054	115	212	345	500	655	788	885

[†] Adapted from Table IV, B, of Helen Walker and J. Lev, Statistical inference. (New York: Holt, Rinehart and Winston, 1953), p. 458, with the kind permission of the authors and publisher.

* 1.0 or approximately 1.0.

Table A4 Distribution of t for Given Probability Levels*

	Level of significance for one-tailed test					
	.10	.05	.025	.01	.005	.0005
	Level of significance for two-tailed test					
df	.20	.10	.05	.02	.01	.001
1	3.078	6.314	12.706	31.821	63.657	636.619
2	1.886	2.920	4.303	6.965	9.925	31.598
3	1.638	2.353	3.182	4.541	5.841	12.941
4	1.533	2.132	2.776	3.747	4.604	8.610
5	1.476	2.015	2.571	3.365	4.032	6.859
6	1.440	1.943	2.447	3.143	3.707	5.959
7	1.415	1.895	2.365	2.998	3.499	5.405
8	1.397	1.860	2.306	2.896	3.355	5.041
9	1.383	1.833	2.262	2.821	3.250	4.781
10	1.372	1.812	2.228	2.764	3.169	4.587
11	1.363	1.796	2.201	2.718	3.106	4.437
12	1.356	1.782	2.179	2.681	3.055	4.318
13	1.350	1.771	2.160	2.650	3.012	4.221
14	1.345	1.761	2.145	2.624	2.977	4.140
15	1.341	1.753	2.131	2.602	2.947	4.073
16	1.337	1.746	2.120	2.583	2.921	4.015
17	1.333	1.740	2.110	2.567	2.898	3.965
18	1.330	1.734	2.101	2.552	2.878	3.922
19	1.328	1.729	2.093	2.539	2.861	3.883
20	1.325	1.725	2.086	2.528	2.845	3.850
21	1.323	1.721	2.080	2.518	2.831	3.819
22	1.321	1.717	2.074	2.508	2.819	3.792
23	1.319	1.714	2.069	2.500	2.807	3.767
24	1.318	1.711	2.064	2.492	2.797	3.745
25	1.316	1.708	2.060	2.485	2.787	3.725
26	1.315	1.706	2.056	2.479	2.779	3.707
27	1.314	1.703	2.052	2.473	2.771	3.690
28	1.313	1.701	2.048	2.467	2.763	3.674
29	1.311	1.699	2.045	2.462	2.756	3.659
30	1.310	1.697	2.042	2.457	2.750	3.646
40	1.303	1.684	2.021	2.423	2.704	3.551
60	1.296	1.671	2.000	2.390	2.660	3.460
120	1.289	1.658	1.980	2.358	2.617	3.373
∞	1.282	1.645	1.960	2.326	2.576	3.291

*This table is abridged from Table III of R. A. Fisher and F. Yates, Statistical Tables for Biological, Agricultural, and Medical Research, New York: Hafner 1974.

Table A5 Critical Values of the Pearson Correlation Coefficient*

df	Level of significance for one-tailed test			
	.05	.025	.01	.005
	Level of significance for two-tailed test			
	.10	.05	.02	.01
1	.988	.997	.9995	.9999
2	.900	.950	.980	.990
3	.805	.878	.934	.959
4	.729	.811	.882	.917
5	.669	.754	.833	.874
6	.622	.707	.789	.834
7	.582	.666	.750	.798
8	.549	.632	.716	.765
9	.521	.602	.685	.735
10	.497	.576	.658	.708
11	.576	.553	.634	.684
12	.458	.532	.612	.661
13	.441	.514	.592	.641
14	.426	.497	.574	.623
15	.412	.482	.558	.606
16	.400	.468	.542	.590
17	.389	.456	.528	.575
18	.378	.444	.516	.561
19	.369	.433	.503	.549
20	.360	.423	.492	.537
21	.352	.413	.482	.526
22	.344	.404	.472	.515
23	.337	.396	.462	.505
24	.330	.388	.453	.496
25	.323	.381	.445	.487
26	.317	.374	.437	.479
27	.311	.367	.430	.471
28	.306	.361	.423	.463
29	.301	.355	.416	.486
30	.296	.349	.409	.449
35	.275	.325	.381	.418
40	.257	.304	.358	.393
45	.243	.288	.338	.372
50	.231	.273	.322	.354
60	.211	.250	.295	.325
70	.195	.232	.274	.303
80	.183	.217	.256	.283
90	.173	.205	.242	.267
100	.164	.195	.230	.254

*Abridged from R. A. Fisher and F. Yates, Statistical Tables for Biological, Agricultural, and Medical Research, New York: Hafner 1974.

Table A6 The 5 (Roman Type) and 1 (Boldface Type) Percent Points for the Distribution of F

n_1 degrees of freedom (for greater mean square)

d.f.	1	2	3	4	5	6	7	8	9	10	11	12	14	16	20	24	30	40	50	75	100	200	500	∞
1	161 **4,052**	200 **4,999**	216 **5,403**	225 **5,625**	230 **5,764**	234 **5,859**	237 **5,928**	239 **5,981**	241 **6,022**	242 **6,056**	243 **6,082**	244 **6,106**	245 **6,142**	246 **6,169**	248 **6,208**	249 **6,234**	250 **6,258**	251 **6,286**	252 **6,302**	253 **6,323**	253 **6,334**	254 **6,352**	254 **6,361**	254 **6,366**
2	18.51 **98.49**	19.00 **99.00**	19.16 **99.17**	19.25 **99.25**	19.30 **99.30**	19.33 **99.33**	19.36 **99.34**	19.37 **99.36**	19.38 **99.38**	19.39 **99.40**	19.40 **99.41**	19.41 **99.42**	19.42 **99.43**	19.43 **99.44**	19.44 **99.45**	19.45 **99.46**	19.46 **99.47**	19.47 **99.48**	19.47 **99.48**	19.48 **99.49**	19.49 **99.49**	19.49 **99.49**	19.50 **99.50**	19.50 **99.50**
3	10.13 **34.12**	9.55 **30.82**	9.28 **29.46**	9.12 **28.71**	9.01 **28.24**	8.94 **27.91**	8.88 **27.67**	8.84 **27.49**	8.81 **27.34**	8.78 **27.23**	8.76 **27.13**	8.74 **27.05**	8.71 **26.92**	8.69 **26.83**	8.66 **26.69**	8.64 **26.60**	8.62 **26.50**	8.60 **26.41**	8.58 **26.35**	8.57 **26.27**	8.56 **26.23**	8.54 **26.18**	8.54 **26.14**	8.53 **26.12**
4	7.71 **21.20**	6.94 **18.00**	6.59 **16.69**	6.39 **15.98**	6.26 **15.52**	6.16 **15.21**	6.09 **14.98**	6.04 **14.80**	6.00 **14.66**	5.96 **14.54**	5.93 **14.45**	5.91 **14.37**	5.87 **14.24**	5.84 **14.15**	5.80 **14.02**	5.77 **13.93**	5.74 **13.83**	5.71 **13.74**	5.70 **13.69**	5.68 **13.61**	5.66 **13.57**	5.65 **13.52**	5.64 **13.48**	5.63 **13.46**
5	6.61 **16.26**	5.79 **13.27**	5.41 **12.06**	5.19 **11.39**	5.05 **10.97**	4.95 **10.67**	4.88 **10.45**	4.82 **10.27**	4.78 **10.15**	4.74 **10.05**	4.70 **9.96**	4.68 **9.89**	4.64 **9.77**	4.60 **9.68**	4.56 **9.55**	4.53 **9.47**	4.50 **9.38**	4.46 **9.29**	4.44 **9.24**	4.42 **9.17**	4.40 **9.13**	4.38 **9.07**	4.37 **9.04**	4.36 **9.02**
6	5.99 **13.74**	5.14 **10.92**	4.76 **9.78**	4.53 **9.15**	4.39 **8.75**	4.28 **8.47**	4.21 **8.26**	4.15 **8.10**	4.10 **7.98**	4.06 **7.87**	4.03 **7.79**	4.00 **7.72**	3.96 **7.60**	3.92 **7.52**	3.87 **7.39**	3.84 **7.31**	3.81 **7.23**	3.77 **7.14**	3.75 **7.09**	3.72 **7.02**	3.71 **6.99**	3.69 **6.94**	3.68 **6.90**	3.67 **6.88**
7	5.59 **12.25**	4.74 **9.55**	4.35 **8.45**	4.12 **7.85**	3.97 **7.46**	3.87 **7.19**	3.79 **7.00**	3.73 **6.84**	3.68 **6.71**	3.63 **6.62**	3.60 **6.54**	3.57 **6.47**	3.52 **6.35**	3.49 **6.27**	3.44 **6.15**	3.41 **6.07**	3.38 **5.98**	3.34 **5.90**	3.32 **5.85**	3.29 **5.78**	3.28 **5.75**	3.25 **5.70**	3.24 **5.67**	3.23 **5.65**
8	5.32 **11.26**	4.46 **8.65**	4.07 **7.59**	3.84 **7.01**	3.69 **6.63**	3.58 **6.37**	3.50 **6.19**	3.44 **6.03**	3.39 **5.91**	3.34 **5.82**	3.31 **5.74**	3.28 **5.67**	3.23 **5.56**	3.20 **5.48**	3.15 **5.36**	3.12 **5.28**	3.08 **5.20**	3.05 **5.11**	3.03 **5.06**	3.00 **5.00**	2.98 **4.96**	2.96 **4.91**	2.94 **4.88**	2.93 **4.86**
9	5.12 **10.56**	4.26 **8.02**	3.86 **6.99**	3.63 **6.42**	3.48 **6.06**	3.37 **5.80**	3.29 **5.62**	3.23 **5.47**	3.18 **5.35**	3.13 **5.26**	3.10 **5.18**	3.07 **5.11**	3.02 **5.00**	2.98 **4.92**	2.93 **4.80**	2.90 **4.73**	2.86 **4.64**	2.82 **4.56**	2.80 **4.51**	2.77 **4.45**	2.76 **4.41**	2.73 **4.36**	2.72 **4.33**	2.71 **4.31**
10	4.96 **10.04**	4.10 **7.56**	3.71 **6.55**	3.48 **5.99**	3.33 **5.64**	3.22 **5.39**	3.14 **5.21**	3.07 **5.06**	3.02 **4.95**	2.97 **4.85**	2.94 **4.78**	2.91 **4.71**	2.86 **4.60**	2.82 **4.52**	2.77 **4.41**	2.74 **4.33**	2.70 **4.25**	2.67 **4.17**	2.64 **4.12**	2.61 **4.05**	2.59 **4.01**	2.56 **3.96**	2.55 **3.93**	2.54 **3.91**
11	4.84 **9.65**	3.98 **7.20**	3.59 **6.22**	3.36 **5.67**	3.20 **5.32**	3.09 **5.07**	3.01 **4.88**	2.95 **4.74**	2.90 **4.63**	2.86 **4.54**	2.82 **4.46**	2.79 **4.40**	2.74 **4.29**	2.70 **4.21**	2.65 **4.10**	2.61 **4.02**	2.57 **3.94**	2.53 **3.86**	2.50 **3.80**	2.47 **3.74**	2.45 **3.70**	2.42 **3.66**	2.41 **3.62**	2.40 **3.60**
12	4.75 **9.33**	3.88 **6.93**	3.49 **5.95**	3.26 **5.41**	3.11 **5.06**	3.00 **4.82**	2.92 **4.65**	2.85 **4.50**	2.80 **4.39**	2.76 **4.30**	2.72 **4.22**	2.69 **4.16**	2.64 **4.05**	2.60 **3.98**	2.54 **3.86**	2.50 **3.78**	2.46 **3.70**	2.42 **3.61**	2.40 **3.56**	2.36 **3.49**	2.35 **3.46**	2.32 **3.41**	2.31 **3.38**	2.30 **3.36**
13	4.67 **9.07**	3.80 **6.70**	3.41 **5.74**	3.18 **5.20**	3.02 **4.86**	2.92 **4.62**	2.84 **4.44**	2.77 **4.30**	2.72 **4.19**	2.67 **4.10**	2.63 **4.02**	2.60 **3.96**	2.55 **3.85**	2.51 **3.78**	2.46 **3.67**	2.42 **3.59**	2.38 **3.51**	2.34 **3.42**	2.32 **3.37**	2.28 **3.30**	2.26 **3.27**	2.24 **3.21**	2.22 **3.18**	2.21 **3.16**

APPENDIX 449

Table A6 Continued

n_1 degrees of freedom (for greater mean square)

d.f.	1	2	3	4	5	6	7	8	9	10	11	12	14	16	20	24	30	40	50	75	100	200	500	∞
14	4.60 8.86	3.74 6.51	3.34 5.56	3.11 5.03	2.96 4.69	2.85 4.46	2.77 4.28	2.70 4.14	2.65 4.03	2.60 3.94	2.56 3.86	2.53 3.80	2.48 3.70	2.44 3.62	2.39 3.51	2.35 3.43	2.31 3.34	2.27 3.26	2.24 3.21	2.21 3.14	2.19 3.11	2.16 3.06	2.14 3.02	2.13 3.00
15	4.54 8.68	3.68 6.36	3.29 5.42	3.06 4.89	2.90 4.56	2.79 4.32	2.70 4.14	2.64 4.00	2.59 3.89	2.55 3.80	2.51 3.73	2.48 3.67	2.43 3.56	2.39 3.48	2.33 3.36	2.29 3.29	2.25 3.20	2.21 3.12	2.18 3.07	2.15 3.00	2.12 2.97	2.10 2.92	2.08 2.89	2.07 2.87
16	4.49 8.53	3.63 6.23	3.24 5.29	3.01 4.77	2.85 4.44	2.74 4.20	2.66 4.03	2.59 3.89	2.54 3.78	2.49 3.69	2.45 3.61	2.42 3.55	2.37 3.45	2.33 3.37	2.28 3.25	2.24 3.18	2.20 3.10	2.16 3.01	2.13 2.96	2.09 2.89	2.07 2.86	2.04 2.80	2.02 2.77	2.01 2.75
17	4.45 8.40	3.59 6.11	3.20 5.18	2.96 4.67	2.81 4.34	2.70 4.10	2.62 3.93	2.55 3.79	2.50 3.68	2.45 3.59	2.41 3.52	2.38 3.45	2.33 3.35	2.29 3.27	2.23 3.16	2.19 3.08	2.15 3.00	2.11 2.92	2.08 2.86	2.04 2.79	2.02 2.76	1.99 2.70	1.97 2.67	1.96 2.65
18	4.41 8.28	3.55 6.01	3.16 5.09	2.93 4.58	2.77 4.25	2.66 4.01	2.58 3.85	2.51 3.71	2.46 3.60	2.41 3.51	2.37 3.44	2.34 3.37	2.29 3.27	2.25 3.19	2.19 3.07	2.15 3.00	2.11 2.91	2.07 2.83	2.04 2.78	2.00 2.71	1.98 2.68	1.95 2.62	1.93 2.59	1.92 2.57
19	4.38 8.18	3.52 5.93	3.13 5.01	2.90 4.50	2.74 4.17	2.63 3.94	2.55 3.77	2.48 3.63	2.43 3.52	2.38 3.43	2.34 3.36	2.31 3.30	2.26 3.19	2.21 3.12	2.15 3.00	2.11 2.92	2.07 2.84	2.02 2.76	2.00 2.70	1.96 2.63	1.94 2.60	1.91 2.54	1.90 2.51	1.88 2.49
20	4.35 8.10	3.49 5.85	3.10 4.94	2.87 4.43	2.71 4.10	2.60 3.87	2.52 3.71	2.45 3.56	2.40 3.45	2.35 3.37	2.31 3.30	2.28 3.23	2.23 3.13	2.18 3.05	2.12 2.94	2.08 2.86	2.04 2.77	1.99 2.69	1.96 2.63	1.92 2.56	1.90 2.53	1.87 2.47	1.85 2.44	1.84 2.42
21	4.32 8.02	3.47 5.78	3.07 4.87	2.84 4.37	2.68 4.04	2.57 3.81	2.49 3.65	2.42 3.51	2.37 3.40	2.32 3.31	2.28 3.24	2.25 3.17	2.20 3.07	2.15 2.99	2.09 2.88	2.05 2.80	2.00 2.72	1.96 2.63	1.93 2.58	1.89 2.51	1.87 2.47	1.84 2.42	1.82 2.38	1.81 2.36
22	4.30 7.94	3.44 5.72	3.05 4.82	2.82 4.31	2.66 3.99	2.55 3.76	2.47 3.59	2.40 3.45	2.35 3.35	2.30 3.26	2.26 3.18	2.23 3.12	2.18 3.02	2.13 2.94	2.07 2.83	2.03 2.75	1.98 2.67	1.93 2.58	1.91 2.53	1.87 2.46	1.84 2.42	1.81 2.37	1.80 2.33	1.78 2.31
23	4.28 7.88	3.42 5.66	3.03 4.76	2.80 4.26	2.64 3.94	2.53 3.71	2.45 3.54	2.38 3.41	2.32 3.30	2.28 3.21	2.24 3.14	2.20 3.07	2.14 2.97	2.10 2.89	2.04 2.78	2.00 2.70	1.96 2.62	1.91 2.53	1.88 2.48	1.84 2.41	1.82 2.37	1.79 2.32	1.77 2.28	1.76 2.26
24	4.26 7.82	3.40 5.61	3.01 4.72	2.78 4.22	2.62 3.90	2.51 3.67	2.43 3.50	2.36 3.36	2.30 3.25	2.26 3.17	2.22 3.09	2.18 3.03	2.13 2.93	2.09 2.85	2.02 2.74	1.98 2.66	1.94 2.58	1.89 2.49	1.86 2.44	1.82 2.36	1.80 2.33	1.76 2.27	1.74 2.23	1.73 2.21
25	4.24 7.77	3.38 5.57	2.99 4.68	2.76 4.18	2.60 3.86	2.49 3.63	2.41 3.46	2.34 3.32	2.28 3.21	2.24 3.13	2.20 3.05	2.16 2.99	2.11 2.89	2.06 2.81	2.00 2.70	1.96 2.62	1.92 2.54	1.87 2.45	1.84 2.40	1.80 2.32	1.77 2.29	1.74 2.23	1.72 2.19	1.71 2.17
26	4.22 7.72	3.37 5.53	2.98 4.64	2.74 4.14	2.59 3.82	2.47 3.59	2.39 3.42	2.32 3.29	2.27 3.17	2.22 3.09	2.18 3.02	2.15 2.96	2.10 2.86	2.05 2.77	1.99 2.66	1.95 2.58	1.90 2.50	1.85 2.41	1.82 2.36	1.78 2.28	1.76 2.25	1.72 2.19	1.70 2.15	1.69 2.13

Table A6 Continued

d.f.	1	2	3	4	5	6	7	8	9	10	11	12	14	16	20	24	30	40	50	75	100	200	500	∞
27	4.21 / 7.68	3.35 / 5.49	2.96 / 4.60	2.73 / 4.11	2.57 / 3.79	2.46 / 3.56	2.37 / 3.39	2.30 / 3.26	2.25 / 3.14	2.20 / 3.06	2.16 / 2.98	2.13 / 2.93	2.08 / 2.83	2.03 / 2.74	1.97 / 2.63	1.93 / 2.55	1.88 / 2.47	1.84 / 2.38	1.80 / 2.33	1.76 / 2.25	1.74 / 2.21	1.71 / 2.16	1.68 / 2.12	1.67 / 2.10
28	4.20 / 7.64	3.34 / 5.45	2.95 / 4.57	2.71 / 4.07	2.56 / 3.76	2.44 / 3.53	2.36 / 3.36	2.29 / 3.23	2.24 / 3.11	2.19 / 3.03	2.15 / 2.95	2.12 / 2.90	2.06 / 2.80	2.02 / 2.71	1.96 / 2.60	1.91 / 2.52	1.87 / 2.44	1.81 / 2.35	1.78 / 2.30	1.75 / 2.22	1.72 / 2.18	1.69 / 2.13	1.67 / 2.09	1.65 / 2.06
29	4.18 / 7.60	3.33 / 5.42	2.93 / 4.54	2.70 / 4.04	2.54 / 3.73	2.43 / 3.50	2.35 / 3.33	2.28 / 3.20	2.22 / 3.08	2.18 / 3.00	2.14 / 2.92	2.10 / 2.87	2.05 / 2.77	2.00 / 2.68	1.94 / 2.57	1.90 / 2.49	1.85 / 2.41	1.80 / 2.32	1.77 / 2.27	1.73 / 2.19	1.71 / 2.15	1.68 / 2.10	1.65 / 2.06	1.64 / 2.03
30	4.17 / 7.56	3.32 / 5.39	2.92 / 4.51	2.69 / 4.02	2.53 / 3.70	2.42 / 3.47	2.34 / 3.30	2.27 / 3.17	2.21 / 3.06	2.16 / 2.98	2.12 / 2.90	2.09 / 2.84	2.04 / 2.74	1.99 / 2.66	1.93 / 2.55	1.89 / 2.47	1.84 / 2.38	1.79 / 2.29	1.76 / 2.24	1.72 / 2.16	1.69 / 2.13	1.66 / 2.07	1.64 / 2.03	1.62 / 2.01
32	4.15 / 7.50	3.30 / 5.34	2.90 / 4.46	2.67 / 3.97	2.51 / 3.66	2.40 / 3.42	2.32 / 3.25	2.25 / 3.12	2.19 / 3.01	2.14 / 2.94	2.10 / 2.86	2.07 / 2.80	2.02 / 2.70	1.97 / 2.62	1.91 / 2.51	1.86 / 2.42	1.82 / 2.34	1.76 / 2.25	1.74 / 2.20	1.69 / 2.12	1.67 / 2.08	1.64 / 2.02	1.61 / 1.98	1.59 / 1.96
34	4.13 / 7.44	3.28 / 5.29	2.88 / 4.42	2.65 / 3.93	2.49 / 3.61	2.38 / 3.38	2.30 / 3.21	2.23 / 3.08	2.17 / 2.97	2.12 / 2.89	2.08 / 2.82	2.05 / 2.76	2.00 / 2.66	1.95 / 2.58	1.89 / 2.47	1.84 / 2.38	1.80 / 2.30	1.74 / 2.21	1.71 / 2.15	1.67 / 2.08	1.64 / 2.04	1.61 / 1.98	1.59 / 1.94	1.57 / 1.91
36	4.11 / 7.39	3.26 / 5.25	2.86 / 4.38	2.63 / 3.89	2.48 / 3.58	2.36 / 3.35	2.28 / 3.18	2.21 / 3.04	2.15 / 2.94	2.10 / 2.86	2.06 / 2.78	2.03 / 2.72	1.98 / 2.62	1.93 / 2.54	1.87 / 2.43	1.82 / 2.35	1.78 / 2.26	1.72 / 2.17	1.69 / 2.12	1.65 / 2.04	1.62 / 2.00	1.59 / 1.94	1.56 / 1.90	1.55 / 1.87
38	4.10 / 7.35	3.25 / 5.21	2.85 / 4.34	2.62 / 3.86	2.46 / 3.54	2.35 / 3.32	2.26 / 3.15	2.19 / 3.02	2.14 / 2.91	2.09 / 2.82	2.05 / 2.75	2.02 / 2.69	1.96 / 2.59	1.92 / 2.51	1.85 / 2.40	1.80 / 2.32	1.76 / 2.22	1.71 / 2.14	1.67 / 2.08	1.63 / 2.00	1.60 / 1.97	1.57 / 1.90	1.54 / 1.86	1.53 / 1.84
40	4.08 / 7.31	3.23 / 5.18	2.84 / 4.31	2.61 / 3.83	2.45 / 3.51	2.34 / 3.29	2.25 / 3.12	2.18 / 2.99	2.12 / 2.88	2.07 / 2.80	2.04 / 2.73	2.00 / 2.66	1.95 / 2.56	1.90 / 2.49	1.84 / 2.37	1.79 / 2.29	1.74 / 2.20	1.69 / 2.11	1.66 / 2.05	1.61 / 1.97	1.59 / 1.94	1.55 / 1.88	1.53 / 1.84	1.51 / 1.81
42	4.07 / 7.27	3.22 / 5.15	2.83 / 4.29	2.59 / 3.80	2.44 / 3.49	2.32 / 3.26	2.24 / 3.10	2.17 / 2.96	2.11 / 2.86	2.06 / 2.77	2.02 / 2.70	1.99 / 2.64	1.94 / 2.54	1.89 / 2.46	1.82 / 2.35	1.78 / 2.26	1.73 / 2.17	1.68 / 2.08	1.64 / 2.02	1.60 / 1.94	1.57 / 1.91	1.54 / 1.85	1.51 / 1.80	1.49 / 1.78
44	4.06 / 7.24	3.21 / 5.12	2.82 / 4.26	2.58 / 3.78	2.43 / 3.46	2.31 / 3.24	2.23 / 3.07	2.16 / 2.94	2.10 / 2.84	2.05 / 2.75	2.01 / 2.68	1.98 / 2.62	1.92 / 2.52	1.88 / 2.44	1.81 / 2.32	1.76 / 2.24	1.72 / 2.15	1.66 / 2.06	1.63 / 2.00	1.58 / 1.92	1.56 / 1.88	1.52 / 1.82	1.50 / 1.78	1.48 / 1.75
46	4.05 / 7.21	3.20 / 5.10	2.81 / 4.24	2.57 / 3.76	2.42 / 3.44	2.30 / 3.22	2.22 / 3.05	2.14 / 2.92	2.09 / 2.82	2.04 / 2.73	2.00 / 2.66	1.97 / 2.60	1.91 / 2.50	1.87 / 2.42	1.80 / 2.30	1.75 / 2.22	1.71 / 2.13	1.65 / 2.04	1.62 / 1.98	1.57 / 1.90	1.54 / 1.86	1.51 / 1.80	1.48 / 1.76	1.46 / 1.72
48	4.04 / 7.19	3.19 / 5.08	2.80 / 4.22	2.56 / 3.74	2.41 / 3.42	2.30 / 3.20	2.21 / 3.04	2.14 / 2.90	2.08 / 2.80	2.03 / 2.71	1.99 / 2.64	1.96 / 2.58	1.90 / 2.48	1.86 / 2.40	1.79 / 2.28	1.74 / 2.20	1.70 / 2.11	1.64 / 2.02	1.61 / 1.96	1.56 / 1.88	1.53 / 1.84	1.50 / 1.78	1.47 / 1.73	1.45 / 1.70

n degrees of freedom (for greater mean square)

APPENDIX 451

Table A6 Continued

n_1 degrees of freedom (for greater mean square)

d.f.	1	2	3	4	5	6	7	8	9	10	11	12	14	16	20	24	30	40	50	75	100	200	500	∞
50	4.03 7.17	3.18 5.06	2.79 4.20	2.56 3.72	2.40 3.41	2.29 3.18	2.20 3.02	2.13 2.88	2.07 2.78	2.02 2.70	1.98 2.62	1.95 2.56	1.90 2.46	1.85 2.39	1.78 2.26	1.74 2.18	1.69 2.10	1.63 2.00	1.60 1.94	1.55 1.86	1.52 1.82	1.48 1.76	1.46 1.71	1.44 1.68
55	4.02 7.12	3.17 5.01	2.78 4.16	2.54 3.68	2.38 3.37	2.27 3.15	2.18 2.98	2.11 2.85	2.05 2.75	2.00 2.66	1.97 2.59	1.93 2.53	1.88 2.43	1.83 2.35	1.76 2.23	1.72 2.15	1.67 2.06	1.61 1.96	1.58 1.90	1.52 1.82	1.50 1.78	1.46 1.71	1.43 1.66	1.41 1.64
60	4.00 7.08	3.15 4.98	2.76 4.13	2.52 3.65	2.37 3.34	2.25 3.12	2.17 2.95	2.10 2.82	2.04 2.72	1.99 2.63	1.95 2.56	1.92 2.50	1.86 2.40	1.81 2.32	1.75 2.20	1.70 2.12	1.65 2.03	1.59 1.93	1.56 1.87	1.50 1.79	1.48 1.74	1.44 1.68	1.41 1.63	1.39 1.60
65	3.99 7.04	3.14 4.95	2.75 4.10	2.51 3.62	2.36 3.31	2.24 3.09	2.15 2.93	2.08 2.79	2.02 2.70	1.98 2.61	1.94 2.54	1.90 2.47	1.85 2.37	1.80 2.30	1.73 2.18	1.68 2.09	1.63 2.00	1.57 1.90	1.54 1.84	1.49 1.76	1.46 1.71	1.42 1.64	1.39 1.60	1.37 1.56
70	3.98 7.01	3.13 4.92	2.74 4.08	2.50 3.60	2.35 3.29	2.23 3.07	2.14 2.91	2.07 2.77	2.01 2.67	1.97 2.59	1.93 2.51	1.89 2.45	1.84 2.35	1.79 2.28	1.72 2.15	1.67 2.07	1.62 1.98	1.56 1.88	1.53 1.82	1.47 1.74	1.45 1.69	1.40 1.62	1.37 1.56	1.35 1.53
80	3.96 6.96	3.11 4.88	2.72 4.04	2.48 3.56	2.33 3.25	2.21 3.04	2.12 2.87	2.05 2.74	1.99 2.64	1.95 2.55	1.91 2.48	1.88 2.41	1.82 2.32	1.77 2.24	1.70 2.11	1.65 2.03	1.60 1.94	1.54 1.84	1.51 1.78	1.45 1.70	1.42 1.65	1.38 1.57	1.35 1.52	1.32 1.49
100	3.94 6.90	3.09 4.82	2.70 3.98	2.46 3.51	2.30 3.20	2.19 2.99	2.10 2.82	2.03 2.69	1.97 2.59	1.92 2.51	1.88 2.43	1.85 2.36	1.79 2.26	1.75 2.19	1.68 2.06	1.63 1.98	1.57 1.89	1.51 1.79	1.48 1.73	1.42 1.64	1.39 1.59	1.34 1.51	1.30 1.46	1.28 1.43
125	3.92 6.84	3.07 4.78	2.68 3.94	2.44 3.47	2.29 3.17	2.17 2.95	2.08 2.79	2.01 2.65	1.95 2.56	1.90 2.47	1.86 2.40	1.83 2.33	1.77 2.23	1.72 2.15	1.65 2.03	1.60 1.94	1.55 1.85	1.49 1.75	1.45 1.68	1.39 1.59	1.36 1.54	1.31 1.46	1.27 1.40	1.25 1.37
150	3.91 6.81	3.06 4.75	2.67 3.91	2.43 3.44	2.27 3.14	2.16 2.92	2.07 2.76	2.00 2.62	1.94 2.53	1.89 2.44	1.85 2.37	1.82 2.30	1.76 2.20	1.71 2.12	1.64 2.00	1.59 1.91	1.54 1.83	1.47 1.72	1.44 1.66	1.37 1.56	1.34 1.51	1.29 1.43	1.25 1.37	1.22 1.33
200	3.89 6.76	3.04 4.71	2.65 3.88	2.41 3.41	2.26 3.11	2.14 2.90	2.05 2.73	1.98 2.60	1.92 2.50	1.87 2.41	1.83 2.34	1.80 2.28	1.74 2.17	1.69 2.09	1.62 1.97	1.57 1.88	1.52 1.79	1.45 1.69	1.42 1.62	1.35 1.53	1.32 1.48	1.26 1.39	1.22 1.33	1.19 1.28
400	3.86 6.70	3.02 4.66	2.62 3.83	2.39 3.36	2.23 3.06	2.12 2.85	2.03 2.69	1.96 2.55	1.90 2.46	1.85 2.37	1.81 2.29	1.78 2.23	1.72 2.12	1.67 2.04	1.60 1.92	1.54 1.84	1.49 1.74	1.42 1.64	1.38 1.57	1.32 1.47	1.28 1.42	1.22 1.32	1.16 1.24	1.13 1.19
1000	3.85 6.66	3.00 4.62	2.61 3.80	2.38 3.34	2.22 3.04	2.10 2.82	2.02 2.66	1.95 2.53	1.89 2.43	1.84 2.34	1.80 2.26	1.76 2.20	1.70 2.09	1.65 2.01	1.58 1.89	1.53 1.81	1.47 1.71	1.41 1.61	1.36 1.54	1.30 1.44	1.26 1.38	1.19 1.28	1.13 1.19	1.08 1.11
∞	3.84 6.64	2.99 4.60	2.60 3.78	2.37 3.32	2.21 3.02	2.09 2.80	2.01 2.64	1.94 2.51	1.88 2.41	1.83 2.32	1.79 2.24	1.75 2.18	1.69 2.07	1.64 1.99	1.57 1.87	1.52 1.79	1.46 1.69	1.40 1.59	1.35 1.52	1.28 1.41	1.24 1.36	1.17 1.25	1.11 1.15	1.00 1.00

Table A7 Table of χ^2*

Degrees of Freedom	P = .99	.98	.95	.90	.80	.70	.50	.30	.20	.10	.05	.02	.01
1	.000157	.000628	.00393	.0158	.0642	.148	.455	1.074	1.642	2.706	3.841	5.412	6.635
2	.0201	.0404	.103	.211	.446	.713	1.386	2.408	3.219	4.605	5.991	7.824	9.210
3	.115	.185	.352	.584	1.005	1.424	2.366	3.665	4.642	6.251	7.815	9.837	11.341
4	.297	.429	.711	1.064	1.649	2.195	3.357	4.878	5.989	7.779	9.488	11.668	13.277
5	.554	.752	1.145	1.610	2.343	3.000	4.351	6.064	7.289	9.236	11.070	13.388	15.086
6	.872	1.134	1.635	2.204	3.070	3.828	5.348	7.231	8.558	10.645	12.592	15.033	16.812
7	1.239	1.564	2.167	2.833	3.822	4.671	6.346	8.383	9.803	12.017	14.067	16.622	18.475
8	1.646	2.032	2.733	3.490	4.594	5.527	7.344	9.524	11.030	13.362	15.507	18.168	20.090
9	2.088	2.532	3.325	4.168	5.380	6.393	8.343	10.656	12.242	14.684	16.919	19.679	21.666
10	2.558	3.059	3.940	4.865	6.179	7.267	9.342	11.781	13.442	15.987	18.307	21.161	23.209
11	3.053	3.609	4.575	5.578	6.989	8.148	10.341	12.899	14.631	17.275	19.675	22.618	24.725
12	3.571	4.178	5.226	6.304	7.807	9.034	11.340	14.011	15.812	18.549	21.026	24.054	26.217
13	4.107	4.765	5.892	7.042	8.634	9.926	12.340	15.119	16.985	19.812	22.362	25.472	27.688
14	4.660	5.368	6.571	7.790	9.467	10.821	13.339	16.222	18.151	21.064	23.685	26.873	29.141
15	5.229	5.985	7.261	8.547	10.307	11.721	14.339	17.322	19.311	22.307	24.996	28.259	30.578
16	5.812	6.614	7.962	9.312	11.152	12.624	15.338	18.418	20.465	23.542	26.296	29.633	32.000
17	6.408	7.255	8.672	10.085	12.002	13.531	16.338	19.511	21.615	24.769	27.587	30.995	33.409
18	7.015	7.906	9.390	10.865	12.857	14.440	17.338	20.601	22.760	25.989	28.869	32.346	34.805
19	7.633	8.567	10.117	11.651	13.716	15.352	18.338	21.689	23.900	27.204	30.144	33.687	36.191
20	8.260	9.237	10.851	12.443	14.578	16.266	19.337	22.775	25.038	28.412	31.410	35.020	37.566
21	8.897	9.915	11.591	13.240	15.445	17.182	20.337	23.858	26.171	29.615	32.671	36.343	38.932
22	9.542	10.600	12.338	14.041	16.314	18.101	21.337	24.939	27.301	30.813	33.924	37.659	40.289
23	10.196	11.293	13.091	14.848	17.187	19.021	22.337	26.018	28.429	32.007	35.172	38.968	41.638
24	10.856	11.992	13.848	15.659	18.062	19.943	23.337	27.096	29.553	33.196	36.415	40.270	42.980
25	11.524	12.697	14.611	16.473	18.940	20.867	24.337	28.172	30.675	34.382	37.652	41.566	44.314
26	12.198	13.409	15.379	17.292	19.820	21.792	25.336	29.246	31.795	35.563	38.885	42.856	45.642
27	12.879	14.125	16.151	18.114	20.703	22.719	26.336	30.319	32.912	36.741	40.113	44.140	46.963
28	13.565	14.847	16.928	18.939	21.588	23.647	27.336	31.391	34.027	37.916	41.337	45.419	48.278
29	14.256	15.574	17.708	19.768	22.475	24.577	28.336	32.461	35.139	39.087	42.557	46.693	49.588
30	14.953	16.306	18.493	20.599	23.364	25.508	29.336	33.530	36.250	40.256	43.773	47.962	50.892

*Table A7 is reprinted from Table III of Fisher: Statistical Methods for Research Workers, New York: Hafner Press, © 1973. For larger values of df, the expression $\sqrt{2\chi^2} - \sqrt{2(df) - 1}$ may be used as a normal deviate with unit standard error.

Table A8 Determination of r_{tet} for Various Values* of bc/ad or ad/bc from a Four-fold Contingency Table†

r_{tet}	$\dfrac{bc}{ad}$ or $\dfrac{ad}{bc}$	r_{tet}	$\dfrac{bc}{ad}$ or $\dfrac{ad}{bc}$	r_{tet}	$\dfrac{bc}{ad}$ or $\dfrac{ad}{bc}$	r_{tet}	$\dfrac{bc}{ad}$ or $\dfrac{ad}{bc}$
0	1.000	.26	1.941–1.993	.51	4.068–4.205	.76	11.513–12.177
.010	1.013–1.039	.27	1.994–2.048	.52	4.206–4.351	.77	12.178–12.905
.02	1.040–1.066	.28	2.049–2.105	.53	4.352–4.503	.78	12.906–13.707
.03	1.067–1.093	.29	2.106–2.164	.54	4.504–4.662	.79	13.708–14.592
.04	1.094–1.122	.30	2.165–2.225	.55	4.663–4.830	.80	14.593–15.574
.05	1.123–1.151	.31	2.226–2.288	.56	4.831–5.007	.81	15.575–16.670
.06	1.152–1.180	.32	2.289–2.353	.57	5.008–5.192	.82	16.671–17.899
.07	1.181–1.211	.33	2.354–2.421	.58	5.193–5.388	.83	17.900–19.287
.08	1.212–1.242	.34	2.422–2.491	.59	5.389–5.595	.84	19.288–20.865
.09	1.243–1.275	.35	2.492–2.563	.60	5.596–5.813	.85	20.866–22.674
.10	1.276–1.308	.36	2.564–2.638	.61	5.814–6.043	.86	22.675–24.766
.11	1.309–1.342	.37	2.639–2.716	.62	6.044–6.288	.87	24.767–27.212
.12	1.343–1.377	.38	2.717–2.797	.63	6.289–6.547	.88	27.213–30.105
.13	1.378–1.413	.39	2.798–2.881	.64	6.548–6.822	.89	30.106–33.577
.14	1.414–1.450	.40	2.882–2.968	.65	6.823–7.115	.90	33.578–37.815
.15	1.451–1.488	.41	2.969–3.059	.66	7.116–7.428	.91	37.816–43.096
.16	1.489–1.528	.42	3.060–3.153	.67	7.429–7.761	.92	43.097–49.846
.17	1.529–1.568	.43	3.154–3.251	.68	7.762–8.117	.93	49.847–58.758
.18	1.569–1.610	.44	3.252–3.353	.69	8.118–8.499	.94	58.759–71.035
.19	1.611–1.653	.45	3.354–3.460	.70	8.500–8.910	.95	71.036–88.964
.20	1.654–1.697	.46	3.461–3.571	.71	8.911–9.351	.96	88.965–117.479
.21	1.698–1.743	.47	3.572–3.687	.72	9.352–9.828	.97	117.480–169.503
.22	1.744–1.790	.48	3.688–3.808	.73	9.829–10.344	.98	169.504–292.864
.23	1.791–1.838	.49	3.809–3.935	.74	10.345–10.903	.99	292.865–923.687
.24	1.839–1.888	.50	3.936–4.067	.75	10.904–11.512	1.00	923.688– ∞
.25	1.889–1.940						

*Values in this table were calculated by Thomas O. Maguire.
†If bc/ad is greater than 1, the value of r_{tet} is read directly from this table. If ad/bc is greater than 1, the table is entered with ad/bc and the value of r_{tet} is negative.

INDEX

A

Alpha, 264, 301–302
Analysis of variance (ANOVA), 350–393
    assumptions in, 365
    degrees of freedom in, 356
    factorial, 372
    mean squares in, 355–356
    null hypothesis in, 357
    one-way, 350–371
    sums of squares in, 354
    test of significance in, 356–358
    two-way, 372–393
Averages, 52

B

Bias, 288–289
Bimodal distribution, 43
Binomial distribution, 278
    mean of, 279
    standard deviation of, 279
Binomial probability, 269–271
Biserial coefficient, 236–240
Bivariate distribution, 163

C

Central tendency, 52
    measures of, 76–77
    selecting a measure of, 74–75
Chi square, 395–413
    assumptions in, 409
    contingency coefficient from, 412
    contingency table in, 302, 406–407
    and continuity correction (for 1 d.f.), 400
    degrees of freedom in, 398, 405
    goodness-of-fit, 396–401
    phi coefficient from, 411–412
    test of independence, 402–409
Class interval, 28–31
    apparent limits, 30
    midpoint of, 29
    number of, 28
    real limits, 30
    size, 29
Coefficient of determination, 186
Combinations, 272
Constant, 9
Contingency coefficient, 248, 412–413
Correlation, 162–163
    assumptions of Pearson, 182–184

Correlation (*cont.*)
  coefficient of, 168
  factors influencing, 184–186
  interpretation of, 186–187
  negative, 166
  Pearson product-moment coefficient of, 173
    computation of, 173–181
  positive, 165
  range of coefficient of, 179
  scatter diagram in, 164–166, 169–170
  strength of, 168
Correlation coefficient, 168, 226
  biserial, 236–240
  contingency, 248, 412
  Pearson product-moment, 173–184
  phi, 244–248, 411–412
  point biserial, 232–234
  rank-order, 227–231
  Spearman rho, 227–231
  tetrachoric, 241–244
Correlated samples, 334–336
Cumulative frequency distribution, 32–33
  polygon, 39–40
Cumulative percentage distribution, 33–34
  polygon, 40–41

D

Decile, 124
Degrees of freedom (d.f.), 315–316
  in analysis of variance, 356, 375
  in chi square, 398, 405
  in a t-distribution, 316
  in a t-test, 327, 338
Descriptive statistics, 4
Deviation score, 90
Difference, distribution of between means, 323–324
  mean of, 323
  standard deviation of, 324
Distribution, bimodal, 43
  binomial probability, 278–279

Distribution, bimodal (*cont.*)
  of a difference in means, 323–324
  frequency, 24–34
  of a mean, 293–294
  normal, 138–160
  skewed, 42
  standard normal, 144
  of t, 315–318
  univariate, 163

E, F

Error, grouping, 28
  of prediction, 203, 214–215
  sampling 291–293
F-ratio, in one-way ANOVA, 356–357
  in two-way ANOVA, 375–376
Factorial analysis of variance, 372–388
  computation in, 374–380
  interaction effect in, 374
  K-factor, 386–387
  main effects in, 374
  null hypothesis in, 376
  partitioning sum of squares in, 374–379
  test of significance, 375
Factorial design, 373–374
Factorial n, 272
Factors, 372
Fisher's exact test, 410
Frequency distribution, 24–34
  grouped, 26–34
  ungrouped, 24–26
Frequency polygon, 37–38

G

Goodness-of-fit, 396–401
Gosset, William ("Student"), 314
Graphs, of frequency distributions, 34–41
Grouping data, 26–34

H

Histogram, 35–36
Homoscedasticity, 184, 218
Hypothesis testing, about a population correlation, 342–343
    about a population mean, using the normal distribution, 295–298
    about a population mean, using a t-distribution, 319–321

I

Independent samples, 330
Inferential statistics, 4–5
Interaction, 372
    interpretation of, 380–383
    second- and third-order, 387
Interval scale, 15–16

K, L

Kurtosis, 143
Least squares criterion, 201, 203
Leptokurtosis, 143
Limits, of class intervals, 30

M

Main effects, 374, 376
Mean, arithmetic, 63
    characteristics of, 65–69
    of a combined set of measures, 67
    compared to median and mode, 70–73
    defined, 63
    of a frequency distribution, 64
    of a population, 63
    of a sample, 63
    sum of deviations from, 66
Mean square, 356, 375
Measurement, scales of, 12–17
Median, computation of, 55–60
    defined, 54

Median (*cont.*)
    properties of, 61–62
Median test, 416–417
Mode, 52–53

N

Nominal scale, 12–13
Nonparametric tests, 395–420
    compared with parametric, 419–420
Normal curve, 138–160
    areas of the, 145–146
    formula for, 141
    properties of, 143
    standard normal curve, 144
Normal curve approximation of the binomial, 278–281
Null hypothesis, 262–263, 299
    and alternative hypotheses, 299–300
    interpreting rejected null, 304–305
    interpreting retained null, 304–305

O

Ogive, 39
One-tailed test, 303
Ordinal scale, 13–15

P

Parameters, 6, 288
    defined, 6
    distinguished from statistics, 7
    symbols for, 7
Parametric tests, 313–388
    assumptions of, 395
Pearson, Karl, 173
Percentile, computation of, 119–123
    defined, 118
Percentile rank, calculation of, 114–116
    defined, 113
    finding percentile rank equivalents for z-scores, 148–150
    finding z-score equivalents for percentile ranks, 152–153

Phi coefficient, 244–248
Platykurtosis, 143
Point-biserial coefficient, 232–234
Population, 6, 287
Power, of a test of significance, 340
Prediction, 196–197
  equation for, 206
  "line of best fit," 203
  using raw scores, 206–208
  of z-scores, 198
Probability, areas under the normal curve and, 278–281
  binomial expansion and, 272–276
  binomial probability formula, 274

Q, R

Quartile, 86–87, 123
Quartile deviation, 86–89
Random assignment, 290
Random numbers, use of table of, 289
Random sampling, 289
Range, 84–85
Rank-order coefficient, 227–231
Ratio scale, 16–17
Regression, 199
  coefficient, 207
  defined, 200
  effect, 200
  equation, 206
  line, 201, 203
  linear, 205
  toward the mean, 200
Rejection of null hypothesis, 304–305
Research hypothesis, 299
Rho (see Correlation coefficient)

S

Sample, biased, 288
  defined, 7
  random, 289
  representative, 7
Sampling error, 291–293

Scales, interval, 15–16
  of measurement, 12–17
  nominal, 12–13
  ordinal, 13–15
  ratio, 16–17
Scattergram, 164–165
Semi-interquartile range, 86
Significance, statistical, 305
Sign test, assumptions of, 413
  computation of, 415
  null hypothesis in, 415
Skewed distributions, 42
Spearman, Charles, 227
Standard deviation, 94–105
  computation from frequency distribution, 99–100
  computation from raw scores, 95–98
  defined, 94
  interpretation of, 103–104
  of a population, 101
  of a sample, 101–102
Standard error, of the difference between two means when observations are correlated, 337
  of the difference between two means when observations are independent, 324
  of estimate, 216–217, 220
  of the mean, 294
    estimated from sample data, 314
Standard scores, 125–134
  z-scores, characteristics of, 127–129
  conversion to other standard scores, 133
  defined, 125
  for given percentile ranks, 152–153
  use of, 130–131
  Z-scores, 132
Stanines, 134
Statistics, defined, 3
  distinguished from parameters, 7, 288
  functions of, 3–4
  symbols for, 7
Student's t, 315
Sum of squares, in analysis of variance, 354–355

Sum of squares (*cont.*)
    computation of, in one-way ANOVA, 359–362
    in two-way ANOVA, 374–379

T

t-Distribution (*see* Distribution, of *t*)
t-Test, 324
    assumptions of, 329
    for correlated samples, 336–339
    error term in, 327
    for independent samples, 330–332
    logic of, 327–328
    null hypothesis in, 326
    of significance of Pearson correlation coefficient, 342–343
Tetrachoric coefficient, 241–244
Transformation, of scores, 113
Two-tailed test, 303
Type I error, 262, 264, 300

Type II error, 262, 264, 300
    effect of numbers on, 307

V

Variables, continuous, 10
    defined, 9
    dependent, 9
    discrete, 9
    independent, 9
    qualitative, 9
    quantitative, 9
Variance, of a population, 92
    of a sample, 101

Y, Z

Yates' correction, 400
z-Scores (*see* Standard scores)